Progress in Molecular and Subcellular Biology 14

AF785832

Series Editors

Ph. Jeanteur, Y. Kuchino,
W.E.G. Müller (*Managing Editor*)
P.L. Paine

W.E.G. Müller H.C. Schröder (Eds.)

Biological Response Modifiers – Interferons, Double-Stranded RNA and 2′,5′-Oligoadenylates

With 68 Figures

Springer-Verlag
Berlin Heidelberg New York
London Paris Tokyo
Hong Kong Barcelona
Budapest

Prof. Dr. W.E.G. Müller

Prof. Dr. Dr. H.C. Schröder

Institut für Physiologische Chemie
Abteilung Angewandte Molekularbiologie
Universität Mainz
Duesbergweg 6
55099 Mainz
Germany

ISSN 0079-6484

ISBN-13: 978-3-642-78551-1 e-ISBN-13: 978-3-642-78549-8
DOI: 10.1007/ 978-3-642-78549-8

This work is subject to copyright. All rights are reserved, whether the whole or part of the material is concerned, specifically the rights of translation, reprinting, reuse of illustrations, recitation, broadcasting, reproduction on microfilm or in any other way, and storage in data banks. Duplication of this publication or parts thereof is permitted only under the provisions of the German Copyright Law of September 9, 1965, in its current version, and permission for use must always be obtained from Springer-Verlag. Violations are liable for prosecution under the German Copyright Law.

© Springer-Verlag Berlin Heidelberg 1994
Softcover reprint of the hardcover 1st edition 1994

The use of general descriptive names, registered names, trademarks, etc. in this publication does not imply, even in the absence of a specific statement, that such names are exempt from the relevant protective laws and regulations and therefore free for general use.

Typesetting: Macmillan India Ltd., Bangalore 25
39/3130/SPS-5 4 3 2 1 0 - Printed on acid-free paper

Preface

The observation that higher organisms do not suffer from more than one viral infection at a time resulted in the discovery of Isaacs and Lindenmann (1957) that after virus infection the cells release a protein which protects them against infection from a broad spectrum of unrelated viruses. The authors termed this factor "interferon". Interferon (IFN) is the body's natural defence against viruses. IFN was found to be stable against acid treatment and proved to be virus-unspecific but species-specific. Later, it was discovered that IFN cannot only be induced by viruses but also by other microorganisms, components of microorganisms and synthetic substances. Among them are also polyanions, e.g. nucleic acids, synthetic polynucleotides and lipopolysaccharides. Several kinds of IFNs were recognized and classified according to IFN-producing cells, inducing agents, and physical/chemical properties of the IFN. They were termed α-IFN (produced from lymphocytes and macrophages) and β-IFN (produced from fibroblasts) which were grouped together as IFN-I, and IFN-II or γ-IFN, which is produced by T-lymphocytes.

In the 1970s it became clear that IFN has multiple biological effects (Gresser 1977). IFN inhibits growth of both normal and tumor cells, alters the cell surface, induces the expression of histocompatibility antigens and modulates immune response, e.g. it inhibits antibody formation and delayed-hypersensitivity reaction, and enhances natural killer cell activity.

A series of proteins and enzyme systems have been identified that are induced after incubation of cells with IFN. Among them are the two enzymes whose activity depends on the presence of dsRNA, the 2′,5′-oligoadenylate synthetase and a specific protein kinase, which inhibits peptide chain initiation (Baglioni 1979; Hovanessian 1979; Williams et al. 1979; Revel et al. 1980; Gordon and Minks 1981; Lengyel 1981; Lebleu and Content 1982). In contrast to these enzymes, other enzymes, e.g. ornithine decarboxylase, are inhibited by IFN.

The overall objective of this book is to provide reasoned assessments of the current knowledge on (1) the molecular and cellular mechanisms which establish the antiviral state of cells. The interaction of IFN with receptors in the target cell membrane is described highlighting the mechanisms involved in signal transduction from the membrane to the nucleus. (2) In a further part, the mechanisms by which IFN inhibits virus replication are described and, finally, and with major emphasis (3) strategies are outlined which apply cell biological and molecular knowledge to therapeutic treatment of selected diseases, including AIDS. The strategies discussed comprise the application of dsRNA and analogues of 2′,5′-oligodenylate.

We would like to express our gratitude to all authors for their cooperation and understanding, particularly during the discussions about the final format of their contributions.

References

Baglioni C (1979) Interferon-induced enzymatic activities and their role in the antiviral state. Cell 17:255–264

Gordon J, Minks MA (1981) The interferon renaissance: molecular aspects of the induction and action. Microbiol Rev 45 : 244–266

Gresser I (1977) On the varied biological effects of interferon. Cell Immunol 34:406–415

Hovanessian AG (1979) Intracellular events in interferon-treated cells. Differentiation 15:139–151

Isaacs A, Lindenmann J (1957) Virus interference. I. The interferon. Proc R Soc Lond B 147:258–273

Lebleu B, Content J (1982) Mechanisms of interferon action: biochemical and genetic approaches. Interferon 4:47–94

Lengyel P (1981) Enzymology of interferon action – a short survey. Methods Enzymol 79:135–148

Revel M, Kimchi A, Shulman L, Fradin A, Schuster R, Yakobson E, Chernajovsky Y, Schmidt A, Shure A, Bendori R (1980) Role of interferon-induced enzymes in the antiviral and antimitogenic effects of interferon. Ann N Y Acad Sci 350:459–472

Williams BRG, Golgher RR, Brown RE, Gilbert CS, Kerr IM (1979) Natural occurrence of 2-5A in interferon-treated EMC virus-infected L cells. Nature 282:582–586

Mainz, Germany W.E.G. Müller
January 1994 H.C. Schröder

Contents

List of Contributors

Addresses are given at the beginning of the respective contribution

Activation of the dsRNA-Dependent Kinase

R.A. Petryshyn, J. Li, and R. Judware[1]

1 Introduction

The double-stranded RNA (dsRNA)-dependent protein kinase, designated here as dsI, but also referred to as DAI, P1 kinase, p68 kinase and P1/eIF-2α protein kinase, is an interferon-induced serine/threonine protein kinase. It is uniquely distinguished from other protein kinases in that its activation to a functional enzyme, and in some cases prevention of its activation, is dependent on RNAs containing accessible double-stranded structures. However, like several other protein kinases, once activated, it elicits a modulating effect that alters cellular processes.

The importance of dsI as an inhibitor of the initiation of protein synthesis and its cellular role in mediating, in part, the antiviral action of interferon has already been well established. Considerable additional interest in dsI has arisen from the realization that several viruses have evolved mechanisms to overcome the antiviral effect of interferon by specifically inhibiting the activation or activity of the kinase. Furthermore, it has become evident from recent studies that this enzyme may be critical for regulating cell growth and differentiation, suppression of tumor growth and in modulating signal transduction by interferon and other signal inducers. With these aspects in mind and within the limitations of this review, we attempt to briefly summarize recent findings regarding the mechanisms that regulate the activation of dsI and its physiological role. Where possible, unresolved questions have been raised. The authors guide the readers to several previous reviews (Pestka et al. 1987; Hershey 1991; Samuel 1992) which provide both a historical perspective and a more detailed description of the biochemistry of dsI.

2 Mechanism of Activation of the dsRNA-Dependent Kinase

dsI is a ribosome-associated protein kinase of 65 (murine)-68 (human) kDa in size, whose synthesis is transcriptionally induced by interferon (Meurs et al. 1990; Saito and Kawakita 1991; Sen and Lengyel 1992). Its conversion from

[1]Department of Biochemistry and Molecular Biology, State University of New York, Health Science Center at Syracuse, 750 East Adams Street, Syracuse, New York 13210, USA

Progress in Molecular and Subcellular Biology, Vol. 14
W.E.G. Müller/H.C. Schröder (Eds.)
© Springer-Verlag Berlin Heidelberg 1994

a latent to active enzyme requires low levels of dsRNA, ATP and millimolar concentrations of Mg^{2+} or Mn^{2+} (Levin et al. 1981; Petryshyn et al. 1983; Galabru and Hovanessian 1985). The kinase is not activated by DNA, DNA:RNA hybrids, or by single-stranded RNAs that are either devoid of or contain an inaccessible extensive secondary structure (Hunter et al. 1975; Bischoff and Samuel 1989). Heparin has also been observed to activate dsI, but the physiological significance of this observation is not known (Galabru and Hovanessian 1985). In response to dsRNA from viral infection or upon addition of dsRNA in vitro, two dsI-mediated phosphorylation steps occur. The first is an autophosphorylation reaction in which the kinase undergoes phosphorylation at multiple serine/threonine residues (Galabru and Hovanessian 1987). It has been shown that maximally phosphorylated dsI is in its most activated state (Petryshyn et al. 1982; Galabru and Hovanessian 1987). In the second step, activated dsI phosphorylates the α-subunit of the eukaryotic initiation factor-2 (eIF-2α). This reaction is independent of dsRNA (Farrell et al. 1 977; Levin and London 1978; Samuel 1979). The phosphorylation of eIF-2α occurs on serine-51 (Pathak et al. 1988; Price et al. 1991) and is directly related to inhibition of protein synthesis (London et al. 1987; Hershey 1991). eIF-2 is a guanine nucleotide binding protein that requires the exchange of GTP for GDP for recycling after one round of initiation of protein synthesis has occurred (London et al. 1987; Hershey 1991). This exchange is catalyzed by the guanine nucleotide exchange factor (eIF-2B) (Konieczny and Safer 1983; Matts et al. 1983; Rowlands et al. 1988). When eIF-2α becomes phosphorylated it forms a tight association with eIF-2B rendering the exchange factor nonfunctional (Matts and London 1984). Since eIF-2B is present in limiting quantities in most cells (Konieczny and Safer 1983; Matts et al. 1983; Hershey 1991), phosphorylation of 20–40% of the total cellular eIF-2 is sufficient to sequester and inactivate all of the available exchange factor, resulting in an inhibition of protein synthesis (Leroux and London 1982; Siekierka et al. 1984). The principal features of the activation and activity of dsI are illustrated in Fig. 1.

Despite our understanding of the activation and activity of dsI, the mechanism by which it interacts with dsRNA is both complex and not fully understood. This complexity is evident from the observations that a diverse number of synthetic dsRNAs and viral RNAs, both double-stranded and single-stranded, activate dsI (Hunter et al. 1975; Farrell et al. 1977; Levin and London 1978; Nilsen et al. 1982; Bischoff and Samuel 1989; Black et al. 1989; Edery et al. 1989; SenGupta and Silverman 1989; Roy et al. 1991; Meurs et al. 1992). Thus, there appears to be no specific sequence requirement, but rather a dependency on the extent of secondary and probably tertiary structure within the RNA. Moreover, viral RNAs such as adenovirus VAI RNA (Kitajewski et al. 1986; Mathews and Shenk 1991) and Epstein-Barr virus EBER-1 RNA (Clarke et al. 1990, 1991) bind but fail to activate the enzyme. Furthermore, these viral RNAs prevent the activation of dsI by other RNAs (Clarke et al. 1991; Mathews and Shenk 1991). It is of interest that the TAR sequence in the 5′ leader region of human immunodeficiency virus-1 mRNA has been shown to both activate (Edery et al.

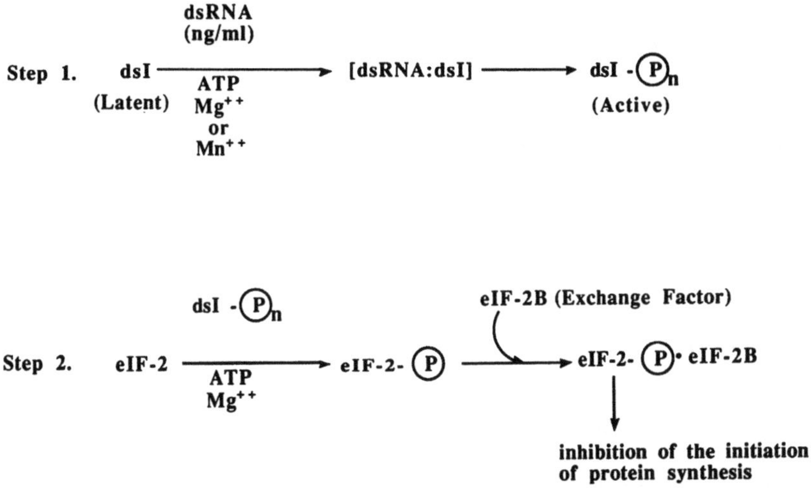

Fig. 1. Activation and activity of dsI. A generalized diagram illustrating the dsRNA-dependent phosphorylation of dsI (*Step 1*). Activated dsI phosphorylates eIF-2 (on the α-subunit), which leads to the sequestration of the guanine nucleotide exchange factor (eIF-2B), resulting in the inhibition of the initiation of protein synthesis (*Step 2*)

1989; SenGupta and Silverman 1989; Roy et al. 1991) and inhibit (Gunnery et al. 1990) dsI. The reason for both observations is not known but may depend on the preparation or level of the TAR RNA and dsI used. Interestingly, we have recently observed that a peptide containing the TAR RNA binding sequence of the tat protein specifically inhibits the activation of dsI by the TAR RNA in vitro (Judware et al. 1993), suggesting the possibility that the activation of dsI by the TAR RNA is regulated during infection with HIV-1. In this regard it is notable that the tat protein has also been shown to block the TAR RNA-mediated activation of the (2′, 5′) oligoadenylate synthetase, another interferon-induced enzyme which is activated by dsRNA (Schröder et al. 1990). In addition, the activation of dsI exhibits a paradoxic dependency on the concentration of the RNA. While activation occurs at low concentrations (ng/ml), high concentrations (μg/ml) of the same RNA prevent activation (Hunter et al. 1975; Farrell et al. 1977; Levin and London 1978; Clarke et al. 1990). Once activated, however, neither high concentrations of RNA nor addition of the inhibitory RNAs affect its activity as an eIF-2α kinase (Hunter et al. 1975; Farrell et al. 1977; Levin and London 1978; Clarke et al. 1990).

It has been established that the activation of dsI proceeds through an ordered reaction mechanism, in that dsRNA binds to the enzyme and this is required for the subsequent binding of ATP and autophosphorylation (Bischoff and Samuel 1985; Galabru and Hovanessian 1987; Judware and Petryshyn 1992). Molecular studies have indicated that the dsRNA binding domain is localized to the amino-terminal portion of the protein (regulatory domain) (Katze et al. 1991;

```
      ..........  ..........  ..........  ....
  1-MAGDLSAGFF  MEELNTYRQK  QGVVLKYQEL  PNSGPPHDRR  FTFQVIIDGR-50

               ***  **********  *****
 51-EFPEGEGRSK  KEAKNAAAKL  AVEILNKEKK  AVSPLLLTTT  NSSEGLSMGN-100

                                    ..  ..........  .......*.*
101-YIGLINRIAQ  KKRLTVNYEQ  CASGVHGPEG  FHYKCKMGQK  EYSIGTGSTK-150

    .*.*.*.*.*  .*.*.*....
151-QEAKQLAAKL  AYLQILSEET  SVKSDYLSSG-180
```

Fig. 2. A summary of the putative dsRNA binding regions within the regulatory domain of dsI. The deduced amino acid sequence of human dsI starting at position 1 is from Meurs et al. (1990). The amino acid sequences reported to be involved in the binding of dsRNA are indicated by the following symbols: ...., amino acids 1–34 and 129–170 (Patel and Sen 1992); ——, amino acids 39–50 and 58–69 (Feng et al. 1992); * * * *, amino acids 58–75 and 148–166 (McCormack et al. 1992)

Feng et al. 1992; McCormack et al. 1992; Patel and Sen 1992), while the carboxyl-terminal portion (catalytic domain) contains the putative ATP binding sites and other motifs found conserved among serine/threonine protein kinases (Meurs et al. 1990; Barber et al. 1991; Thomis et al. 1992). The serine/threonine residues, where autophosphorylation occurs, as well as the ribosome and eIF-2α recognition sites, have yet to be identified. It is likely that the binding of dsRNA at the regulatory domain induces a conformational change which is transmitted to the catalytic domain to facilitate ATP binding and autophosphorylation. Attempts to delineate the precise regions responsible for the binding of dsRNA have met with varied results and these are summarized in Fig. 2. Regions of importance may include (1) amino acids 58–75 and a homologous repeat of this sequence extending from amino acids 148–166, which exhibit significant homology to motifs in a number of other RNA binding proteins; (McCormack et al. 1992) (2) amino acids 1–34 and 129–170, which when deleted by truncation were shown to be necessary for the binding of dsRNA to the amino-terminal region of the protein (Patel and Sen 1992); and (3) amino acids 39–50 and 58–69, which were reported to be necessary for dsRNA binding and for growth inhibition in yeast (Feng et al. 1992). A summary of the available data suggests that dsI contains at least two critical dsRNA binding regions delineated by the amino acid sequences 58–69 and 148–168.

The precise mechanism by which dsRNA interacts with the binding sites to facilitate both activation and inhibition of activation of dsI remains to be determined. For example, questions remain as to whether dsRNAs bind at both sites with equal affinity and whether one or two dsRNA molecules bind to the enzyme during activation or inhibition. Earlier studies (Galabru et al. 1989) have suggested that dsI contains two dsRNA binding sites: a high-affinity site

($K_d = 0.35$ nM) responsible for the activation of dsI at low concentrations of dsRNA, and a low-affinity binding site ($K_d = 1$ nM) which mediates the inhibition of activation in the presence of high concentrations of dsRNA. A recent study by Manche et al. (1992), using short synthetic dsRNAs, has provided a more complex view for the interaction of dsRNA with dsI. In this study it was observed that dsRNA of ~ 30 bp (three helical turns) could bind and activate a single molecule of dsI. Moreover, activation increased as the length of the dsRNA was increased to a maximum size of ~ 80 bp (seven to eight helical turns). It was suggested that a single molecule of dsRNA of ~ 30–80 bp in length was sufficient to span both of the RNA binding sites on the enzyme and form a bivalent interaction which resulted in a conformational change necessary for activation. On the other hand, it was observed that low concentrations of dsRNAs of ~ 11–30 bp in length formed unstable complexes and failed to activate dsI. At high concentrations, however, these short duplexes blocked the activation of dsI by longer dsRNAs, presumably by interference with one or both of the dsRNA binding sites. The authors also postulated that long dsRNAs, which activated dsI at low concentrations, prevented activation at high concentrations because the two binding sites form complexes with separate RNA duplexes which prevented the conformational change needed for activation. Clearly the interaction of dsRNA with dsI is complex and further investigation will be necessary to delineate the nature of the binding between dsRNA and activation of dsI.

Another question concerning the activation of dsI is whether the autophosphorylation occurs via an intra- or intermolecular mechanism. Both mechanisms have been proposed, with evidence to support each conclusion (Galabru et al. 1989; Kostura and Mathews 1989). The reason for this seemingly contradictory finding is unknown. It may be due to the vastly different procedures used to obtain the enzyme in each study, and it is possible that both intra- and intermolecular phosphorylation may occur depending on the preparation used. For example, Langland and Jacobs (1992) have identified two forms of dsI: a latent unphosphorylated monomeric form, and a partially phosphorylated soluble form consisting of a dsI:dsI dimer of 140–160 kDa. The dimer form contains 2–3 mol PO_4/mol monomer of dsI and this level could be increased by incubation with [γ-^{32}P] ATP. Moreover, the dimer could be converted to monomer form by treatment with alkaline phosphatase, suggesting that phosphorylation was necessary for dimer formation. Whether intra- or intermolecular autophosphorylation of the dimer, remains to be determined.

However, it is important to note that in a recent study it was shown that co-expression of wild-type and a nonfunctional mutant form of the human kinase resulted in a marked reduction in the autophosphorylation of the wild-type enzyme (Koromilas et al. 1992). It was proposed that the dominant negative effect of the mutant protein was due to its association with the wild-type enzyme which prevented its autophosphorylation and activation (Koromilas et al. 1992). These findings provide strong evidence that an intermolecular autophosphorylation mechanism is critical for the activation of dsI as was originally proposed (Kostura and Mathews 1989).

3 Regulation of the dsRNA-Dependent Kinase

The role of dsI in mediating the antiviral activity of interferon has been well documented (Nilsen et al. 1982; Pestka et al. 1987; Schneider and Shenk 1987; Hovanessian 1989; Mathews and Shenk 1991; Samuel 1991, 1992; Meurs et al. 1992). Its importance in this regard is further demonstrated by the observations that a number of viruses have developed specialized molecules that inhibit dsI and circumvent its antiviral activity (Schneider and Shenk 1987). These include viral RNAs (Kitajewski et al. 1986; Clarke et al. 1990, 1991; Ghadge et al. 1991; Mathews and Shenk 1991) and viral-associated proteins (Whitaker-Dowling and Youngner 1984; Imani and Jacobs 1988; Watson et al. 1991; Davis et al. 1992) which prevent the binding of activating dsRNA to dsI, and other activities which cause its degradation (Hovanessian et al. 1987; Black et al. 1989; Roy et al. 1990). For brevity, the source and nature of these inhibitory molecules have been listed in Table 1. In addition to these viral inhibitors, there have been several reports of cellular inhibitors of the kinase. The best characterized of these is the dsI-regulatory factor (dRF) which has been purified from mouse 3T3-F442A

Table 1. Summary of viral and cellular factors which regulate dsI

Inhibitor		Source	Nature		Reference
Viral					
	VAI	Adenovirus	RNA		Kitajewski et al. (1986) Mathews and Shenk (1991) Ghadge et al. (1991)
	EBER-1	Epstein-Barr virus	RNA		Clarke et al. (1990, 1991)
	SKIF/E3L	Vaccinia virus	Protein-25	kDa	Whitaker-Dowling and Youngner (1984) Watson et al. (1991)
	K3L	Vaccinia virus	Protein-11	kDa	Davis et al. (1992)
	$\sigma 3$	Reovirus	Protein-?	kDa	Imani and Jacobs (1988)
Cellular					
	dRF	Mouse 3T3-F442A cells	Protein-15	kDa	Judware and Petryshyn (1991, 1992)
	p58	Bovine MDBK cells	Protein-58	kDa	Lee et al. (1990, 1992)
	p160	Human FL cells	Protein-160	kDa	Saito and Kawakita (1991)
		Rabbit reticulocyte	Protein-?	kDa	Petryshyn et al. (1982) Szyszka et al. (1989)
phosphoprotein phosphatases		p21v-ras transformed-BALB/c-3T3 fibroblasts	Protein-?	kDa	Mundschan and Faller (1992)

cells (Judware and Petryshyn 1991, 1992), where its activity is markedly elevated in cells cultured under conditions which suppress differentiation (Judware and Petryshyn 1991). dRF is a 15-kDa protein, which does not bind or degrade dsRNA, nor does it function to dephosphorylate or degrade dsI (Judware and Petryshyn 1991, 1992). It does appear to inhibit the activation by preventing the binding of dsRNA to dsI through a direct effect on the enzyme (Judware and Petryshyn 1992). Other cellular inhibitors of dsI have been isolated from bovine (Lee et al. 1990, 1992) and human cells (Saito and Kawakita 1991). The exact mechanism of action of the latter two inhibitors awaits further characterization, but they appear to be distinct from dRF. It is also likely that dsI may be regulated by cellular protein phosphatases. There have been a few reports describing putative dsI-specific phosphatases (Petryshyn et al. 1982; Szyszka et al. 1989; Mundschan and Faller 1992) but their full characterization and physiological significance remains to be determined.

The role of cellular inhibitors is less obvious than those of the virus-associated inhibitors. Their existence, however, suggests that dsI has a physiological function in noninfected cells and that its activity must be precisely modulated.

4 Biological Significance of the dsRNA-Dependent Kinase

Biochemical and molecular studies have shown that enzymatic activities possessing the characteristics of dsI are highly conserved in a number of species and animal cells. Protein kinases which phosphorylate eIF-2α have been characterized in mammalian cells (Petryshyn et al. 1983, Kostura and Mathews 1989; Samuel 1991), in yeast (Chen et al. 1991; Chong et al. 1992; Devers et al. 1992), and in a plant tissue (Crum et al. 1988). The molecular cloning of the human (Meurs et al. 1990) and murine (Feng et al. 1992) cDNA for dsI has provided further support for a universal role of this and other dsI-like kinases in regulating the initiation of protein synthesis and for a role of dsI in mediating the antiviral action of interferon (Meurs et al. 1992). For example, it has been shown that the human sequence is 61% identical in its overall nucleotide sequence with the murine cDNA (Feng et al. 1992) and 38% identical in the catalytic domain with the yeast GCN2 eIF-2α kinase (Chen et al. 1991; Feng et al. 1992). Moreover, the human sequence has 42% nucleotide sequence identity with the cDNA sequence of the heme-regulated eIF-2α kinase from rabbit reticulocytes (Chen et al. 1991). The availability of cDNA clones for dsI is now beginning to provide detailed information about the functionally important domains of this kinase and how it is regulated (Barber et al. 1991; Katze et al. 1991; Feng et al. 1992; Koromilas et al. 1992; McCormack et al. 1992; Meurs et al. 1992; Patel and Sen 1992; Thomis et al. 1992).

While it is well established that dsI regulates the initiation of protein synthesis by phosphorylating eIF-2α, the full physiological consequence of this regulation on cellular processes remains to be elucidated. For example, it is possible that activated dsI may lead to a complete inhibition of protein synthesis, while at other times it may contribute to a selective effect on translation of specific mRNAs (DeBenedetti and Baglioni 1984; Kaufman et al. 1989; Mathews and Shenk 1991). These effects are likely to depend on the amount of dsI available, its cellular localization, the structure and amount of the activating RNA available, and on other cellular factors which regulate its activation and activity. It is of interest that in yeast, under conditions of amino acid starvation, eIF-2α becomes phosphorylated by the dsI-like protein kinase GCN2 (Devers et al. 1992). This limited phosphorylation is associated with an increased frequency of ribosomal initiation at a downstream AUG codon specific for the GCN2 open reading frame and results in synthesis of the GCN4 protein. The GCN4 protein is responsible for the transcriptional activation of genes for amino acid biosynthesis (Devers et al. 1992). Whether such a mechanism for selection of protein initiation start sites by phosphorylation of eIF-2α occurs in other cells in response to depletion of amino acids or other nutritional components remains to be determined. It is also possible that limited phosphorylation of eIF-2α by dsI may preferentially limit the availability of short-lived proteins which may function as positive or negative regulators of translation or transcription and thus modify the fundamental behavior of cells. This possibility remains to be examined.

Regardless of its exact mechanism of action, the kinase has been implicated in a number of cellular events in addition to its antiviral action. These include regulation of growth and differentiation of 3T3-F442A preadipocyte fibroblasts where autocrine interferon induces dsI in a transient manner specific to stages of growth and adipose differentiation (Petryshyn et al. 1984, 1988). In these cells, it has also been shown that dsI becomes activated in response to a specific cellular RNA which has been partially cloned and characterized (Li and Petryshyn 1991 J. Li and R. Petryshyn unpubl. observ. 1992;). Moreover, under culture conditions which suppress adipose differentiation, the activation of dsI becomes inhibited by a cellular protein which specifically prevents the binding of the activating RNA to dsI (Judware and Petryshyn 1991, 1992). Since cessation of cell growth is normally required for subsequent differentiation, dsI may have a critical function in limiting normal cell growth. Recent studies using other biological systems support such a role for dsI. Birnbaum et al. (1990) have reported that dsI is transiently expressed in rat myoblasts and is elevated at a time immediately preceding cessation of growth and myotube formation. Additional studies using the cloned cDNA for human dsI have led to two important observations: First, the expression of the wild-type but not a nonfunctional mutant form of the gene in yeast resulted in a phenotype exhibiting growth suppression which correlated with an active kinase and phosphorylation of eIF-2α (Feng et al. 1992). Second, Koromilas et al. (1992) have shown that transformation and overexpression of nonfunctional mutant forms of human kinase in NIH-3T3 cells gave rise to transfectants that were characterized by increased proliferation rates. These transfectants also developed tumor cell morphology, exhibited sustained growth on soft agar plates, and developed tumors at an accelerated rate in nude mice. These observations are significant because they indicate that dsI may normally function as an inhibitor of cell proliferation and as a tumor suppressor. Thus, it is possible that mutations leading to expression of a nonfunctional enzyme, or to an altered expression or function of other cellular components involved in regulating dsI, may contribute to the development of tumor cells.

dsI has also been implicated, albeit indirectly, as a component of the signal transduction pathways mediated by several biological effectors including β- and γ-interferon (Zinn et al. 1988; Bandyopadhyay and Sen 1992), NF-κB (Ghosh and Baltimore 1990), and platelet-derived growth factor (PDGF) (Zullo et al. 1985). Recent studies have pointed to a direct involvement of dsI in the PDGF signal transduction pathway. PGDF specifically induces transcription of a number of growth-related genes and triggers a number of intracellular activities which are blocked in cells expressing v-ras genes (Quiñones et al. 1991). Mundschau and Faller (1992) have shown that in BALB/c-3T3 fibroblasts, oncoprotein p21v-ras inhibited dsI activity by inducing or activating a cytoplasmic protein with dsI-specific phosphatase activity. This inhibition was specific to p21v-ras since it was not seen in cells transformed by the oncogenes v-src, v-mos, or v-abl. dsI activity was restored in these cells by elevation of intracellular cyclic AMP levels or by expression of the Krev-1 gene, both of which specifically reversed transformation by p21v-ras. These findings raise the possibility that dsI may be an important regulator of second messenger systems.

In addition, dsI has been reported to be involved in the heat shock response, where its activity has been reported to be reduced as a consequence of denaturation (Dubois et al. 1991). It has been postulated that inactivation of the kinase during heat shock may be important to the cell in order to recover from the inhibition of protein synthesis (Dubois et al. 1991).

5 Conclusion and Prospective

Substantial advances have recently been made in understanding the molecular biology and biochemistry of dsI, its effect on the initiation of protein synthesis, and how this is manifested during viral infection. However, the exact consequences of the activity of dsI on cellular processes in non-virally infected cells, either as a direct result of altered protein translation or otherwise, remain to be fully elucidated. Moreover, at present, relatively little is known as to the existence and the spectrum of both agonists and antagonists of this important enzyme in noninfected cells. Questions remain as to how dsI becomes activated in normal cells. For example: What is the nature and scope of the cellular RNAs and other cellular factors involved in regulating its activation and activity under physiological conditions? Are these cellular components altered during tumorigenesis or other diseased states? What is the role of protein phosphatases in regulating dsI? Do additional substrates exist for dsI other than eIF-2, which may be involved in the signal transduction pathway of interferon or other inducers?

Beyond its role in mediating the antiviral effects of interferon, dsI appears to have additional important effects on those cellular processes modulated by interferon. Studies to address these and other questions may further advance our understanding of the physiological role of dsI under normal and pathological conditions.

Acknowledgments. Research in the authors' laboratory is supported in part by grant number 91-R-3 from the Children's Miracle Network Telethon and grant number 001278-9-RG from the American Foundation for AIDS Research.

References

Bandyopadhyay SK, Sen GC (1992) Role of protein phosphorylation in activation of interferon-stimulated gene factors. J Biol Chem 267:6389–6395

Barber GN, Tomita J, Hovanessian AG, Meurs E, Katze MG (1991) Functional expression and characterization of the interferon-induced double-stranded RNA activated P68 kinase from *Escherichia coli*. Biochemistry 30:10356–10361

Birnbaum M, Trink B, Shainberg A, Salzberg S (1990) Activation of the interferon system during myogenesis in vitro. Differentiation 45:138–145

Bischoff JR, Samuel CE (1985) Mechanism of interferon action: the interferon-induced phosphoprotein P_1 possesses a double-stranded RNA-dependent ATP-binding site. J Biol Chem 260:8237–8239

Bischoff JR, Samuel CE (1989) Mechanism of interferon action. Activation of the human PI/eIF-2α protein kinase by individual reovirus s-class mRNAs: SI mRNA is a potent activator relative to S4 mRNA. Virology 172:106–115

Black TL, Safer B, Hovanessian A, Katze MG (1989) The cellular 68,000-M_r protein kinase is highly autophosphorylated and activated yet significantly degraded during poliovirus infection: implications for translational regulation. J Virol 63:2244–2251

Chen J-J, Throop MS, Gehrke L, Kuo I, Pal JK, Brodsky M, London IM (1991) Cloning of the cDNA of the heme-regulated eukaryotic initiation factor 2α (eIF-2α) kinase of rabbit reticulocytes: homology to yeast GCN2 protein kinase and human double-stranded RNA-dependent eIF-2α kinase. Proc Natl Acad Sci USA 88:7729–7733

Chong KL, Feng L, Schappert K, Meurs E, Donahue TF, Friesen JD, Hovanessian AG, Williams BG (1992) Human p68 kinase exhibits growth suppression in yeast and homology to the translational regulator GCN2. EMBO J 11:1553–1562

Clarke PA, Sharp NA, Clemens MJ (1990) Translational control by the Epstein-Barr virus small RNA EBER-1: reversal of the double-stranded RNA-induced inhibition of protein synthesis in reticulocyte lysates. Eur J Biochem 193:635–641

Clarke PA, Schwemmle M, Schickinger J, Hilse K, Clemens MJ (1991) Binding of Epstein-Barr virus small RNA EBER-1 to the double-stranded RNA-activated protein kinase DAI. Nucleic Acids Res 19:243–248

Crum JC, Hu J, Hiddinga HJ, Roth DA (1988) Tobacco mosaic virus infection stimulates the phosphorylation of a plant protein associated with double-stranded-RNA-dependent protein kinase activity. J Biol Chem 263:13440–13442

Davis M, Elroy-Stein O, Jagus R, Moss B, Kaufman RJ (1992) The vaccinia virus K3L gene product potentiates translation by inhibiting double-stranded-RNA-activated protein kinase and phosphorylation of the alpha subunit of eukaryotic initiation factor 2. J Virol 66:1943–1950

DeBenedetti A, Baglioni C (1984) Inhibition of mRNA binding to ribosomes by localized activation of dsRNA-dependent protein kinase. Nature 311:79–81

Devers TE, Feng L, Wek RC, Cigan MA, Donahue TF, Hinnebusch AG (1992) Phosphorylation of initiation factor 2α by protein kinase GCN2 mediates gene-specific translational control of GCN4 in yeast. Cell 68:585–596

Dubois MF, Hovanessian AG, Bensaude O (1991) Heat-shock-induced denaturation of proteins. J Biol Chem 266:9707–9711

Edery I, Petryshyn R, Sonenberg N (1989) Activation of double-stranded RNA-dependent kinase (dsI) by the TAR region of HIV-1 mRNA: a novel translational control mechanism. Cell 56:303–312

Farrell PJ, Balkow K, Hunt T, Jackson RJ, Trachsel H (1977) Phosphorylation of initiation factor eIF-2 and control of reticulocyte protein synthesis. Cell 11:187–200

Feng GS, Chong K, Kumar A, Williams BRG (1992) Identification of double-stranded RNA-binding domains in the interferon-induced double stranded RNA-activated p68 kinase. Proc Natl Acad Sci USA 89:5447–5451

Galabru J, Hovanessian AG (1985) Two interferon-induced proteins are involved in the protein kinase complex dependent on double-stranded RNA. Cell 43:685–694

Galabru J, Hovanessian A (1987) Autophosphorylation of the protein kinase dependent upon double-stranded RNA. J Biol Chem 262:15538–15544

Galabru J, Katze MG, Robert N, Hovanessian AG (1989) The binding of double-stranded RNA and adenovirus VAI RNA to the interferon-induced protein kinase. Eur J Biochem 178:581–589

Ghadge GD, Swaminathan S, Katze MG, Thimmappaya B (1991) Binding of the adenovirus VAI RNA to the interferon-induced 68-kDA protein kinase correlates with function. Proc Natl Acad Sci USA 88:7140–7144

Ghosh S, Baltimore D (1990) Activation in vitro of NF-κB by phosphorylation of its inhibitor IκB. Nature 344:678–682

Gunnery S, Rice AP, Robertson HD, Mathews MB (1990) Tat-responsive region RNA of human immunodeficiency virus I can prevent activation of the double-stranded RNA-activated protein kinase. Proc Natl Acad Sci USA 87:8687–8691

Hershey JWB (1991) Translational control in mammalian cells. Annu Rev Biochem 60:717–755

Hovanessian AG (1989) The double-stranded RNA-activated protein kinase induced by interferon: dsRNA-PK. J Interferon Res 9:641–647

Hovanessian AG, Galabru J, Meurs E, Buffet-Janvresse C, Svab J, Robert N (1987) Rapid decreases in the levels of the double-stranded RNA-dependent protein kinase during virus infections. Virology 159:126–136

Hunter T, Hunt T, Jackson RJ, Robertson HD (1975) The characteristics of inhibition of protein synthesis by double-stranded ribonucleic acid in reticulocyte lysates. J Biol Chem 250:409–417

Imani F, Jacobs BL (1988) Inhibitory activity for the interferon-induced protein kinase is associated with the reovirus serotype 1σ3 protein. Proc Natl Acad Sci USA 85:7887–7891

Judware R, Petryshyn R (1991) Partial characterization of a cellular factor that regulates the double-stranded RNA-dependent eIF-2α kinase in 3T3-F442A fibroblasts. Mol Cell Biol 11:3259–3267

Judware R, Petryshyn R (1992) Mechanism of action of a cellular inhibitor of the dsRNA-dependent protein kinase from 3T3-F442A cells. J Biol Chem 267:21685–21690

Judware R, Li J, Petryshyn R (1993) Inhibition of the dsRNA-dependent protein kinase by a peptide derived from the human immunodeficiency virus type I tat protein. J Interferon Res 13:153–160

Katze MG, Wambach M, Wong M-L, Garfinkel M, Meurs E, Chong K, Williams BRG, Hovanessian AG, Barber GN (1991) Functional expression and RNA binding analysis of the interferon-induced, double-stranded RNA-activated, 68,000-M_r protein kinase in a cell-free system. Mol Cell Biol 11:5497–5505

Kaufman RJ, Davies MV, Pathak VK, Hershey JWB (1989) The phosphorylation state of eukaryotic initiation factor 2 alters translational efficiency of specific mRNAs. Mol Cell Biol 9:946–958

Kitajewski J, Schneider RJ, Safer B, Munemitsu SM, Samuel CE, Thimmappaya B, Shenk T (1986) Adenovirus VAI RNA antagonizes the antiviral action of interferon by preventing activation of the interferon-induced eIF-2α kinase. Cell 45:195–200

Konieczny A, Safer B (1983) Purification of the eukaryotic initiation factor 2-eukaryotic initiation factor 2B complex and characterization of its guanine nucleotide exchange activity during protein synthesis initiation. J Biol Chem 258:3402–3408

Koromilas AE, Roy S, Barber GN, Katze MG, Sonenberg N (1992) Malignant transformation by a mutant of the IFN-inducible dsRNA-dependent protein kinase. Science 257:1685–1689

Kostura M, Mathews MG (1989) Purification and activation of the double-stranded RNA-dependent eIF-2 kinase DAI. Mol Cell Biol 9:1576–1586

Langland JO, Jacobs BL (1992) Cytosolic double-stranded RNA-dependent protein kinase is likely a dimer of partially phosphorylated M_r = 66,000 subunits. J Biol Chem 267:10729–10736

Lee TG, Tomita J, Hovanessian AG, Katze MG (1990) Purification and partial characterization of a cellular inhibitor of the interferon-induced protein kinase of M_r 68,000 from influenza virus-infected cells. Proc Natl Acad Sci USA 87:6208–6212

Lee TG, Tomita J, Hovanessian AG, Katze MG (1992) Characterization and regulation of the 58,000-dalton cellular inhibitor of the interferon-induced, dsRNA-activated protein kinase. J Biol Chem 267:14238–14243

Leroux A, London IM (1982) Regulation of protein synthesis by phosphorylation of eukaryotic initiation factor 2α in intact reticulocytes and reticulocyte lysates. Proc Natl Acad Sci USA 79:2147–2151

Levin D, London IM (1978) Regulation of protein synthesis: activation by double-stranded RNA of a protein kinase that phosphorylates eukaryotic initiation factor 2. Proc Natl Acad Sci USA 75:1121–1125

Levin DH, Petryshyn R, London IM (1981) Characterization of purified double-stranded RNA-activated eIF-2α kinase from rabbit reticulocytes. J Biol Chem 256:7638–7641

Li J, Petryshyn R (1991) Activation of the double-stranded RNA-dependent eIF-2α kinase by a cellular RNA from 3T3-F442A cells. Eur J Biochem 195:41–48

London IM, Levin DH, Matts RL, Thomas NSB, Petryshyn R, Chen J-J (1987) Regulation of protein synthesis. In: Boyer PD, Krebs EG (ed) The enzymes, vol 18. Academic Press, New York, pp 359–380

Manche L, Green SR, Schmedt C, Mathews MB (1992) Interaction between double-stranded RNA regulators and the protein kinase DAI. Mol Cell Biol 12:5238–5248

Mathews MB, Shenk T (1991) Adenovirus virus-associated RNA and translational control. J Virol 65:5657–5662

Matts RL, London IM (1984) The regulation of initiation of protein synthesis by phosphorylation of eIF-2(α) and the role of reversing factor in the recycling of eIF-2. J Biol Chem 259:6708–6711

Matts RL, Levin DH, London IM (1983) Effect of phosphorylation of the α-subunit of eukaryotic initiation 2 on the function of reversing factor in the initiation of protein synthesis. Proc Natl Acad Sci USA 80:2559–2563

McCormack SJ, Thomis DC, Samuel CE (1992) Mechanism of interferon action: identification of a RNA binding domain within the N-terminal region of the human RNA-dependent P1/eIF-2α protein kinase. Virology 188:47–56

Meurs E, Chong K, Galabru J, Thomas NSB, Kerr IM, Williams BRG, Hovanessian AG (1990) Molecular cloning and characterization of the human double-stranded RNA-activated protein kinase induced by interferon. Cell 62:379–390

Meurs EF, Watanabe Y, Kadereit S, Barber GN, Katze MG, Chong K, Williams BRG, Hovanessian AG (1992) Constitutive expression of human double-stranded RNA-activated p68 kinase in murine cells mediates phosphorylation of eukaryotic initiation factor 2 and partial resistance to encephalomyocarditis virus growth. J Virol 66:5805–5814

Mundschau LJ, Faller DV (1992) Oncogenic ras induces an inhibitor of double-stranded RNA-dependent eukaryotic initiation factor 2α-kinase activation. J Biol Chem 267:23092–23098

Nilsen TW, Maroney PA, Baglioni C (1982) Inhibition of protein synthesis in reovirus-infected HeLa cells with elevated levels of interferon-induced protein kinase activity. J Biol Chem 257:14593–14596

Patel RC, Sen GC (1992) Identification of the double-stranded RNA-binding domain of the human interferon-inducible protein kinase. J Biol Chem 267:7671–7676

Pathak VK, Schindler D, Hershey JWB (1988) Generation of a mutant form of protein synthesis initiation factor eIF-2 lacking the site of phosphorylation by eIF-2 kinases. Mol Cell Biol 8:993–995

Pestka S, Langer JA, Zoon KC, Samuel CE (1987) Interferons and their actions. Annu Rev Biochem 56:727–777

Petryshyn R, Levin DH, London IM (1982) Regulation of double-stranded RNA-activated eukaryotic initiation factor 2α kinase by type 2 protein phosphatase in reticulocyte lysates. Proc Natl Acad Sci USA 79:6512–6516

Petryshyn R, Levin DH, London IM (1983) Purification of the double-stranded RNA-dependent protein kinase. Methods Enzymol 99:346–362

Petryshyn R, Chen J-J, London IM (1984) Growth related expression of a dsRNA-dependent eIF-2α kinase in 3T3 cells. J Biol Chem 259:14736–14742

Petryshyn R, Chen J-J, London IM (1988) Detection of activated double-stranded RNA-dependent protein kinase in 3T3-F442A cells. Proc Natl Acad Sci USA 85:1427–1431

Price NT, Welsh GI, Proud CG (1991) Phosphorylation of only serine-51 in protein synthesis initiation factor-2 is associated with inhibition of peptide-chain initiation in reticulocyte lysates. Biochem Biophys Res Commun 176:993–999

Quiñones MA, Mundschau LJ, Rake JB, Faller DV (1991) Dissociation of platelet-derived growth factor (PDGF) receptor autophosphorylation from other PDGF-mediated second messenger events. J Biol Chem 266:14055–14063

Rowlands AG, Panniers R, Henshaw EC (1988) The catalytic mechanism of guanine nucleotide exchange factor action and competitive inhibition by phosphorylation eukaryotic initiation factor 2. J Biol Chem 263:5526–5553

Roy R, Katze MG, Parkin NT, Edery I, Hovanessian AG, Sonenberg N (1990) Control of the interferon-induced 68-kilodalton protein kinase by the HIV-1 tat gene product. Science 247:1216–1219

Roy S, Agy M, Hovanessian AG, Sonenberg N, Katze M (1991) The integrity of the stem structure of human immunodeficiency virus type 1 tat-responsive sequence RNA is required for interaction with the interferon-induced 68,000-M_r protein kinase. J Virol 65:632–640

Saito S, Kawakita M (1991) Inhibitor of interferon-induced double-stranded RNA-dependent protein kinase and its relevance to alteration of cellular protein kinase activity level in response to external stimuli. Microbiol Immunol 35:1105–1114

Samuel CE (1979) Mechanism of interferon action: phosphorylation of protein synthesis initiation factor eIF-2 in interferon-treated human cells by a ribosome-associated kinase possessing site specificity similar to hemin-regulated rabbit reticulocyte kinase. Proc Natl Acad Sci USA 76:600–604

Samuel CE (1991) Antiviral actions of interferon: interferon regulated cellular proteins and their surprisingly selective antiviral activities. Virology 83:1–11

Samuel CE (1992) RNA-dependent P1/eIF-2α protein kinase. In: Baron S, Coppenhaver DH, Dianzani F, Fleischmann WR, Hughes TK, Klimpel GR, Niesel DW, Stanton GJ, Tyring SK (eds) Interferon principles and medical application, 1st edn. UTMB, Galveston, pp 237–249

Schneider RJ, Shenk T (1987) Impact of virus infection on host cell protein synthesis. Annu Rev Biochem 56:317–332

Schröder HC, Ugarkovic'D, Wenger R, Reuter P, Okamoto T, Müller WEG (1990) Binding of tat protein to TAR region of human immunodeficiency virus type I blocks TAR-mediated activation of (2′–5′) oligoadenylate synthetase. AIDs Res Hum Retrovir 6:659–672

Sen GS, Lengyel P (1992) The interferon system. J Biol Chem 267:5017–5020

SenGupta DN, Silverman RH (1989) Activation of interferon-regulated, dsRNA-dependent enzymes by human immunodeficiency virus-1 leader RNA. Nucleic Acids Res 17:969–978

Siekierka J, Manne V, Ochoa S (1984) Mechanism of translational control by partial phosphorylation of the α subunit of eukaryotic initiation factor 2. Proc Natl Acad Sci USA 81:352–356

Szyszka R, Kudlicki W, Kramer G, Hardesty B, Galabru J, Hovanessian AG (1989) A type I phosphoprotein phosphatase active with phosphorylated $M_r = 68{,}000$ initiation factor 2 kinase. J Biol Chem 264:3827–3831

Thomis DC, Doohan JP, Samuel CE (1992) Mechanism of interferon action: cDNA structure, expression, and regulation of the interferon-induced, RNA-dependent P1/eIF-2α protein kinase from human cells. Virology 188:33–46

Watson JC, Chang HW, Jacobs BL (1991) Characterization of a vaccinia virus-encoded double-stranded RNA-binding protein that may be involved in inhibition of a double-stranded RNA-dependent protein kinase. Virology 185:206–216

Whitaker-Dowling P, Youngner JS (1984) Characterization of a specific kinase inhibitory factor produced by vaccinia virus which inhibits the interferon-induced protein kinase. Virology 137:171–181

Zinn K, Keller A, Whittemore L-A, Maniatis T (1988) 2-Aminopurine selectively inhibits the induction of β-interferon, c-fos and c-myc gene expression. Science 240:210–213

Zullo JN, Cochran BH, Huang AS, Stiles CD (1985) Platelet-derived growth factor and double-stranded ribonucleic acids stimulate expression of the same genes in 3T3 cells. Cell 43:793–800

Double-Stranded RNAs as Gene Activators

S. Landolfo[1,2], M. Gariglio[1], G. Gribaudo[1], and G. Garotta[3]

1 Introduction

In 1963 Gomatos and Tamm discovered the first natural dsRNA, namely the genomic RNA of reoviruses. A few years later, Lampson et al. (1967) established that dsRNA is the active agent in an extract of a *Penicillium funiculosum* preparation which, following injection into rabbits, induces interferon (IFN) in serum. They also showed that various natural (i.e. reovirus genomic RNA) or synthetic (i.e. poly rI:rC) dsRNAs induce IFN-α/β in animals and cultured animal cells (Field et al. 1967). These findings support the hypothesis that virus-infected cells stimulate IFN synthesis of surrounding cells by releasing some dsRNA formed during viral replication. Although dsRNAs induce IFN synthesis, several observations show that immunomodulatory and inflammatory activities of dsRNA are achieved through activation of a defined set of genes (O'Malley et al. 1979; Stewart 1979; De Maeyer and De Maeyer-Guignard 1988; Haines et al. 1991). As a consequence, cells synthesize new proteins that change their growth capacity and modulate their functions.

2 Double-Stranded RNA as Gene Inducers

Many synthetic dsRNAs have been studied for their capacity to activate IFN-α/β synthesis and induce the antiviral state (De Maeyer and De Maeyer-Guignard 1988). Using a transcription assay, Raji and Pitha (1983) demonstrated that the synthesis of IFN-β-specific RNA in nuclei becomes detectable after induction of a human fibroblast cell line by poly rI:rC. Recently, dsRNA has been proven to stimulate the IFN-γ production by those lymphocytes that are not retained by a nylon wool column (Tamura-Nishimura and Sasakawa 1989).

All the active dsRNAs share common properties: (1) a stable secondary structure (poly rI:rC is a better IFN inducer than poly AU); (2) relatively strong resistance to ribonucleases; (3) high molecular weight, the minimum effective

[1]Institute of Microbiology, Medical School, University of Torino, Italy
[2]Immunogenetics and Histocompatibility Center, CNR, Torino, Italy
[3]Hoffmann-La Roche, Inc., Basle, Switzerland

Progress in Molecular and Subcellular Biology, Vol. 14
W.E.G. Müller/H.C. Schröder (Eds.)
© Springer-Verlag Berlin Heidelberg 1994

chain length being about 100 residues. In addition, the ribonuclease-resistant thiophosphate derivative of poly AU is a much more effective IFN inducer than poly AU, and the activity of poly IC is greatly enhanced when it is complexed with polycations such as DEAE-dextran, methylated albumin, protamine or histone (Field et al. 1970; Dianzani et al. 1971; Lampson et al. 1981).

Exposure of macrophages to poly rI:rC is accompanied by massive stimulation of the synthesis of the C-components, factors B and C3, and concomitant inhibition of the synthesis of the lysosomal enzyme β-glucuronidase (Riches et al. 1988). Moreover, treatment of different animal cell types, namely fibroblasts, macrophages and endothelial cells, induces the simultaneous production of cytokines, such as IL-1, IL-6 and various colony stimulating factors (e.g. GM-CSF, G-CSF, M-CSF) (Fibbe et al. 1988; Souvannavong and Adam 1990). As far as IFNs stimulate macrophage and endothelial cell release of IL-1, IL-6 and CSF, the production of these cytokines could be secondary to the secreted IFN (Landolfo and Garotta 1991; Table 1).

Zullo et al. (1985) demonstrated that a set of between 10 to 30 genes is rapidly induced when BALB/c-3T3 cells are exposed to PDGF or poly rI:rC. These genes, defined as "competence" genes, comprise some well-characterized genes promoting cell division, such as c-myc or c-fos, others that function as feedback inhibitors of mitogenic response, such as the (2′,5′)-oligoadenylate synthetase, and finally others still functionally unknown, such as JE, KC and a fos-related gene termed r-fos. Directly related to this finding, Vilcek et al. (1987) demonstrated that in human diploid foreskin fibroblasts poly rI:rC does have mitogenic activity, and that this can be amplified either by the addition of antiserum to neutralize IFN-β, or by inhibiting its synthesis with dexamethasone.

In addition to the induction of nuclear proto-oncogenes, dsRNA affects the expression of membrane-linked proto-oncogenes, namely ras genes (Maran et al. 1990). Addition of poly rI:rC to NIH-3T3 cells, indeed, increases the RNA levels of c-Ha-ras. The dsRNA configuration is required to increase mRNA levels, since heat-denaturated dsRNA and ssRNA had no effect.

Although the majority of the studies on gene induction by natural or synthetic dsRNA were conducted in vitro, few studies have been made on the ability of dsRNA to increase the expression of various genes in vivo (Flenniken et al. 1988; Gariglio et al. 1991, 1992; Terao et al. 1992). The injection of dsRNA is always followed by a rapid appearance of significant IFN-α/β serum titers. Moreover, a few hours after injection, poly rI:rC induced the appearance in the spleen and bone marrow of mRNAs coded for by the 202, (2′,5′)oligoadenylate synthetase (2-5 OAse) and class I histocompatibility antigen genes (Gariglio et al. 1991, 1992). The 202 encodes a cytoplasmic protein of unknown function, which translocates into the nucleolus upon IFN or dsRNA treatment (Choubey and Lengyel 1992), whereas the 2-5 OAse encodes an enzyme system involved in the antiviral action of IFNs (Stewart 1979). Similar observations have been made by Terao et al. (1992): injection of poly rI:rC in CD-1 mice induces in several tissues the appearance of the mRNA coding for the xanthine oxidoreductase system,

Table 1. dsRNA-induced genes and encoded protein functions

Designation	Encoded protein	Biological function recognized[a]	Reference
IFN-α/β	Cytokine	AV, AP, IM, CD	De Maeyer and De Maeyer-Guignard (1988)
IFN-γ	Cytokine	AV, AP, IM, CD	De Maeyer and De Mayer Guignard (1988)
IL-1	Cytokine	IM, CD	Souvannavong and Adam (1990)
IL-6	Cytokine	AV, AP, IM, CD	Milhaud et al. (1987)
TNF	Cytokine	AV, AP, IM, CD	North et al. (1991)
CSF (G-, GM-, M-)	Cytokines	Hematopoiesis regulation	Fibbe et al. (1988)
Factor B	C-component	C cascade, IM	Riches et al. (1988)
C3	C-component	C cascade, IM	Riches et al. (1988)
Competence genes	DNA-binding, proteins, enzymes, unknown proteins	Cell growth regulation	Zullo et al. (1985)
MHC	Histocompatibility antigens	Self-recognition, IM	Gariglio et al. (1991)
2-5-OAse	Synthetase	AV, AP	Gariglio et al. (1991)
202	Mouse cytoplasmic protein (56 kDa)	None	Gariglio et al. (1991)
Xanthine oxidase	Oxidoreductase	Uric acid synthesis	Terao et al. (1992)
IFI-56K	Human cytoplasmic protein (56 kDa)	None	Wathelet et al. (1988)
IFI-54K	Human cytoplasmic protein (54 kDa)	None	Wathelet et al. (1988)
IRF-1	DNA-binding protein	IFN gene regulation	Miyamoto et al. (1988)

[a]AV: antiviral activity; AP: antiproliferative activity; IM: immunomodulatory activity; CD: cell differentiation activity.

which catalyzes the oxidation of hypoxanthine to xanthine and subsequently to uric acid. Eventually, the injection of dsRNA in vivo caused the regression of a murine sarcoma (North et al. 1991), apparently mediated by the induction of a cytokine, namely tumor necrosis factor (TNF), and concomitant activation of cytotoxic T-cells.

Many of the dsRNA-inducible genes are directly induced by IFNs. As far as dsRNAs are potent IFN inducers (Stevenson et al. 1985), their ability to activate target genes could be mediated by IFNs. Support for this hypothesis comes from the finding that addition of anti-IFN-α/β antibodies inhibit gene induction as well as biological action of RNA copolymers (Vilcek et al. 1987; Maran et al. 1990; Yoshida and Marcus 1990; Gariglio et al. 1991; Pyo et al. 1991). In a few cases, however, IFN involvement could be totally excluded. Wathelet et al. (1988) observed, indeed, that induction of IFI-56K and IFI-54K genes by poly rI:rC was not impaired in the presence of IFN-neutralizing antibodies. In human GM2767 cells, 2-aminopurine inhibited induction of mRNAs 561 and 6-16 by dsRNA but not by IFN-α/β, suggesting that there are remarkable differences in the two processes (Tiwari et al. 1988). Eventually, Memet et al. (1991), in addition to the finding that gene induction occurred in fibroblasts in spite of the presence of cycloheximide and/or anti-IFN-α/β antibodies, demonstrated that, in IFN-resistant Daudi cells, ISG15 and IP-10 genes were still inducible by dsRNA.

Taken as a whole, these results suggest that dsRNAs can exploit two pathways of gene induction, the first one activated upon their direct interaction with target cells, the second one mediated by the induction of cytokines such as IFNs that in turn activate target genes.

3 Interaction of dsRNA with Target Cells

The molecular events leading to the induction of cell functions by dsRNA are still poorly defined. Poly rI:rC adsorbed rapidly to the cells, but the induction of biological activity did not correlate with rapid binding. Many attempts have been made over several years to determine whether poly rI:rC attached to solid matrices can induce antiviral activity without entering the cells (Johnston et al. 1976; Stewart 1979). The main problem with such studies has been "leaking" of polynucleotides from the supporting matrix, or immobilization in an inactive configuration. Thus, in all these experiments some polynucleotide was either available to cells in a soluble form, or was unable to induce antiviral activity. Indirect evidence for the presence of specific receptors for polynucleotides came from studies by Yoshida and Azuma (1985). The PR-RK cell line, derived from the parental NRK cells by repeated treatment with a copolymer of poly rI:rC, lost its ability to bind the polynucleotide and became resistant to its cytotoxic effect. Surprisingly, however, they did not attempt to directly measure the binding of the radiolabeled copolymer to the cell surface. Different conclusions have been reached by Milhaud et al. (1987) in the attempt to determine whether the intracellular penetration of poly rI:rC is a prerequisite for cell lysis. Treatment of murine 1022 cells with enzymes, such as proteases, glycosidases or phospholipases, did not prevent dsRNA uptake by target cells. By contrast, poly rI:rC internalization was blocked when sodium azide and 2-deoxyglucose were added simultaneously, suggesting that dsRNA is endocytosed without the need for specific membrane receptors. In any event, the initial interaction of polynucleotide, binding nonspecifically and superficially, can be neutralized by antibodies to double-stranded polynucleotides. Following brief incubation at 37 °C, polynucleotides bound to cells become resistant to ribonuclease and antibodies (Stewart 1979).

The multistep process from the interaction of the dsRNA with the cell surface to the initiation of gene transcription was partially clarified by studies on IFN gene activation. The length of the period between the addition of poly rI:rC and activation of IFN genes suggested the importance of discrete "second messengers" for the transmission of signals from the cell membrane to the nucleus. In this regard some distinct signal transduction mechanisms have been proposed (Berridge 1987; Alkon and Rasmussen 1988): one is mediated via the diacylglycerol-protein kinase C, a second involves the calcium-calmodulin-mediated cellular processes, a third leads to activation of adenylate cyclase with a subsequent increase in the level of cellular cAMP, whereas the last involves phosphorylation of tyrosine residues on intracellular domains of surface receptors. Significant enhancement of IFN-γ production by peripheral blood lymphocytes upon stimulation with poly rI:rC was observed in the presence of the phorbol ester, 12-O-tetradecanoylphorbol-13-acetate (TPA, a PKC activator). By contrast, addition of the PKC inhibitor H-7, or the cAMP-dependent PK inhibitor H-8, abrogated IFN-γ production (Tamura-Nishimura and Sasakawa 1989). These results as a whole suggest that the mechanisms of signal transduction activated by poly rI:rC involve PKC along with the cAMP-dependent protein kinase. Thacore et al. (1990) reached similar conclusions studying the effect of PKC inhibitors on IFN-β production by viral and non-viral (poly rI:rC) inducers. In particular their studies supported an integral role for membrane-associated PKC in nonviral induction of IFN-β in human, simian and mouse fibroblasts. With the aim to study the direct antiproliferative action of mismatched dsRNA, Hubbell et al. (1991) used specific inhibitors of cAMP-dependent protein kinase or protein kinase C. Direct measurement of adenylate cyclase activity showed that the cAMP system is utilized by mismatched dsRNA as an early signal transduction mechanism for growth control. Finally, the possibility that dsRNA and IFN-α induce 2-5 OAse, 6-16, IP-10 and ISG15 mRNAs by triggering different protein kinases (PK), was demonstrated by the finding that 2-aminopurine (an inhibitor of various PKs) inhibited gene activation by dsRNA, but not by IFN-α (Tiwari et al. 1988; Memet et al. 1991). Similar conclusions were reached by Hall et al. (1989). These investigators observed that both PDGF and poly rI:rC activated transcription initiation of the JE gene (a cytokine-like protein synthesizing gene). The JE response to dsRNA stimulation, however, did not appear to be channeled through the PDGF signal transduction pathway that needs the hydrolysis of PI bisphosphate and the activation of PKC.

What, then, is the nature of the signal generated by dsRNA that is capable of activating gene expression? It has been demonstrated that dsRNA, such as viral RNA or poly rI:rC, is capable of activating specific protein kinases in vivo and in vitro (Morrow et al. 1985; O'Malley et al. 1986). One of these, pk 68, phosphorylates the α-subunit of the eIF2, resulting in a block of protein synthesis (O'Malley et al. 1986). On the other hand, another dsRNA-dependent protein kinase appears to be induced during stimulation of 3T3 cell growth (Petryshyn et al. 1984). These kinases may be activated in vivo by poly rI:rC and in turn generate the signal for increased gene expression (Table 2).

Table 2. Involvement of second messengers in the dsRNA-induced response

Function analyzed	Modulation of activity of	Modulation of activity by	References
IFN-γ induction	PKC PKA	TPA, H7, H8	Tamura-Nishimura and Sasakawa (1984)
IFN-β induction	PKC	K252a, H7	Thacore et al. (1990)
JE gene induction	PK?	--	Hall et al. (1989)
2-5A, ISG15 induction	PK?	2-Aminopurine	Tiwari et al. (1988)
6-16	PK?	2-Aminopurine	Memet et al. (1991)
Cell growth arrest	PKA	H7, HA1004, 3-isobutyl-1-methylxanthine	Hubbell et al. (1991)

4 dsRNAs as Gene Activators

Many viral and cellular genes require enhancer elements for normal levels of expression. These elements can act at considerable distances upstream or downstream of promoters irrespective of their orientation, and on heterologous genes (Maniatis et al. 1987; Mitchell and Tjian 1989).

The human IFN-α/β genes are highly inducible in fibroblasts by virus or dsRNA. Between the initial interaction of dsRNA with cells and the appearance of IFN, there is a period of 60 to 90 min depending on the cell system. This induction, designated as "lag period", was the black box to which interferonologists addressed their efforts for some 20 years until the middle of the 1980s (Stewart 1979). On the basis of the results accumulated so far, we can divide activation of the IFN system by dsRNA into two phases (Fig. 1). The first one, including the former lag period, takes place a few minutes after treatment, is mediated by regulatory genes, is CHX-resistant, and involves the modification of existing cellular factors (Taniguchi 1988; Levy and Darnell 1990). In the second phase, transcription of structural genes increases gradually a few hours after dsRNA treatment, requires active protein synthesis, and terminates with the secretion of IFN proteins (Lengyel 1982). Mammalian cell transfection systems have been used in conjunction with in vitro mutagenesis procedures to identify DNA sequences involved in IFN gene regulation (Ohno and Taniguchi 1983; Ragg and Weissmann 1983; Zinn et al. 1983). Although there is general agreement that regulatory sequences are located in the 5′-flanking sequences of the gene, the boundaries of the sequences established in different laboratories are not the same. Using a stable transfection assay, Fujita et al. (1985, 1987) demonstrated that the 5′ boundary of the human IFN-β gene required for maximal induction by Newcastle disease virus (NDV), poly rI:rC, and Sendai

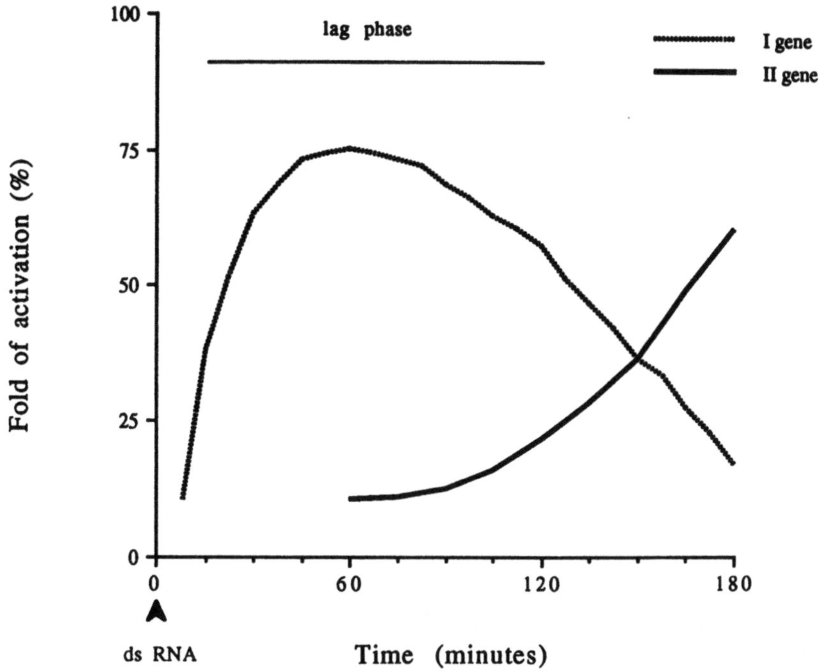

Fig. 1. The primary and secondary response of IFN genes to dsRNA

virus lies between −117 and −95 in mouse L929 cells (Taniguchi 1988). On the other hand, Zinn et al. (1983) by analyzing the transient expression of IFN-β genes introduced into mouse C127 cells, identified another element, whose 5′ boundary lies between −77 and −73. Initial delineation of this regulatory sequence identified a 14-bp element (−77 to −64) capable of restoring some inducibility by poly rI:rC to the minimal promoter of the IFN-β gene (Zinn et al. 1983). However, further examination of the 5′ region revealed additional sequence elements related to this segment (Goodbourn et al. 1985). The region basically consists of repeated hexanucleotide sequences (seven times) between −109 and −65, where the consensus sequence was deduced as AAGTGA. All these sequences display a high degree of homology and are present in both murine and human IFN-α/β genes.

Based on these results Maniatis and collaborators (Goodbourn et al. 1986; Zinn and Maniatis 1986) proposed that the enhancer of the IFN-β gene consists of a "constitutive transcription element", and a "negative regulatory element" that blocks an otherwise functional, constitutive enhancer prior to induction. Thus, transcription of the IFN-β gene might be regulated by three elements: (1) inducible elements; (2) a constitutive element whose function requires two of the inducible elements; and (3) a negative element. These results raise the question whether it is activation or depression or both that mainly determine the induction of the IFN-β gene. In this connection it has been observed that the SV40 enhancer upstream of four-tandem copies of the consensus sequence (AAGTGA), failed to exert its activity on a distal reporter gene (CAT) unless the recipient cells were virus-induced (Dinter and Hauser 1987; Fujita et al. 1987). Induction does not require de novo protein synthesis (Weissman and Weber 1986), and nuclear factors are present which specifically bind to the consensus motifs both in uninduced and induced cells, raising the intriguing question whether such factors can act as repressors in the former and as activators in the latter (Zinn and Maniatis 1986). Another possibility was that in the uninduced cells the hexamer motifs are bound by a given factor, whereas in the induced cells they are bound by a different one. To investigate these two possibilities, the factors binding the AAGTGA motifs have been identified and cloned (Fujita et al. 1988). In cells treated with inducing agents, such as viruses, dsRNA or IFN itself (priming phenomenon), a factor designated as Interferon Regulatory Factor 1 (IRF-1) (Miyamoto et al. 1988) appears to translocate into the nucleus and bind to the hexamer motif. This binding is accompanied by activation of the transcription of the IFN-β gene. IRF-1 is a protein of 329 amino acids in length. The amino-terminal half, spanning down to around amino acid 140, is rich in lysine and arginine. This region shows strong hydrophobicity, and is very likely primarily responsible for the binding to the specific DNA sequences. The rest of the molecule (i.e. end carboxyl-terminal half) shows a relative abundance of aspartic acid, glutamic acid, serine and threonine. Many Ser and Thr residues appear to form clusters that could represent phosphorylation sites implicated in the activation of the IRF-1 protein. It is worth noting that IRF-1 also manifests affinities to regulatory sequences of genes other than the IFN-β gene, suggesting that induction of IRF-1 may be involved in the regulation of a set of genes in various cell types. This observation could explain the ability of dsRNA to augment the expression of a heterogeneous set of genes sharing a consensus sequence for transcription factors activated by dsRNA.

Since viruses or other agents, such as dsRNA, induce a transient expression of the IFN-β gene, one can argue that other regulatory molecules modulate its expression. Subsequent studies have, in fact, revealed a second gene encoding a different regulatory factor, termed IRF-2 (Harada et al. 1989), showing a marked homology with IRF-1 (Miyamoto et al. 1988). Unlike IRF-1, IRF-2 does not stimulate transcription, but displays a repressor-like activity on the repeated AAGTGA sequence that functions either as an inducible enhancer, or as a silencing element, to juxtaposed viral enhancers in induced or uninduced cells, respectively. One can therefore hypothesize the following events in IFN-β gene regulation: IRF-2 is bound to the IFN genes in uninduced cells to the same sequence elements as IRF-1; upon induction, IRF-2 is replaced by IRF-1, which would cooperate with other factors to promote transcription.

In the case of the IFN-β gene, it has indeed been observed that another factor, namely NFkB, binds to a decamer sequence at -66 to -57 position (designated as PRDII site) upstream from the starting site and it appears to play a role in inducing transcription in cooperation with IRF-1 (Fujita et al. 1989; Hiscott et al. 1989; Lenardo et al. 1989). Previous studies have shown that dsRNA and viruses activate adenylate cyclase and increase the level of cellular cAMP, suggesting that the cAMP signal is involved in the transduction signal (Hubbell et al. 1991). It is interesting from this point of view that NFkB can be activated in vitro by protein kinase A (Shirakawa and Mizel 1989), and by dsRNA or virus induction (Fujita et al. 1989; Hiscott et al. 1989; Lenardo et al. 1989). It is, therefore, conceivable to conclude that dsRNA or viruses induce NFkB binding to the PRDII site through activation of protein kinase A. It will be interesting to test this possibility by stimulating with cAMP the expression of a construct containing multiple copies of PRDII driving a reporter gene.

Virus induction of the human IFN-β gene appears also to depend on other activating transcription elements binding to sequences upstream of the DNA segments recognized by IRF-1 and NFkB. Du and Maniatis (1992) have reported the characterization of a distinct regulatory element of the human IFN-β gene promoter, designated PRDIV (positive regulatory domain IV). This sequence appears to be a binding site for a protein of the activating transcription factor/cAMP response element binding protein (ATF/CREB) family of transcription factors. The overall picture shows that at least four positive and one negative regulatory domains are involved in the regulation of the HuIFN-β gene (Fig. 2).

5 Conclusions

A picture is slowly emerging which shows that individual dsRNA-induced proteins have distinct biochemical activities leading to discrete physiological

Fig. 2. The organization of the HuIFN-β gene promoter

changes in dsRNA-treated cells. As a consequence, a given dsRNA-induced protein might dramatically modulate a particular cell function (e.g. secretion of C3 and B factors) and at the same time have no detectable effects on other functions. By contrast, other dsRNA-inducible genes (e.g. competence genes) encode proteins that regulate some basic metabolic pathways, thereby affecting cellular functions such as protein synthesis or cell growth. This differential gene control appears to be related to the type of target cell along with the complexity of the mechanisms that regulate the activation of the inducible genes at transcriptional and post-transcriptional levels. It is, therefore, reasonable to assume that the dsRNA action should be considered the sum of biological activities of a large number of induced proteins acting in concert.

Acknowledgements. Some of the studies discussed in this review were supported by grants from the National Research Council (PF "Biotechnologie e Biostrumentazione", PF "A.C.R.O.") and from A.I.R.C. to S.L. We wish to thank Mrs. Monica Raviola for typing the manuscript.

References

Alkon DL, Rasmussen H (1988) A spatial-temporal model of cell activation. Science 239:998–1004
Berridge MJ (1987) Inositol lipids and cell proliferation. Biochim Biophys Acta 907:33–45
Choubey D, Lengyel P (1992) Interferon action: nuclear and nucleoplasmic localization of the interferon-inducible 72-kD protein that is encoded by the Ifi 204 gene from the gene 200 cluster. J Cell Biol 116:1333–1341
De Maeyer E, De Maeyer-Guignard J (1988) Interferons and other regulatory cytokines. Wiley, New York
Dianzani F, Baron S, Buckler CE (1971) Mechanisms of DEAE-dextran enhancement of polynucleotide induction of interferon. Proc Soc Exp Biol Med 136:1111–1114
Dinter H, Hauser H (1987) Cooperative interaction of multiple DNA elements in the human interferon-β promoter. Eur J Biochem 166:103–110
Du W, Maniatis T (1992) An ATF/CREB binding site protein is required for virus induction of the human interferon β gene. Proc Natl Acad Sci USA 89:2150–2154
Fibbe WE, Van Damme J, Billiau A, Duinkerken N, Lurvink E, Ralpha P, Altrock BW, Kaushansky K, Willemze R, Falkenburg JH (1988) Human fibroplasts produce granulocyte-CSF, macrophage-CSF, and granulocyte-macrophage-CSF following stimulation by interleukin-1 and poly (rI), poly(C). Blood 72:869–866
Field AK, Tytell AA, Lampson GP, Hilleman MR (1967) Inducers of interferon and host resistance. II. Multistranded synthetic polynucleotide complexes. Proc Natl Acad Sci USA 58:1004–1010
Field AK, Tytell AA, Lampson GT, Nemes MM, Hilleman MR (1970) Double-stranded polynucleotides as interferon inducers. J Gen Physiol 57:90S
Flenniken AM, Galabru J, Rutherford MN, Hovanessian AG, Williams BR (1988) Expression of interferon-induced genes in different tissues of mice. J Virol 62:3077–3083
Fujita T, Ohno S, Yasumitsu H, Taniguchi T (1985) Delimitation and properties of DNA sequences required for the regulated expression of human interferon-β gene. Cell 41:489–498
Fujita T, Shibuya H, Hotta H, Yamanishi K, Taniguchi T (1987) Interferon-β gene regulation: tandemly repeated sequences of a synthetic 6 bp oligomer function as a virus-inducible enhancer. Cell 49:357–366
Fujita T, Sakakibara J, Sudo Y, Miyamoto M, Kimura Y, Taniguchi T (1988) Evidence for a nuclear factor(s), IRF-1, mediating induction and silencing properties to human IFN-β gene regulatory elements. EMBO J 7:3397–3405

Fujita T, Miyamoto M, Kimura Y, Hammer J, Taniguchi T (1989) Involvement of a *cis*-element that binds an H2TF-1/NF-kB like factors(s) in the virus-induced interferon-β gene expression. Nucleic Acids Res 17:3335–3346

Gariglio M, Cinato E, Panico S, Cavallo G, Landolfo S (1991) Activation of interferon-inducible genes in mice by poly rI:rC or alloantigens. J Immunother 10:20–27

Gariglio M, Panico S, Cavallo G, Divaker C, Lengyel P, Landolfo S (1992) Impaired transcription of the poly rI:rC- and interferon-activatable 202 gene in mice and cell lines from the C57BL/6 strain. Virology 187:115–123

Gomatos PJ, Tamm I (1963) The secondary structure of reovirus RNA. Proc Natl Acad Sci USA 49:707–714

Goodbourn S, Zinn K, Maniatis T (1985) Human β-interferon gene expression is regulated by an inducible enhancer element. Cell 41:509–520

Goodbourn S, Burstein H, Maniatis T (1986) The human β-interferon gene enhancer is under negative control. Cell 45:601–610

Haines DS, Strauss KI, Gillespie DH (1991) Cellular response to double-stranded RNA. J Cell Biochem 46:2–20

Hall DJ, Jones SD, Kaplan DR, Whitman M, Rollins BJ, Stiles CD (1989) Evidence for a novel signal transduction pathway activated by platelet-derived growth factor and by double-stranded RNA. Mol Cell Biol. 9:1705–1713

Harada H, Fujita T, Miyamoto M, Kimura Y, Maruyama M, Furia A, Miyata T, Taniguchi T (1989) Structurally similar but functionally distinct factors, IRF-1 and IRF-2, bind to the same regulatory elements of IFN and IFN-inducible genes. Cell 58:729–739

Hiscott J, Alper D, Cohen L, Leblac J-F, Sportza L, Wong A, Xanthoudakis S (1989) Induction of human interferon gene expression is associated with a nuclear factor that interacts with the NF-kB site of the human immunodeficiency virus enhancer. J Virol 63:2557–2566

Hubbell HR, Boyer JE, Roane P, Burch RM (1991) Cyclic AMP mediates the direct antiproliferative action of mismatched double-stranded RNA. Proc Natl Acad Sci USA 88:906–910

Johnston MT, Atherton KT, Hutchinson DW, Burke DC (1976) The binding of poly I-poly C to human fibroblasts and the induction of interferon. Biochim Biophys Acta 435:69–73

Lampson GP, Tytel AA, Field AK, Nemes MM, Hilleman MR (1967) Inducers of interferon and host resistance. I. Double stranded RNA from extracts of *Penicillium funiculosum.* Proc Natl Acad Sci USA 58:782–789

Lampson GP, Field AK, Tytell AA, Hilleman MR (1981) Poly I:C/poly-L-lysine: potent inducer of interferons in primates. J Interferon Res 1:539–549

Landolfo S, Garotta G (1991) IFN-gamma, a lymphokine that modulates immunological and inflammatory responses. J Immunol Res 3:81–94

Lenardo MJ, Fan CM, Maniatis T, Baltimore D (1989) The involvement of NF-kB in β-interferon gene regulation reveals its role as widely inducible mediator of signal transduction. Cell 57:287–294

Lengyel P (1982) Biochemistry of interferons and their actions. Annu Rev Biochem 51:251–282

Levy D, Darnell JE Jr (1990) Interferon-dependent transcriptional activation: signal transduction without second messenger involvement? New Biol 2:923–928

Maniatis T, Goodbourn S, Fischer JA (1987) Regulation of inducible and tissue-specific gene expression. Science 236:1237–1244

Maran A, Goldberg ID, Steinberg BM (1990) Induction of c-Ha-ras gene expression by double-stranded RNA and interferon requirement. Mol Cell Biol 10:4424–4426

Memet S, Besancon F, Bourgeade MF, Thang MN (1991) Direct induction of interferon-γ- and interferon α/β-inducible genes by double-stranded RNA. J Interferon Res 11:131–141

Milhaud PG, Silhol M, Salehzada T, Lebleu B (1987) Requirement for endocytosis of poly(rI)·poly (rC) to generate toxicity on interferon-treated LM cells. J Gen Virol 68:1125–1134

Mitchell PJ, Tjian R (1989) Transcriptional regulation in mammalian cells by sequence-specific DNA binding proteins. Science 245:371–378

Miyamoto M, Fujiita T, Kimura Y, Maruyama M, Harada H, Sudo Y, Miyata T, Taniguchi T (1988) Regulated expression of a gene encoding a nuclear factor, IRF-1, that specifically binds to IFN-β gene regulatory elements. Cell 54:903–913

Morrow CD, Gibbons GF, Dasgupta A (1985) The host protein required for in vitro replication of poliovirus is a protein kinase that phosphorylates eukaryotic initiation factor-2. Cell 40:913–921

North RJ, Dunn PL, Havell EA (1991) A role for tumor necrosis factor in poly(I–C)-induced hemorrhagic necrosis and T-cell-dependent regression of a murine sarcoma. J Interferon Res 11:333–340

Ohno S, Taniguchi T (1983) The 5′ flanking sequence of human interferon-β1 gene is responsible for viral induction of transcription. Nucleic Acids Res 11:5403–5412

O'Malley JA, Leong SS, Horoszewicz JS, Carter WA, Alderfer JL, Ts'o POP (1979) Polyinosinic acid-polycytidilic acid and its mismatched analogues: differential effects on human cell function. Mol Pharmacol 15:165–173

O'Malley RP, Mariano TM, Siekierka J, Mathews MN (1986) A mechanism for the control of protein synthesis by adenovirus VA RNA. Cell 44:391–400

Petryshyn R, Chen JJ, London IM (1984) Growth-related expression of a double-stranded RNA-dependent protein kinase in 3T3 cells. J Biol Chem 259:14736–14742

Pyo S, Gangemi JD, Ghaffar A, Mayer EP (1991) Poly-I:C-induced anti-herpes simplex virus type-1 activity in inflammatory macrophages is mediated by induction of interferon-β. J Leuk Biol 50:479–487

Ragg H, Weissmann C (1983) Not more than 117 base pairs of 5′ flanking sequence are required for inducible expression of a human IFN-α gene. Nature 303:439–442

Raji NBK, Pitha PM (1983) Two levels of regulation of β-interferon gene expression in human cells. Proc Natl Acad Sci USA 80:3923–3927

Riches DW, Henson PM, Remigio LK, Catterall JF, Strunk RC (1988) Differential regulation of gene expression during macrophage activation with a polyribonucleotide. The role of endogenously derived IFN. J Immunol 141:180–188

Shirakawa F, Mizel SB (1989) In vitro activation and nuclear translocation of NF-kB catalyzed by cyclic AMP-dependent protein kinase and protein kinase C. Mol Cell Biol 9:2424–2430

Souvannavong V, Adam A (1990) Macrophages from C3H/HeJ mice require an additional step to produce monokines: synergistic effects of silica and poly(I:C) in the release of interleukin 1. J Leuk Biol 48:183–192

Stevenson HC, Dekaban GA, Miller PJ, Benyajati C, Pearson ML (1985) Analysis of human blood monocyte activation at the level of gene expression. Expression of alpha interferon genes during activation of human monocytes by poly IC/LC and muramyl dipeptide. J Exp Med 161:503–513

Stewart II WE (1979) The interferon system. Springer, Berlin Heidelberg New York

Tamura-Nishimura M, Sasakawa S (1989) The roles of protein kinase C and cyclic nucleotide dependent kinase in signal transduction in human interferon gamma induction by poly I:poly C. FEBS Lett 248:73–77

Taniguchi T (1988) Regulation of cytokine gene expression. Annu Rev Immunol 6:439–464

Terao M, Cazzaniga G, Ghezzi P, Bianchi M, Falciani F, Perani P, Garattini E (1992) Molecular cloning of a cDNA coding for mouse liver xanthine dehydrogenase. Biochem J 283:863–870

Thacore HR, Lin H-Y, Davis PJ, Schoenl M (1990) Effect of protein kinase C inhibitors on interferon-β production by viral and non-viral inducers. J Gen Virol 71:2833–2839

Tiwari RK, Kusari J, Kumar R, Sen GC (1988) Gene induction by interferons and double-stranded RNA: selective inhibition by 2-aminopurine. Mol Cell Biol 8:4289–4294

Vilcek J, Kohase M, Henriksen-DeStefano D (1987) Mitogenic effect of double-stranded RNA in human fibroblasts: role of autogenous interferon. J Cell Physiol 130:37–43

Wathelet MG, Clauss IM, Content J, Huez GA (1988) Regulation of two interferon-inducible human genes by interferon, poly(rI):poly(rC) and viruses. Eur J Biochem 174:323–329

Weissman C, Weber H (1986) The interferon genes. Prog Nucleic Acid Res 33:251–268

Yoshida I, Azuma M (1985) Adsorption of poly rI:rC on cell membrane participating and nonparticipating in interferon induction. J Interferon Res 5:1–10

Yoshida I, Marcus PI (1990) Interferon induction by viruses. 20. Acid-labile interferon accounts for the antiviral effect induced by poly(rI)·poly(rC) in primary chick embryo cells. J Interferon Res 10:461–468

Zinn K, Maniatis T (1986) Detection of factors that interact with the human β-interferon regulatory region in vivo by DNAse footprinting. Cell 45:611–620

Zinn K, DiMaio D, Maniatis T (1983) Identification of two distinct regulatory regions adjacent to the human β-interferon gene. Cell 34:865–879

Zullo JN, Cochran BH, Huang AS, Stiles CD (1985) Platelet-derived growth factor and double-stranded ribonucleic acids stimulate expression of the same genes in 3T3 cells. Cell 43:793–800

Viral-Dependent Phosphorylation of a dsRNA-Dependent Kinase

D.A. Roth and X. He[1]

1 Introduction

Plant viruses and viroids are minimal pathogens and depend upon interaction with key host components during pathogenesis. The temporal and spatial regulation of host factors during initial host-pathogen interactions is initiated by host recognition of viral signals, and disease development is conditioned by the subsequent response to these signals. The identity and function of host proteins involved in the recognition and response to pathogen signal molecules as well as the mechanisms by which pathogens manipulate host plant proteins are unknown. Phosphorylation and dephosphorylation are important posttranslational events which alter target protein conformation and function in response to specific metabolic, developmental, and environmental signals. Thus, these processes serve to regulate and coordinate metabolic responses. Phosphorylation-mediated regulation offers a potential mechanism by which plant-virus/viroid interactions may be governed. We have characterized a plant-encoded protein kinase (p68) whose phosphorylation is induced in viral and viroid infected plants and in noninfected tissue extracts supplemented with dsRNA. p68 phosphorylation is dependent upon signal concentration and occurs in the initial phases of pathogenesis. To our knowledge p68 is the only plant protein associated with dsRNA-stimulated protein kinase activity.

2 Plant-Virus Interactions

The interaction between viruses and their plant hosts is governed by a precise set of temporally and spatially expressed regulatory factors. Viral pathogenesis depends upon the efficient regulation of key host components to complement and extend the limited genomic capacity of the virus. Although significant progress has been made on elucidating plant virus genomic structure and protein products, very little is known regarding signal recognition and response (Culver et al. 1991).

The majority of plant viruses are composed of single-stranded, positive-sense RNA encapsidated by protein (Matthews 1991). The structure and nucleic acid

[1]Dept. Plant, Soil and Insect Sciences, University of Wyoming, Laramie, Wyoming 82071, USA

Progress in Molecular and Subcellular Biology, Vol. 14
W.E.G. Müller/H.C. Schröder (Eds.)
© Springer-Verlag Berlin Heidelberg 1994

sequence of many viruses have been determined and have provided valuable information related to potential functions (Francki et al. 1985; Matthews 1991). In general, the limited plant viral genome encodes proteins with primary functions associated with replication, transport and encapsidation, although it is conceivable that virus components serve multifunctional roles during interactions with the host. With few exceptions plant viruses are obligate wound pathogens (Matthews 1991). Specific interactions begin when the host recognizes and responds to pathogen signal molecules. Evidence suggests that penetration, uncoating, and initial translation events of plus-sense RNA viruses involve nonspecific interaction with plant cells. Pseudovirions are capable of penetration, uncoating, and translation but are incapable of replication (Wilson et al. 1990). These studies and the demonstration of a plant-encoded subunit in the viral-specific replicase (Hayes and Buck 1990) suggest that specific recognition between pathogen and host begins at replication.

Significant homology exists between plant and mammalian RNA viruses (Argos et al. 1984; Franssen et al. 1984; Haseloff et al. 1984). Although individual viruses may be morphologically distinct, amino acid sequence similarities between nonstructural proteins suggest real relationships between members. Gene arrangement similarities further suggest a common origin. Thus, even though significant differences exist between plant and animal cells it is conceivable that similar mechanisms are operable during host-virus interactions. Plant cell-virus interactions provide ideal models to define this complex and dynamic process. In addition, these systems are useful in the dissection of normal cellular function. Many plants are amendable to stable introduction and expression of foreign genes and some, such as *Arabidopsis*, have small genomes with few redundant genes which simplifies molecular genetic analyses. Further, differences exist between pathogen strains and host cultivars, resulting in an array of interactions and resultant symptoms.

The identification of plant proteins induced by viral infection has been studied in only a few systems. Host-encoded pathogenesis-related (PR) proteins which are induced by infection in hypersensitive-responding cultivars have been identified and, in part, characterized (reviewed in Linthorst 1991). Although these proteins are expressed during the resistant reaction and many have known enzymatic activity, the functional significance, if any, of PR proteins in the resistance response is not clear. Recently, the cucumber mosaic virus RNA-dependent RNA polymerase (replicase) has been solubilized and purified (Hayes and Buck 1990). Characterization shows that it is composed of two viral peptides and an Mr 50000 host-encoded peptide, supporting prior models which suggested host-encoded protein involvement in plant viral replication (Hall et al. 1982). The identity and function of the host-encoded subunit are not known. Finally, our studies have shown that virus and viroid infection induces the phosphorylation of a plant-encoded protein associated with dsRNA-stimulated protein kinase activity (Crum et al. 1988a; Hiddinga et al. 1988).

3 Plant-Protein Phosphorylation

Nucleotide binding and protein phosphorylation are fundamental events involved in energy balance and regulation of cellular metabolism (Edelman et al. 1987). These key posttranslational events provide an efficient mechanism to amplify the response rate of various extracellular signals. Phosphorylation-dephosphorylation is of demonstrated importance in numerous virus-host interactions (Mohr et al. 1987; Grose et al. 1989; Hovanessian 1989; Ransone and Dasgupta 1989; Lees-Miller et al. 1990) and provides a regulatory mechanism consistent with pathogen needs for a dynamic, timely, and energy efficient system.

The impact of phosphorylation has only recently been recognized as an important process in the regulation of plant metabolism and gene expression (Ranjeva and Boudet 1987; see Randall and Blevins 1990; Bennett 1991; Klimczak et al. 1992). Phosphorylation also may have a central role in the way in which plants perceive and transduce various signals generated during pathogenesis or in response to wounding (Dixon and Lamb 1990; Jin et al. 1990; Bajar et al. 1991; Ryan and Farmer 1991).

It is too early in the characterization of plant kinases to accurately determine the degree of homology between structure and function of plant and animal kinases. However, the deduced polypeptide sequences encoded by putative kinase cDNAs of maize, rice and beans have catalytic domain motifs similar to animal protein kinases (Hanks et al. 1988; Lawton et al. 1989; Biermann et al. 1990). The conservation of catalytic domains suggests a common role for eukaryotic protein kinases in extracellular signal transduction. Regulatory regions between plant and animal kinases appear to be divergent. Plant casein kinase I, II, and $P34^{cdc2}$ kinases are biochemically closely related to the analogous animal kinases. In plants, the predominate stimulatory molecule affecting kinase activity is Ca^{2+}, although other effector molecules are known (see Ranjeva and Boudet 1987; Randall and Blevins 1990). In a significant departure from other eukaryotic systems, no cAMP-dependent protein kinases have been documented from plants.

4 Detection of Viral-Induced Phosphorylation

In view of the demonstrated importance of phosphorylation-dephosphorylation in the regulation of normal and abnormal metabolic function, we investigated changes in the nucleotide binding and phosphoprotein profiles in plants following viral infection using tobacco-tobacco mosaic virus (TMV) as a model. TMV has long served as a model for studying protein-nucleic acid structure-function relationships and host-virus interactions. It is a single-stranded, plus-sense RNA virus with a relatively small 6.4-kb genome (Goulet et al. 1983). The genome has been completely sequenced and translation products have been identified and, in part, characterized (Matthews 1991).

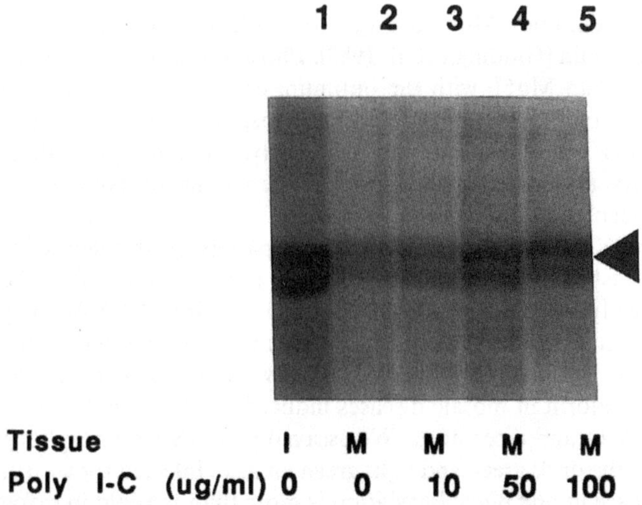

Fig. 1. Phosphorylation of an Mr 68 000 protein in response to viral infection or dsRNA. Homogenates from TMV-infected (*I*) or mock-inoculated (*M*) tissues were subject to in vitro phosphorylation in the absence (*lanes 1, 2*) or presence (*lanes 3, 4, 5*) of dsRNA. This is the autoradiogram of the SDS-PAGel. *Arrowhead* indicates Mr 68 000

Evans et al. (1985) provided the first evidence that TMV infection induces significant changes in the phosphoprotein and nucleotide binding protein profiles of tobacco. A crude soluble fraction was prepared from TMV-infected and mock-inoculated tobacco leaves. Photoaffinity labeling of dialyzed aliquots using ($\gamma^{32}P$) 8-N_3ATP revealed that numerous quantitative and qualitative differences in the nucleotide binding and protein phosphorylation profiles are induced by TMV infection. In particular, an apparent doublet at Mr 68 000 (p68) was labeled in extracts of infected tobacco plants in both the presence and absence of photoactivating light. The 8-azido analog of ATP is an effective mimic of the parent compound in both phosphotransferase reactions and nucleotide binding (Potter and Haley 1982). Label incorporation from ($\gamma^{32}P$) 8-N_3ATP in the presence and absence of photoactivating UV light suggests that p68 is a phosphoprotein or both a phosphoprotein and a nucleotide binding protein. Crum et al. (1988a) showed that a low basal level of ^{32}P incorporation into p68 occurred when homogenates from mock-inoculated tissues were incubated with ($\gamma^{32}P$) 8-N_3ATP. However, an eight- to-ten fold increase was observed in in vitro phosphorylation of p68 in extracts from infected compared to mock-inoculated tissue (Fig. 1). Further studies showed that p68 is a very low abundance, host-encoded protein. The doublet is due, in part, to partial proteolytic digestion during fraction preparation (particularly when using tobacco leaf tissue). Also, the phosphorylation conditions used influences the relative amount of ^{32}P label in the doublet bands. Amino acid analysis indicates that the doublet peptides have identical compositions. Phosphorylation of p68 is dependent upon Mn^{2+} or Mg^{2+} but independent of cAMP, cGMP, calcium or calmodulin (Hiddinga et al. 1988). Phosphorylation is greater in the presence of Mn^{+} than Mg^{2+} with the optimum concentration being 1 mM. Inclusion of Mn^{2+} and Mg^{2+} at 1 mM in the phosphorylation mix provides the optimum stimulation. Phosphoamino acid analysis demonstrates that phosphorylation occurs at serine residues, but not threonine or tyrosine. Also, GTP is not a phosphate donor.

TMV induces a mosaic symptom pattern in infected leaves. The pattern is established very early in the infection process. Dark green areas in the pattern typically do not contain infectious virus, viral dsRNAs, or proteins, whereas adjacent yellow or light-green/yellow areas contain high concentrations of viral RNAs, proteins, and virions (Matthews 1991). This phenomenon is consistent and uniform in mosaic diseases induced by structurally diverse viruses in both monocot and dicot hosts. We assayed p68 in vitro phosphorylation in extracts from the dark green and light green areas of infected tobacco and found that the increase in p68 phosphorylation is more than ten-fold in extracts from the light green areas compared to extracts from dark-green areas or in extracts from healthy controls. This further confirms the association of p68 phosphorylation with the presence of infectious virus components.

Phosphorylation of plant p68 is also induced by infection in other diverse host-pathogen interactions including potato spindle tuber viroid infection of tomato (Hiddinga et al. 1988), brome mosaic virus-infection of the monocot host barley, and cucumber mosaic virus infection of cucumber. In all cases, the response of p68 phosphorylation to infection is similar to that documented in the TMV-tobacco system.

5 Characterization of Viral-Induced Phosphorylation

5.1 Stimulatory Molecules

We evaluated p68 phosphorylation in extracts from healthy tissues in response to exogenous amendments of potentially stimulatory molecules such as cAMP, cGMP, Ca, Ca-calmodulin, DNA, fatty acids, and dsRNA. Phosphorylation of p68 in homogenates from mock-inoculated tissues is elevated to levels similar to infected tissues by dsRNA as poly (I)-poly (C) (Fig. 1), poly (A)-poly (U), or viral dsRNA purified from infected tissues. The response is dependent upon the concentration of exogenous dsRNA; with maximal p68 phosphorylation achieved at 10–50 μg/ml dsRNA in crude tissue extracts. The stimulatory influence of TMV dsRNA is observed at lower concentrations than with poly (I)-poly (C). Inhibition of p68 phosphorylation is observed at poly (I)-poly (C) concentrations > 50–100 μg/ml. Poly (I) and mRNA from mock-inoculated tissue (Fig. 2), as well as poly (C), ribosomal RNA, and salmon sperm DNA have no effect on p68 phosphorylation. DsRNA (10-20 μg/ml) added to infected tissue homogenates did not further elevate p68 kinase activity above that due to

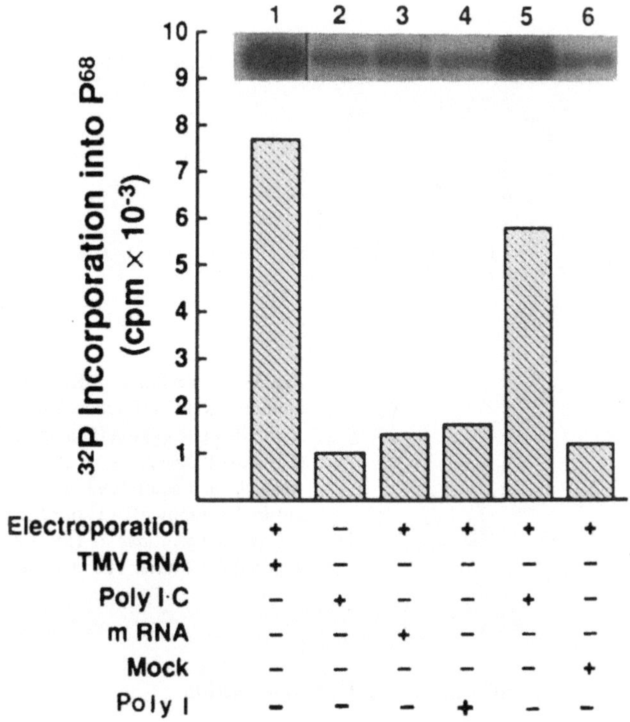

Fig. 2. Stimulation of p68 phosphorylation by dsRNA or viral infection. Protoplasts were infected by electroporation with TMV RNA, mock-inoculated and extracts were subjected to in vitro phosphorylation or electroporated with poly (I)-poly (C), poly I or mRNA

infection. Because these experiments were done using crude extracts, the precise concentration effect often varied between experiments, and precise measurement of the effect of dsRNA will depend upon obtaining purified plant p68.

Further evidence of the dsRNA binding capability of p68 is provided by experiments showing that p68 selectively binds to dsRNA-agarose. A crude 65% ammonium sulfate fraction from a plant tissue extract was in vitro phosphorylated and incubated with poly (I)-poly (C) agarose for 1h at 4°C. The poly (I)-poly (C) agarose was then washed twice with 10 vol of 30 mM Pipes, pH 7.5, 3 mM $Mg(OAc)_2$, 1 mM DTT, 10% glycerol containing 120 mM KC1. The matrix was then solubilized in SDS gel electrophoresis sample buffer, boiled for 4 min and analyzed by SDS-PAGE and autoradiography (Evans et al. 1985). Only one band was observable on the Coomassie-stained gel and it was at Mr 75000. This band was not observable on the autoradiogram even after a 12-day exposure. The only radiolabeled band on the autoradiogram was at Mr 68000 (Fig. 3).

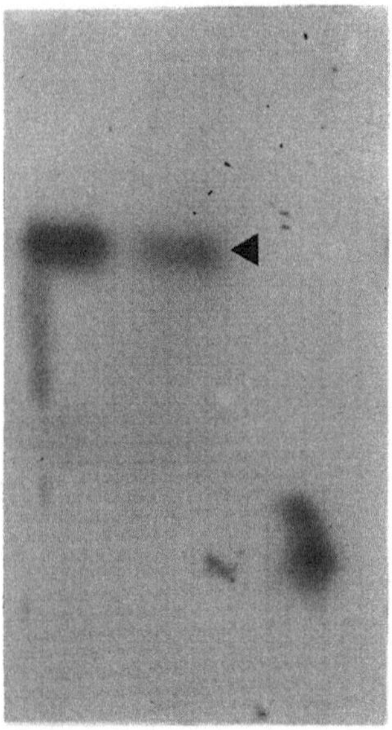

Fig. 3. P68 binding to dsRNA-agarose. 25 µl of a crude 65% ammonium sulfate fraction from a plant extract was subjected to in vitro phosphorylation and incubated with poly (I)-poly (C) agarose (*lane 1*) for 1 h at 4°C. The matrix was extensively washed, then solubilized and analyzed by SDS-PAGE and autoradiography. HPLC-purified p68 previously labeled with ^{32}P is shown in *lane 2*. This autoradiogram of the gel was exposed for 2 days. The *arrowhead* indicated Mr 68 000

5.2 Temporal Pattern of Phosphorylation

Pathogenesis is a dynamic process and key events in the virus-plant interaction are temporally regulated. Our initial studies were done using leaf tissue 14-day postinoculation which showed severe systemic mosaic symptoms to TMV. Temporal events are difficult to analyze using whole plant leaves where it is not possible to obtain synchronous infection of a high percentage of the cells. Further, infection effects are often masked in tissues because only approximately 0.1% of the cells are infected by TMV by mechanical inoculation (Matthews 1991). However, these disadvantages can be overcome using plant protoplasts derived from mesophyll cells where greater than 70% one-step infection is achievable. In order to analyze the temporal induction of p68 phosphorylation we isolated protoplasts and inoculated them with TMV (Hu and Roth 1991). Protoplasts from healthy tobacco were infected with TMV virions or with TMV-RNA via electroporation and incubated for intervals of 0–72 h postinoculation (Fig. 4). Stimulation of p68 in vitro phosphorylation first was observed in homogenates of protoplasts harvested between 5–6 h postinoculation period and progressively increased to a peak stimulation four to five fold greater than in mock-inoculated protoplasts at 12 h postinfection. Electroporation of artificial dsRNA or TMV dsRNA into uninfected protoplasts induced similar levels of p68 phosphorylation to those observed with infected protoplasts. Phosphorylation of p68 subsequently decreased after 12 h postinoculation but was still greater than in extracts from mock-inoculated protoplasts up to 72 h postinfection.

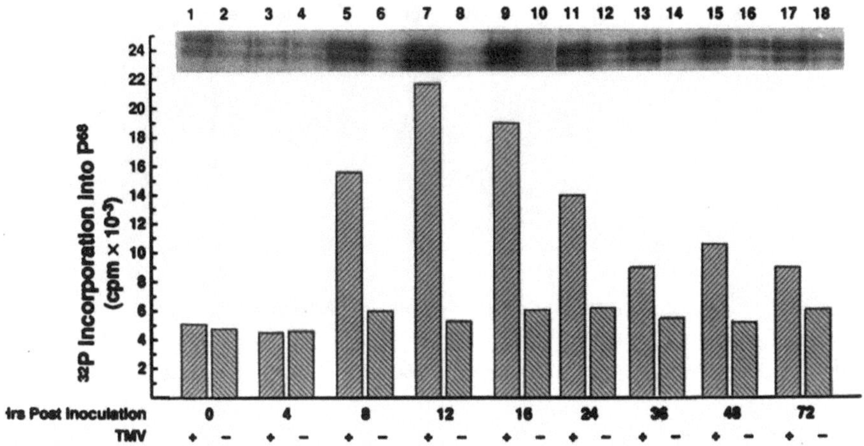

Fig. 4. Regulation of p68 phosphorylation by virus infection. Tobacco protoplasts were synchronously infected with TMV or mock-inoculated and incubated for various times. Fractions enriched for p68 were in vitro phosphorylated subjected to SDS-PAGE and autoradiography

No differences were found in protoplast viability between infected and mock-inoculated treatments. Viability was monitored by fluoroscein diacetate staining and was between 95-98% from 0–26 h postinoculation.

In synchronously infected protoplasts, TMV RNAs, including dsRNAs, are detectable 3 h postinoculation and replication activity significantly increases at 6–9 h postinoculation (Watanabe et al. 1984; Watanabe and Okada 1986). The level of these viral products is observable until at least 24 h postinoculation, although at a substantially lower level. Although synthesis of viral proteins is initially detectable 2–4 h postinoculation, peak synthesis of proteins implicated in viral-specific replication occurs 6–9 h postinoculation and then declines. This time frame positively correlates with maximum stimulation of p68 kinase activity.

5.3 In Vivo Phosphorylation of p68

In an effort to demonstrate that the p68 phosphorylation observed in in vitro phosphorylation assays occurs in vivo in plant cells in response to viral infection, protoplasts were pulse-labeled with $H_3{}^{32}PO_4$ following infection with TMV (Hu and Roth 1991). Following specific incubation periods, protoplasts were fractionated and aliquots were subjected to SDS-PAGE and autoradiography. Phosphorylation of p68 was elevated in vivo after 4-h infection and maximum p68 phosphorylation occurred between 9–15 h (greater than eight fold more in extracts from infected protoplasts compared to mock-inoculated

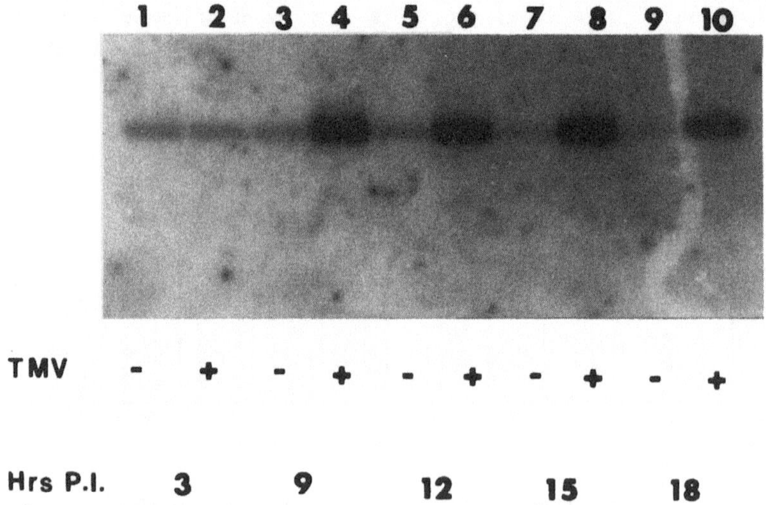

Fig. 5. In vivo phosphorylation of p68 is induced by virus infection. Protoplasts were mock-inoculated or TMV-infected and pulse-labeled with $[^{32}P]H_3PO_4$ for 2 h. Following 3 (*lanes 1, 2*), 9 (*lanes 3, 4*), 12 (*lanes 5, 6*), 15 (*lanes 7, 8*), and 18 (*lanes 9, 10*) h incubation protoplasts were lysed and p68-enriched fractions were subjected to SDS-PAGE and autoradiography

controls) (Fig. 5). A decrease in p68 phosphorylation was observed after 15 h post-infection. These in vivo results support data from in vitro phosphorylation assays.

5.4 Characterization of p68 Protein Kinase Activity

Since initial photoaffinity labeling results using $(\gamma^{32}P)8\text{-}N_3ATP$ indicated that p68 may contain a nucleotide binding site, further studies were done using $(\alpha^{32}P)8\text{-}N_3ATP$. p68 in both mock and infected tissue homogenates was photolabeled with $(\alpha^{32}P)8\text{-}N_3ATP$ or $(\alpha^{32}P)2\text{-}N_3ATP$ in a concentration-dependent fashion. The degree of photoaffinity labeling of p68 was four to six fold greater in homogenates from infected tissues than from mock-inoculated tissues (Fig. 6). No difference in photoinsertion was found using the $8\text{-}N_3ATP$ or $2\text{-}N_3ATP$ analogs. The azido group at C-2 results in minimal structural changes from the parent compound which favors an anti-conformation, whereas the azido group at C-8 allows the syn-anti equilibrium ratio to favor the syn-conformation (Czarnecki et al. 1982).

Specificity of p68 nucleotide binding was suggested by saturation effects and competition for the binding site by the parent compound at physiologically relevant concentrations. Saturable binding of $(\alpha^{32}P)8\text{-}N_3ATP$ was accomplished by photolysis of reaction mixtures containing p68 and varying

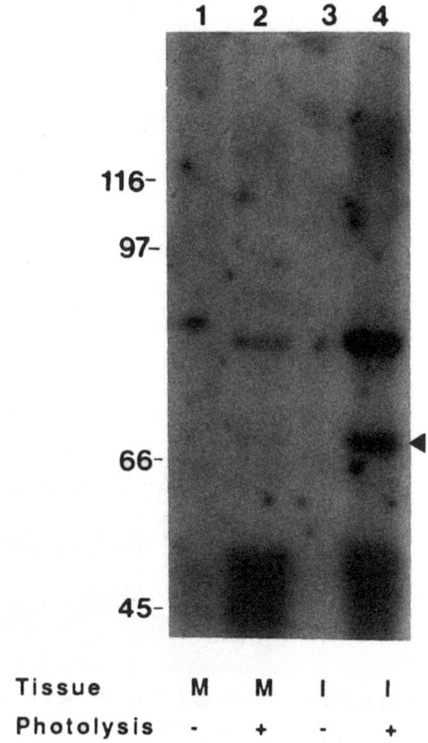

Fig. 6. Nucleotide photoaffinity labeling of p68. Homogenates from mock-inoculated (*M*) or infected (*I*) tissues were incubated with 20 µM [α^{32}P]8-N_3ATP in the presence (*lanes 2, 4*) or absence (*lanes 1, 3*) of UV irradiation. Mixtures were subjected to SDS-PAGE and autoradiography. Molecular weight markers $\times 10^3$ are located to the *left of lane* 1; *arrowhead* indicates position of p68

concentrations of the photoaffinity probe. Saturation experiments indicate an apparent K_d (dissociation constant) of 14 µm (± 3 µM). Further evidence for labeling specificity was provided by competition effects which showed that the nucleotide binding site of p68 is protected only by 200 µM ATP but not other nucleotides including CTP, UTP, and GTP (Fig. 7), as well as, dATP, adenosine, cAMP, or glucose 1,6 diphosphate. Photolabeling was dependent upon Mg^{2+} but not Ca^{2+} and incubation of extracts with (α^{32}P)8-N_3ATP in the absence of photolysis did not result in detectable ^{32}P incorporation into p68. These results clearly demonstrate that 8-N_3ATP is interacting with an ATP binding site in p68 consistent with kinase activity.

Further evidence that p68 is a protein kinase is provided by renaturation-phosphorylation experiments (Fig. 8). Partially purified p68 fractions from *Arabidopsis* and tobacco leaves were subjected to SDS-PAGE. Separated

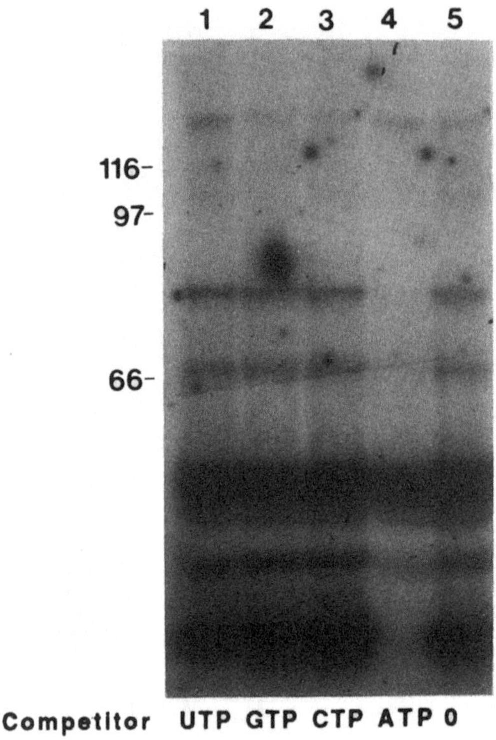

Fig. 7. Protection of p68 photolabeling only by ATP. Photolabeling of TMV-infected homogenates using [α^{32}P]8-N_3ATP was done in the presence or absence of 200 μM of various competitors. Mixtures were subjected to SDS-PAGE and autoradiography

proteins were electroblotted onto PVDF membranes from gels prior to washing and incubation with (γ^{32}P)ATP essentially according to Ferrell and Martin (1990). Figure 8, which is an autoradiogram of the blot, shows that renatured p68 is capable of autophosphorylation. Lane 1 is a control consisting of phosphorylated and HPLC-purified p68 from tobacco extracts.

5.5 Immunological Similarity with the Human p68

The biochemical properties described above for the plant p68 are similar to those described for autophosphorylating dsRNA-dependent protein kinases from mammalian systems (reviewed in Hovanessian 1989). The similarity of the plant and mammalian p68 proteins is further suggested by immunoprecipitation studies using a polyclonal antiserum to the human dsRNA-dependent protein kinase (Berry and Samuel 1985). This antiserum bound to a protein at Mr 68 000 in extracts from infected plant tissues which were previously subject to phosphorylation (lanes 1, 2, 3) or photolabeling with (α^{32}P)8-N_3ATP

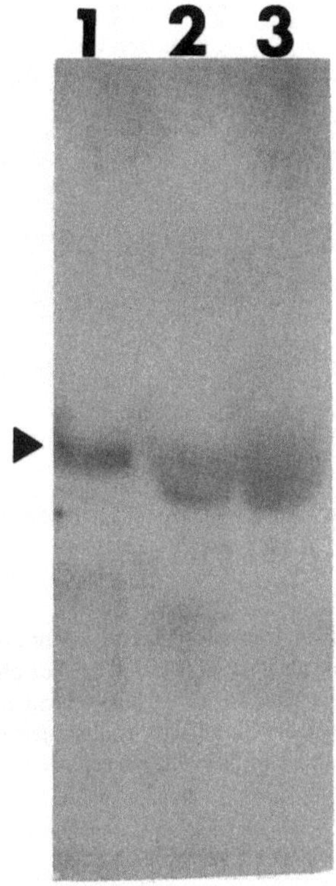

Fig. 8. Detection of p68 protein kinase activity. Partially purified fractions containing p68 were subjected to electrophoresis on 10% SDS-polyacrylamide gels containing BSA, and then washed thoroughly according to Ferrell and Martin (1990). The gel was electroblotted onto PVDF membranes and the blot was washed and subjected to phosphorylation. This is the autoradiogram of the dried blot exposed for 12 h. *Lane 1* HPLC-purified, phosphorylated p68; *lane 2* sample from *Arabidopsis* leaf tissue; *lane 3* sample from tobacco leaf tissue. *Arrowhead* indicates position of p68

(lanes 4, 5, 6) prior to immunoprecipitation (Fig. 9). The ability of plant p68-containing immunocomplexes to catalyze the phosphorylation of p68 was analyzed by performing immunoprecipitation reactions using extracts from infected (Fig. 10, lanes 1, 2, 3) and mock-inoculated (lanes 4, 5, 6) tissues followed by incubation of the thoroughly washed complex with (γ^{32}P)ATP. Immunocomplexes containing p68 were able to catalyze the incorporation of ^{32}P from (γ^{32}P)ATP into p68. These experiments support the concept that p68 is an autophosphorylating protein kinase. We have recently generated polyclonal antiserum to plant p68 which will be useful in the further elucidation of the immunological similarity between the plant and mammalian p68.

5.6 Peptide Sequencing

The purification of the plant p68 has proven difficult. We have adapted techniques used in the purification of the human p68 kinase (Samuel et al. 1985; Kostura and Mathews 1989) where appropriate, although the protocols are not entirely useful for the plant system. We obtained significant purification by

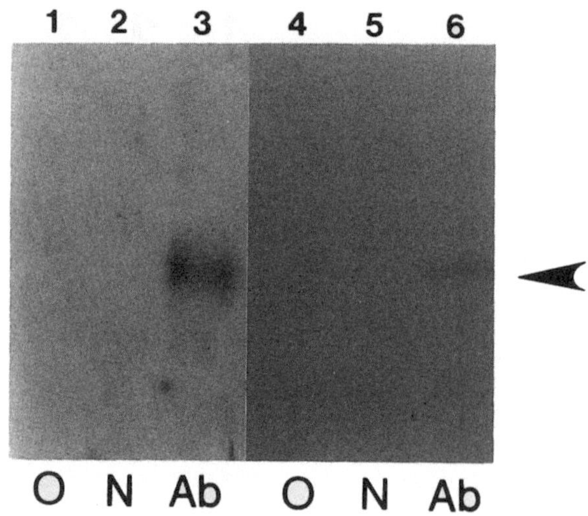

Fig. 9. Immunoprecipitation of p68 from infected tissue. Homogenates from infected tissues were phosphorylated (*lanes 1, 2, 3*) or photolabeled (*lanes 4, 5, 6*) and incubated with either 0, normal, or antiserum to a dsRNA-dependent kinase (Berry and Samuel 1985). Immunoprecipitates were washed, solubilized, and subjected to SDS-PAGE and autoradiography. *Arrowhead* indicates position of p68; *0* no serum, *N* normal serum, *Ab* antiserum

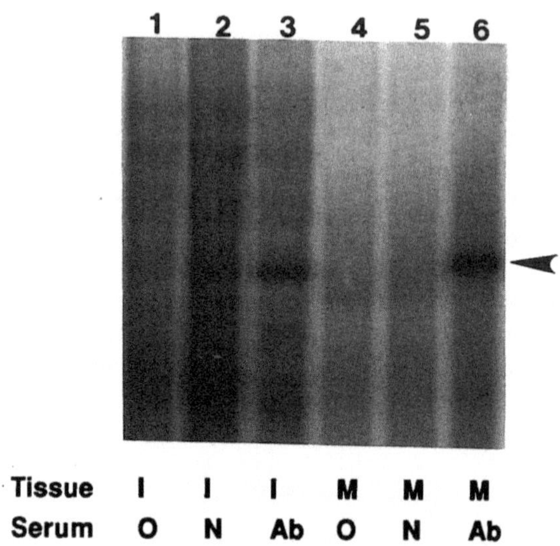

Fig. 10. Protein kinase activity associated with p68 immunocomplexes. Homogenates from mock-inoculated (*M*) or infected (*I*) tissue were immunoprecipitated and the washed complexes were phosphorylated and subjected to SDS-PAGE and autoradiography. *Arrowhead* indicates position of p68; *:0* no serum, *N* normal serum, *Ab* antiserum

initial differential centrifugations to yield a soluble fraction, followed by precipitation at 65% ammonium sulfate. This was followed by DEAE chromatography and ion exchange chromatography using a Mono S column with an FPLC system. The pI of the phosphorylated form (the only form which, at this stage of purification, we can measure) of p68 is 8.2. We expect that the pI of the unphosphorylated form may be in the range of 9 + . In fact, this agrees with data on the human p68, where the pI of the phosphorylated and unphosphorylated forms are 7.8–8.4 and 9.6 (based upon the deduced amino acid sequence), respectively (Meurs et al. 1990).

In order to obtain amino acid sequence information, the partially purified p68 fraction was subjected to further purification by SDS-PAGE. The PAGE step is necessary to eliminate several higher abundance proteins copurifying with p68 up to this step. p68 bands from multiple lanes (up to 300, each with 30 μg total protein) were electroeluted from gels and washed to decrease the SDS and salt concentrations. Residual SDS and contaminants from the gel interfered with subsequent protease digestion and sequence analysis and were minimized using inverse gradient HPLC (Simpson et al. 1987). In this procedure, proteins from SDS-PAGE electroelutions are retained on microbore C-18 columns when loaded at a high concentration of propanol. SDS and gel-related artifacts are not retained and elute at the solvent front. p68, washed free of contaminating SDS, was recovered in an inverse gradient of TFA (0.2% v/v) and propanol. The major peak eluted at 44% propanol, 0.2% TFA and contains uniformly phosphorylated p68 while a trailing shoulder is comprised of p68 phosphorylated to varying degrees. Lyophilized p68 fractions were solubilized in 6 M guanidine HCL (ph 7) and incubated 15 min in a boiling water bath. The guanidine concentration was decreased to 1 M with 25 mM Tris-EDTA, pH 7.7 prior to digestion.

Aliquots of the major radiolabeled peak were completely digested, using Lys-C protease, into distinct bands and peptides were recovered by reverse phase HPLC using a C-4 column with a propanol/TFA gradient. Using this protocol we have purified and sequenced several peptides from the plant p68. Analysis of these peptide sequences (NBRF database) did not reveal statistically significant homologies with the human p68.

5.7 Transgenic Plant Studies

One of the fundamental advantages of using a plant system to study this protein is the ability to generate transgenic plants. Stable integration and expression of foreign genes are now routine for numerous plant species. The regulation of target gene expresssion by the introduction of antisense sequences to endogenous genes has been useful to probe the biological function of specific gene products. We reasoned that if TMV replication is inhibited, dsRNA production would be decreased concomitantly with a decrease in symptom expression. This would provide a useful system to dissect the influence of p68 phosphorylation.

Using in vitro translation, we demonstrated that short DNA and RNA fragments, complementary to two 5′-proximal regions of TMV genomic RNA block the expression of the p126-p183 cistron which is essential for viral replication (Crum et al. 1988b). These DNA fragments were then cloned into the binary vector pMON530 behind the cauliflower mosaic virus 35S RNA promoter, followed by *Agrobacterium tumefaciens*-mediated transformation of *Nicotiana tabacum* cv. *Xanthi* in plants (Horsch et al. 1985). Seed from progeny transgenic plants was challenged with TMV (Nelson et al. unpubl.). Plants expressing the antisense orientation of a 51-bp sequence showed resistance to infection when inoculated with TMV concentrations which produced severe symptoms in control plants. Systemic accumulation of TMV RNA and progeny virus was decreased 15- to 30-fold in these plants. Several of the antisense lines differed in symptom expression as well as viral RNA and protein accumulation. Transgenic plants were inoculated at the three- to four-leaf stage with 0.01 and 0.1 μg/ml TMV, incubated for 5 and 14 days and p68 fractions subject to in vitro phosphorylation, SDS-PAGE, and autoradiography. Initial results showed that p68 phosphorylation in "sense" lines (expressing sense RNA to TMV replicase genes) inoculated with 0.1 and 0.01 μg/ml TMV and incubated for 5 days was equal to that found in TMV-inoculated control plants at the same inoculum levels but four- to six-fold greater than that found in mock-inoculated control plants. p68 phosphorylation in extracts from AS 5 and 6 lines (expressing antisense RNA to TMV replicase genes) was 25 and 75%, respectively, less than that found in sense plants. These data correlated with symptom expression levels and viral concentrations found in the respective lines.

6 Homology with Mammalian Kinases

In mammalian cells an analogous Mr 68 000 dsRNA-dependent kinase (p68) has been identified and characterized (reviewed in Hovanessian 1989). The level of this kinase is significantly increased over basal levels by interferon treatment. p68 undergoes an initial autophosphorylation reaction which is dependent upon dsRNA. Following intramolecular phosphorylation, p68-PO_4 is activated and capable of subsequent phosphotransferase activity. One important substrate for subsequent phosphorylation is the α-subunit of eIF-2. Phosphorylation suppresses protein synthesis by interfering with α-subunit recycling from the eIF-2/GTP/Met tRNA ternary complex necessary for mRNA translation initiation (reviewed in Moldave 1985; London et al. 1987). These events contribute to the interferon-induced, cellular antiviral phenomenon, although numerous strategies have been evolved by animal viruses to overcome this antiviral effect as well as to utilize the mechanism to specifically regulate viral protein synthesis (Katze et al. 1987, 1988; Akkaraju et al. 1989; Black et al. 1989; Edery et al. 1989; Dubois and Hovanessian 1990; Roy et al. 1991).

Recently, the human p68 cDNA sequence has been reported (Meurs et al. 1991). The deduced amino acid sequence from a 2.5-kb mRNA shows 14 of 15 conserved motifs within the nine subdomains of p68 characteristic of protein kinases as described by Hanks et al. (1988). Although there is not a consensus RNA binding site, the hydropathy plot shows that the protein has highly hydrophilic regions (possibly the dsRNA binding regions) with one major hydrophobic region near the C-terminus.

The mammalian and plant kinases have significant biochemical similarities. Both have ATP binding sites, protein kinase activities and are phosphorylated on serine residues. Phosphorylation is dependent upon Mn^{2+} or Mg^{2+} but independent of cAMP or Ca^{2+}. Plant and mammalian p68 phosphorylation occurs in a concentration-dependent fashion in response to exogenous viral dsRNA (purified from infected cells) or synthetic dsRNA as poly (I)-poly (C). The phosphorylation is signal-specific; addition of mRNA, rRNA, viral ssRNA, or DNA does not stimulate phosphorylation of p68. The concentration of exogenous dsRNA required for stimulation of plant p68 (10–50 μg/ml) is considerably higher than the concentration stimulatory for mammalian p68. This may be due to the crude nature of the plant tissue fractions. Phosphorylation of p68 in both systems is inhibited at dsRNA concentrations above the optimum. The plant-encoded p68 is recognized by a polyclonal antiserum against the mammalian p68 dsRNA-dependent protein kinase. Analysis of partial peptide sequences from the plant p68 does not reveal significant homology with the mammalian p68 (Meurs et al. 1990), however, further data are necessary to evaluate this aspect.

7 Conclusions and Future Directions

Mechanisms involved in plant disease development and the contribution of host proteins to pathogenesis in virus and viroid systems have been difficult to dissect and account, in part, for the lack of information on viral/viroid signal-host recognition interactions. It is our hypothesis that dsRNA viral/viroid intermediates or intramolecularly based paired RNA serve as regulatory molecules to induce p68 activity. These RNAs are produced during early plant-virus/viroid interactions and may serve as signals which mediate early plant cell recognition and response to pathogen attack. The relative degree of phosphorylation (multiple sites are present in plant p68) or the specific serine residues phosphorylated may reflect an important regulatory mechanism and be conditioned by the effector concentration. Initial phosphorylation of p68 may activate a subsequent phosphorylation-mediated cascade.

The mammalian p68 is induced by interferon treatment and is involved in host resistance to viral infection. The existence of a phyto-interferon has been questioned for some time and there is no evidence that an analogous system is operable in plant cells (Pierpoint 1983). The function of plant p68 kinase activity in healthy and in virus- or viroid-infected plants is unclear. Viral/viroid-directed changes in the function of key substrates may be necessary for successful completion of the viral/viroid cycle and it is conceivable that the altered function of the substrate conditions disease expression. Alternately, dsRNA-mediated host recognition of pathogen infection may be an initial step in the development of a resistance response. Further characterization of the functional significance of p68 phosphorylation in healthy as well as infected cells may help answer how viruses and viroids interact with and manipulate host metabolic processes. At this time we have insufficient data to determine the contribution of p68 phosphorylation in the host-virus/viroid interaction. These questions are of fundamental importance. This will require the study of a better defined system using more highly purified components, specific antibodies, and cDNA/genomic clones. To this end we have synthesized degenerate oligonucleotides from p68 peptide sequences and polyclonal antiserum according to Ried et al. (1992) to use as probes for further characterization and cloning studies.

In particular, the findings in the viroid system may have important implications. Viroids are plant pathogenic RNAs of between 250–400 nucleotides capable of inciting severe disease in plant hosts (see Diener 1987). The RNA is highly based paired, single-stranded, and lacks a protective protein coat. Importantly, they do not encode any proteins; thus, all plant cell-viroid interactions are mediated by viroid RNA (Keese and Symons 1987).

Although we cannot yet exclude the possibility that p68 kinase activity is only a secondary consequence of pathogenesis, several factors support the hypothesis that activity is, at least, a direct result of pathogen infection. Most importantly, p68 activity during infection is specifically stimulated and is maximum during the periods of peak viral protein translation, decreased host mRNA translation, and peak viral replication. Regulatory roles in protein synthesis and in virus replication are possible based upon analogous systems. Plant viruses do not cause a shutdown of host protein synthesis, however, Fraser and Gerwitz (1980) found that protein synthesis in TMV-infected tissues can be significantly inhibited during the period of virus replication with subsequent recovery to levels comparable with uninfected plants following virus accumulation. No change in host-polyadenylated mRNAs was found, suggesting that inhibition of host protein synthesis is at the posttranscriptional level. Although dramatic changes in overall protein synthesis do not occur in protoplasts in response to viral infection (Siegal et al. 1978), it is conceivable that regulation of the synthesis of specific host proteins occurs and that regulation impacts not only pathogenesis but also subsequent symptom development. In addition, Morrow et al. (1985) have shown that an Mr 68 000 host protein which acts in concert with the poliovirus-encoded replicase proteins is a phosphoprotein associated with dsRNA-dependent protein kinase activity. Although poliovirus replicase proteins have significant homology with plant virus proteins, further in-depth characterization of holoenzymes specific for plant RNA virus replication will be required to clarify this possibility.

Acknowledgments. Special thanks are extended to Boyd Haley for advice and help. This research was supported by NSF grants RII-8610680, BMB-8611424, and DCB-8816085.

References

Akkaraju GR, Whitaker-Dowling P, Youngner JS, Jagus R (1989) Vaccinia specific kinase inhibitory factor prevents translational inhibition by double-stranded RNA in rabbit reticulocyte lysates. J Biol Chem 264:10321–10325

Argos P, Kamer G, Nicklin MJH, Wimmer E (1984) Similarity in gene organization and homology between proteins of animal picornaviruses and a plant comovirus suggest common ancestry of these virus families. Nucleic Acids Res 12:7251–7267

Bajar A, Podila GK, Kolattukudy PE (1991) Identification of a fungal cutinase promoter that is inducible by a plant signal via a phosphorylated *trans*-acting factor. Proc Natl Acad Sci USA 88:8208–8212

Bennett J (1991) Protein phosphorylation in green plant chloroplasts. Annu Rev Plant Physiol Plant Mol Biol 42:281–311

Berry MJ, Samuel CE (1985) Mechanisms of interferon action: production and characterization of monoclonal and polyclonal antibodies to the interferon induced phosphoprotein P1, Biochem Biophys Res Commun 133:168–175

Biermann B, Johnson EM, Feldman CJ (1990) Characterization and distribution of a maize cDNA encoding a peptide similar to the catalytic region of second messenger dependent protein kinases. Plant Physiol 94:1609–1615

Black TL, Safer B, Hovanessian AG, Katze M (1989) The cellular 68 000-Mr protein kinase is highly autophosphorylated and activated yet significantly degraded during poliovirus infection: implications for translational regulation. J Virol 63:2244–2251

Crum CJ, Hu J, Hiddinga HJ, Roth DA (1988a) Tobacco mosaic virus infection stimulates the phosphorylation of a plant protein associated with double-stranded RNA-dependent protein kinase activity. J Biol Chem 263:13440–13443

Crum CJ, Johnson JD, Nelson A, Roth DA (1988b) Complementary oligodeoxynucleotide mediated inhibition of tobacco mosaic virus RNA translation in vitro. Nucleic Acids Res 16:4569–4581

Culver JN, Lindbeck AGC, Dawson WO (1991) Virus-host interactions: induction of chlorotic and necrotic responses in plants by Tobamoviruses. Annu Rev Phytopathol 29:193–217

Czarnecki JJ, Abbott MS, Selman BR (1982) Photoaffinity labeling with 2-azidoadenosine diphosphate of a tight nucleotide binding site on chloroplast coupling factor 1. Proc Natl Acad Sci USA 79:7744–7748

Diener TO (1987) Viroids and viroid diseases. Wiley, New York

Dixon RA, Lamb CJ (1990) Molecular communication in interactions between plants and microbial pathogens. Annu Rev Plant Physiol Plant Mol Biol 41:339–367

Dubois MF, Hovanessian AG (1990) Modified subcellular localization of interferon-induced p68 kinase during encephalomyocarditis virus infection. Virology 179:591–598

Edelman AM, Blumenthal DK, Krebs EG (1987) Protein serine/threonine kinases. Annu Rev Biochem 56:567–613

Edery I, Petryshyn R, Soneberg N (1989) Activation of double-stranded RNA-dependent protein kinase (dsI) by the TAR region of HIV-1 mRNA: a novel translational control mechanism. Cell 56:303–312

Evans RK, Haley BE, Roth DA (1985) Photoaffinity labeling of a viral induced protein from tobacco. J Biol Chem 260:7800–7804

Ferrell JE, Martin GS (1990) Identification of a 42-kilodalton phosphotyrosyl protein as a serine (threonine) protein kinase by renaturation. Mol Cell Biol 10:3020–3026

Francki RIB, Milne RG, Hatta T (1985) Atlas of plant viruses, vols I, II. CRC Press, Boca Raton

Franssen H, Leunissen J, Goldbach R, Lomonossoff G, Zimmern D (1984) Homologous sequences in non-structural proteins from cowpea mosaic virus and picornaviruses. EMBO J 3:855–861

Fraser RSS, Gerwitz A (1980) Tobacco mosaic virus infection does not alter the polyadenylated messenger RNA content of tobacco leaves. J Gen Virol 46:139–148

Goulet P, Lomonossoff GP, Butler PJG, Akam ME, Gait MJ, Karn J (1983) Nucleotide sequence of tobacco mosaic virus RNA. Proc Natl Acad Sci USA 79:5818–5822

Grose C, Jackson W, Traugh JA (1989) Phosphorylation of Varicella-Zoster virus glycoprotein gpI by mammalian casein kinase II and casein kinase I. J Virol 63:3912–3918

Hall TC, Miller WA, Bujarski JJ (1982) Enzymes involved in the replication of plant viral RNAs. In: Ingram D, Williams P (eds) Advances in plant pathology, vol 1. Academic Press, London, pp 179–211

Hanks SK, Quinn AM, Hunter T (1988) The protein kinase family: conserved features and deduced phylogeny of the catalytic domains. Science 341:42–52

Haseloff J, Goelet P, Zimmern D, Ahlquist P, Dasgupta R, Kaesberg P (1984) Striking similarities in amino acid sequence among non-structural proteins encoded by RNA viruses that have dissimilar genomic organization. Proc Natl Acad Sci USA 81:4358–4362

Hayes RJ, Buck KW (1990) Complete replication of an eukaryotic virus RNA in vitro by a purified RNA-dependent RNA polymerase. Cell 63:363–368

Hiddinga HJ, Crum CJ, Hu J, Roth DA (1988) Viroid-induced phosphorylation of a host protein related to a dsRNA-dependent protein kinase. Science 241:451–453

Horsch RB, Fry JE, Hoffman NL, Eichholtz R, Rogers SG, Fraley RT (1985) A simple and general method for transferring genes into plants. Science 227:1229–1231

Hovanessian AG (1989) The double-stranded RNA-activated protein kinase induced by interferon: dsRNA-PK. J Interferon Res 9:641–647

Hovanessian AG, Galabru J, Meurs E, Buffet-Janvresse C, Svab J, Robert N (1987) Rapid decrease in the levels of the double-stranded RNA-dependent protein kinase during virus infection. Virology 159:126–136

Hu J, Roth DA (1991) Temporal regulation of tobacco mosaic virus induced phosphorylation of a host encoded protein. Biochem Biophys Res Commun 179:229–235

Jin S, Roitsch T, Ankenbauer RG, Gordon M, Nester EW (1990) The virA protein of *Agrobacterium tumefaciens* is autophosphorylated and is essential for vir gene regulation. J Bacteriol 172:525–530

Katze MG, DeCorato D, Safer B, Galabru J, Hovanessian AG (1987) Adenovirus VAI RNA complexes with the 68 000 Mr protein kinase to regulate its autophosphorylation and activity. EMBO J 6:689–697

Katze MG, Tomita J, Black T, Krug RM, Safer B, Hovanessian AG (1988) Influenza virus regulates protein synthesis during infection by repressing autophosphorylation and activity of the cellular 68 000-Mr protein kinase. J Virol 62:3710–3717

Keese P, Symons RH (1987) The structure of viroids and virusoids. In: Semancik JS (ed) Viroid and viroid-like pathogens. CRC Press, Boca Raton, pp 1–49

Klimczak LJ, Schindler V, Cashmore AR (1992) DNA binding activity of the *Arabidopsis* G-box binding factor BGF1 is stimulated by phosphorylation by casein kinase II from broccoli. Plant Cell 4:87–98

Kostura M, Mathews MB (1989) Purification and activation of the double-stranded RNA-dependent eIF-2 kinase DAI. Mol Cell Biol 9:1576–1586

Lawton MA, Yamamoto RT, Hanks SK, Lamb CJ (1989) Molecular cloning of plant transcripts encoding protein kinase homologs. Proc Natl Acad Sci USA 86:3140–3144

Lees-Miller SP, Chen Y-R, Anderson CW (1990) Human cells contain a DNA-activated protein kinase that phosphorylates simian virus 40 T antigen, mouse p53 and human Ku autoantigen. Mol Cell Biol 10:6472–6481

Linthorst HJM (1991) Pathogenesis-related proteins of plants. Crit Rev Plant Sci 10:123–150

London IM, Levin DH, Matts RL, Thomas NSF, Petryshyn R, Chen JJ (1987) Regulation of protein synthesis. Enzymes 18:359–380

Matthews REF (1991) Plant virology, 3rd edn. Academic Press, New York

Meurs E, Chong K, Galabru J, Thomas NSF, Kerr IM, Williams BRG, Hovanessian AG (1990) Molecular cloning and characterization of the human double-stranded RNA-activated protein kinase induced by interferon. Cell 62:379–390

Mohr IJ, Stillman B, Gluzman Y (1987) Regulation of SV40 DNA replication by phosphorylation of T antigen. EMBO J 6:153–160
Moldave K (1985) Eukaryotic protein synthesis. Annu Rev Biochem 54:1109–1149
Morrow CD, Gibbons GF, Dasgupta A (1985) The host protein required for in vitro replication of poliovirus is a protein kinase that phosphorylates eukaryotic initiation factor 2. Cell 40:913–921
Pierpoint WS (1983) Is there a phyto-intereferon? Trends Biochem Sci 8:5–7
Potter RL, Haley BE (1982) Photoaffinity labeling of nucleotide binding sites with 8-azidoadenosine analogs: applications and techniques. Methods Enzymol 91:613–633
Randall DD, Blevins DG (eds) (1990) Plant protein phosphorylation, protein kinases, calcium and calmodulin. Current topics in plant biochemistry and physiology, vol 9. Univ Missouri Press, Columbia
Ranjeva R, Boudet AM (1987) Phosphorylation of proteins in plants: regulatory effects and potential involvement in stimulus/response coupling. Annu Rev Plant Physiol 38:73–93
Ransone LJ, Dasgupta A (1989) Multiple isoelectric forms of poliovirus RNA-dependent RNA polymerase: evidence for phosphorylation. J Virol 63:4563–4568
Ried JL, Walker-Simmons MK, Everard JD, Diani J (1992) Production of polyclonal antibodies in rabbits is simplified using perforated plastic golf balls. Biotechniques 12:660–666
Roy S, Agy MB, Hovanessian AG, Sonenberg N, Katze MG (1991) The integrity of the stem structure of human immunodeficiency virus type 1 Tat-responsive sequence RNA is required for interaction with the interferon induced 68000 Mr protein kinase. J Virol 65:632–640
Ryan CA, Farmer EE (1991) Oligosaccharide signals in plants: a current assessment. Annu Rev Plant Physiol Plant Mol Biol 42:651–674
Samuel CE, Knutson GS, Berry MJ, Atwater JA, Lasky SR (1985) Purification of double-stranded RNA-dependent protein kinase from mouse fibroblasts. Methods Enzymol 119:499–516
Siegel A, Hari V, Kolacz K (1978) The effect of tobacco mosaic virus infection on host and virus-specific protein synthesis in protoplasts. Virology 85:494–503
Simpson RJ, Moritz RC, Nice EE, Grego B (1987) A high-performance liquid chromatography procedure for recovering subnanomole amounts of protein from SDS-gel electroeluates for gas-phase sequence analysis. Eur J Biochem 165:21–29
Watanabe Y, Okada Y (1986) In vitro viral RNA synthesis by a subcellular fraction of TMV-inoculated tobacco protoplasts. Virology 149:64–73
Watanabe Y, Emori Y, Ooshika I, Meshi T, Ohno T, Okada Y (1984) Synthesis of TMV-specific RNAs and proteins at the early stage of infection in tobacco protoplasts: transient expression of the 30K protein and its mRNA. Virology 133:18–24
Wilson TMA, Plaskitt KA, Watts JW, Osburn JK, Watkins PAC (1990) Signals and structures involved in early interactions between plants and viruses and pseudoviruses. In: Fraser RSS (ed) Recognition and response in plant-virus interactions. NATO ASI Series 41. Springer, Berlin Heidelberg New York, pp 123–145

Cellular Inhibitors of the Interferon-Induced, dsRNA-Activated Protein Kinase

T.G. Lee and M.G. Katze[1]

1 Introduction

Interferon can modulate multiple cellular functions related to growth, differentiation, and immunomodulation as well as antiviral activity (for reviews, see Petska et al. 1987; Sen and Lengyel 1992). In this review, we will discuss the regulation of the double-stranded (ds) RNA-activated protein kinase which is one of more than 30 proteins induced by interferon. The dsRNA-activated protein kinase will be referred to as the P68 kinase (Katze et al. 1987; Meurs et al. 1990) based on its molecular weight of 68 000 in human cells, although it also has been referred to as dsl (Lengyel 1982), DAI (Mellits et al. 1990), dsRNA-PK (Hovanessian 1989), elF-2α-PKds (Akkaraju et al. 1989), P1/elF-2 kinase (Samuel 1991), and PKR (Clemens 1992).

The P68 kinase is a serine/threonine protein kinase characterized by two distinct kinase activities: the first involves an autophosphorylation (activation) reaction by dsRNA and the second a protein kinase activity on exogenous substrates (Galabru and Hovanessian 1987; Hovanessian 1989). The kinase is activated by low concentration of dsRNA (in the nanogram per milliliter range), or by heparin in vitro in the presence of Mg^{+2} or Mn^{+2} (Hovanessian and Galabru 1987; Meurs et al. 1990). Curiously, activation of the kinase is inhibited by high concentrations of dsRNA (1–10 μg/ml). Once activated, the P68 kinase phosphorylates its natural substrate, the α-subunit of eukaryotic protein synthesis initiation factor 2, elF-2. Phosphorylation of the elF-2 α-subunit blocks the elF-2B (GEF)-mediated exchange of GDP in the inactive elF-2-GDP complex, with GTP required for catalytic utilization of elF-2 (Hershey 1991; Merrick 1992). These events lead to limitation in functional elF-2, which is an essential component of protein synthesis and is normally required to bind initiator Met-tRNA (via the ternary complex elF-2-GTP-Met-tRNA) to the initiating ribosomal subunit before mRNA is bound (Jagus et al. 1981; Hershey 1991). Thus, activation of the kinase triggers a series of events that culminates in an inhibition of protein synthesis initiation. Another protein kinase, the heme-regulated inhibitor (HRI), also can specifically phosphorylate the elF-2

[1] Department of Microbiology, SC-42, School of Medicine, University of Washington, Seattle, Washington 98195, USA

Progress in Molecular and Subcellular Biology, Vol. 14
W.E.G. Müller/H.C. Schröder (Eds.)
© Springer-Verlag Berlin Heidelberg 1994

α-subunit, resulting in the inhibition of protein synthesis in rabbit reticulocytes (Chen et al. 1991). Complementary DNAs for the human P68 kinase and murine P65 kinase have now been cloned (Meurs et al. 1990; Icely et al. 1991; Feng et al. 1992), and the corresponding proteins show 61% identity (Feng et al. 1992). The yeast elF-2 kinase, GCN2, which is thought to be the homologue of the mammalian P68 kinase, displays 38% identity with its human counterpart in the catalytic domains (Ramirez et al. 1991; Dever et al. 1992). HRI also exhibits sequence identity (42%) with P68 (Chen et al. 1991).

Since virus-specific RNAs synthesized during infection have the potential to activate the P68 kinase (Katze et al. 1988; Maran and Mathews 1988; Bischoff and Samuel 1989; Black et al. 1989; Edery et al. 1989; SenGupta and Silverman 1989; Roy et al. 1991), the kinase is likely to mediate some of the antiviral effects of interferon via the inhibition of both host and viral protein synthesis. However, a number of eukaryotic viruses have evolved mechanisms to downregulate the kinase activity in effective and varied ways (for reviews, see Sonenberg 1990; Samuel 1991; Katze 1992, 1993). These viruses include adenovirus (reviewed in Mathews and Shenk 1991), influenza virus (Katze et al. 1986; Lee et al. 1990, 1992), poliovirus (Black et al. 1989, 1993), reovirus (Imani and Jacobs 1988; Lloyd and Shatkin 1992), vaccinia virus (Whitaker-Dowling and Youngner 1984; Akkaraju et al. 1989; Chang et al. 1992; Davies et al. 1992), and human immunodeficiency virus (Edery et al. 1989; Gunnery et al. 1990; Roy et al. 1990).

There is accumulating evidence that the kinase, which is constitutively expressed in cells, plays a pivotal role in the regulation of cellular gene expression in the absence of virus infection and interferon induction. Petryshyn and colleagues have reported that the kinase may be important in controlling growth arrest prior to differentiation into adipocytes (Petryshyn et al. 1984, 1988) and partially purified a 15-kDa inhibitor of the kinase (Judware and Petryshyn 1991, 1992). A cellular inhibitor has also been partially purified from human FL cells (Saito and Kawakita 1991) with a molecular weight of greater than 160 kDa. Furthermore, it has been suggested that the P68 kinase may play a role in the stress-heat shock response (Dubois et al. 1989, 1991), or in the transcriptional regulation of the β-interferon gene or proto-oncogenes such as c-*fos* and c-*myc* (Tiwari et al. 1988; Zinn et al. 1988; Wathelet et al. 1989). Recent studies have shown that introduction of transforming *ras* gene induces a heat-sensitive inhibitor of the P68 kinase in BALB/c-3T3 fibroblasts (Mundschau and Faller 1992). It should also be mentioned that a specific phosphatase, e.g. the type 1 phosphatase, also may regulate kinase activity, although this has not been directly demonstrated by in vivo analysis (Szyszka et al. 1989). Recently, expression of a functionally defective mutant of the human P68 kinase in NIH 3T3 cells resulted in malignant transformation, suggesting that the P68 kinase may function as a suppressor of cell proliferation and tumorigenesis (Koromilas et al. 1992; Meurs et al. 1993). In the first half of this review, we will examine how the P68 kinase is downregulated by a cellular factor during influenza virus as well as poliovirus infection. Following this, we discuss other cellular inhibitors of the P68 kinase which have been identified in a variety of systems.

2 A Cellular Inhibitor of the P68 Kinase from Influenza Virus-Infected Cells

2.1 Downregulation of the P68 Kinase During Influenza Virus Infection

Our laboratory first determined that influenza virus downregulated the P68 kinase activity by analyzing cells doubly infected with influenza virus and the adenovirus VAI RNA mutant dl331 (Katze et al. 1986). In cells infected by dl331 alone there was a dramatic reduction in the levels of both viral and cellular protein synthesis (Thimmappaya et al. 1982). This was due to excessive phosphorylation of the eIF-2 α-subunit by an activated P68 protein kinase which cannot be downregulated due to the absence of the adenovirus-encoded VAI RNA. When dl331-infected cells were superinfected with influenza virus, a dramatic suppression of the protein kinase activity normally detected during dl331 infection was observed (Katze et al. 1986). We measured the activity of the kinase after immunopurifying the P68 kinase from these cells utilizing the high-affinity monoclonal antibody which recognized an epitope located at the very amino terminus of the human kinase (Laurent et al. 1985; Katze et al. 1991). The level of autophosphorylation and activity of kinase on the eIF-2 α-decreased approximately two- and four-fold, respectively (Katze et al. 1988). These data provided the first clues that influenza virus either encoded or activated a gene product that, analogous to adenovirus VAI RNA, inhibited the P68 kinase and any resultant shutdown of protein synthesis. Most importantly, we subsequently observed this repression of kinase activity in cells infected by influenza virus alone and this correlated with a decrease in endogenous levels of eIF-2 α phosphorylation (Katze et al. 1988). This decrease was not due to enhanced P68 degradation as measured by either Western blot or immunoprecipitation analysis. When virus gene expression was restricted to primary transcription by carrying out infections in the presence of the protein synthesis inhibitor anisomycin, P68 kinase activity was not downregulated but actually increased over control levels (Katze et al. 1988). This indicated that viral replication was necessary for kinase suppression and influenza viral RNAs may indeed be able to activate the protein kinase. Based on these series of observations, we next set out to identify what was controlling P68 kinase activity in influenza virus-infected cells and determine how this was occurring at the molecular level.

2.2 Purification and Characterization of a Cellular Inhibitor of the P68 Kinase from Influenza Virus-Infected Cells

To identify the repressor of the protein kinase in influenza virus-infected cells, we first developed an in vitro phosphorylation assay that quantitatively measured inhibitory activity (Lee et al. 1990). Preliminary data showed that the inhibitor of the kinase was most likely a protein based on the decrease in the inhibitory activity after heat or trypsin treatment. Using this assay and 2×10^{10} influenza virus-infected MDBK (Madin-Darby bovine kidney) cells as the starting material, the inhibitor was purified to near homogeneity. The final purified product, which had an apparent molecular weight of 58 000 daltons, inhibited both P68 autophosphorylation and eIF2 α phosphorylation in a dose-dependent manner (Fig. 1). We now know that the purified inhibitor does not function as a protease, phosphatase, ribonuclease, or ATPase (Lee et al. 1990, 1992). We also showed that the 58-kDa protein does not act by sequestering RNA activators as described for other viral gene products. The inhibitor appears to act on the P68 kinase in a stoichiometric manner, although a direct interaction has not yet been demonstrated. As stated previously, P68 kinase has two activities: the kinase first activates itself via an autophosphorylation reaction and then catalyzes the phosphorylation of eIF-2 α (Hovanessian 1989). The best-characterized kinase inhibitor, VAI RNA, can block kinase activation and activity only if it is present in the reaction prior to dsRNA addition (Katze et al. 1987; Galabru et al. 1989). However, the 58-kDa protein blocked the phosphorylation of the eIF-2 α by an activated kinase as well. We speculate that both types of inhibitory activity provide additional safeguards to the cell (and the virus) in the event that P68 autophosphorylation cannot be totally prevented. These data show that the cellular inhibitor works differently than the adenovirus VAI RNA and the reovirus and vaccinia virus-encoded inhibitors.

2.3 Identification of the 58-kDa Protein and a Specific Anti-Inhibitory Activity in Uninfected MDBK Cells

An important remaining question was whether the purified inhibitor was an influenza virus-encoded protein. Western blot analysis using virus-specific antibodies indicated that the purified repressor was a cellular and not a viral protein (Lee et al. 1990). As a first step towards understanding the molecular events controlling the regulation of the kinase by the 58-kDa protein, we obtained peptide microsequence information on the purified inhibitor (Lee et al. 1992). Using the peptide antibody we confirmed that indeed the 58-kDa protein was cellular in origin and further that the inhibitor was present in identical levels in uninfected and influenza virus-infected cells (Fig. 2). This raised the interesting question of how the cellular 58-kDa inhibitor was activated during influenza virus infection because the inhibitory activity was not previously found in crude, uninfected cell extracts. A clue was obtained from the observation that kinase inhibitory activity was measurable after these crude extracts were subjected to ammonium sulfate fractionation. We first tested whether the inhibitory activity recovered after ammonium sulfate fractionation from uninfected cell extracts was similar to that originally described for infected cells by following the purification protocol originally described. The inhibitor purified from uninfected cells had identical chromatographic properties as that from virus-infected cells, including sedimentation on a glycerol gradient (Lee et al. 1992). We speculated that another factor (which we referred to as an anti-inhibitor), which normally functioned to inhibit the inhibitor, may have been dissociated from the

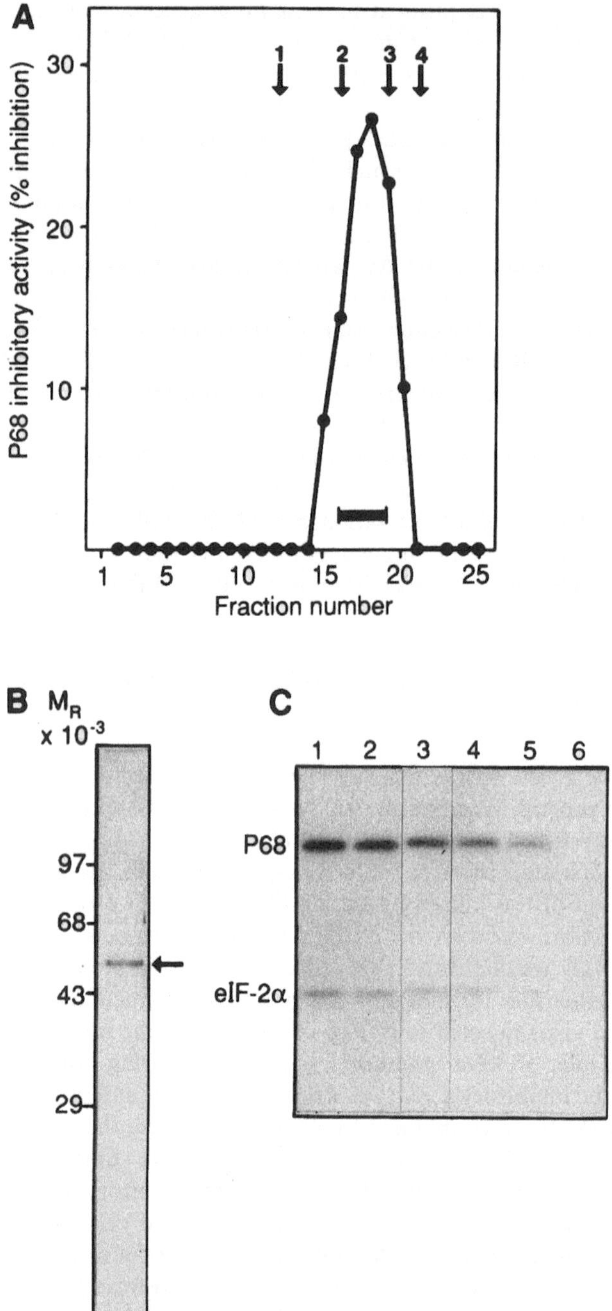

Fig. 1. Glycerol gradient analysis of the P68 inhibitor. **A** Glycerol gradient profile of the Mono S fraction layered onto a 10–30% glycerol gradient and spun at 49 000 rpm for 21 h at 4 °C. Fractions were assayed for inhibition of P68 autophosphorylation utilizing purified P68 and the

58-kDa protein during the high salt treatment. An alternative explanation was that ammonium sulfate treatment altered the inhibitor in some way. To distinguish between these possibilities, we performed mixing experiments with the different ammonium sulfate fractions. In this series of experiments, we observed that addition of increasing amounts of the 60–80% fraction partially repressed the kinase inhibitory activity of the 40–60% fraction which contained the strongest inhibitory activity (Lee et al. 1992). From these results, we concluded that a specific anti-inhibitory factor(s) was present in uninfected cells, which could regulate activity of the 58-kDa inhibitor and consequently the kinase itself.

2.4 Model of P68 Kinase Regulation

Based on the data accumulated so far, we present a hypothetical model for the P68 kinase regulation in influenza virus-infected cells (Fig. 3). Figure 3A depicts the scenario in the absence of any regulation when the kinase is activated by dsRNAs, and as a result protein synthesis initiation is inhibited Figure 3B shows the results of P68 downregulation; after virus infection and synthesis of viral-specific dsRNAs, the 58-kDa inhibitor (P58), which had been in an inactive complex with its own inhibitor (the anti-inhibitor referred to as I-P58), becomes dissociated and can then block the autophosphorylation and/or activity of the kinase possibly through a direct interaction with the kinase. The end result is that protein synthetic rates are not compromised. The work on the cellular 58-kDa inhibitor and its regulation should provide novel insights into the regulation of the kinase not only in virus-infected cells but also in uninfected cells. If the kinase does regulate cell growth and differentiation as reported (Judware and Petryshyn 1991), it is possible that the 58-kDa inhibitor would become dissociated or associated with its inhibitor in response to a specific environmental signal or stimulus to control kinase activity levels and overall protein rates as needed. Regulation of the kinase by activators and inhibitors is unlikely to be an all or none phenomenon. A more subtle regulation is more probable given that small changes in kinase activity may have large effects on protein synthesis.

in vitro assay. The positions of molecular weight standards analyzed on a parallel gradient are shown at *top*: *1* gamma globulin (M_r 158 000); *2* bovine serum albumin (M_r 68 000); *3* ovalbumin (M_r 44 000); *4* myoglobin (M_r 17 000). Active fractions were pooled as shown by the *brackets*. **B** SDS-14% PAGE analysis of pooled active gradient fractions as detected by silver staining. The *arrow* indicates the 58-kDa polypeptide with the position of molecular weight standards shown on the *left*.C Activity of the purified 58-kDa polypeptide. The gradient peak was analyzed for inhibitory activity utilizing purified P68 and eIF-2. Purified P68 was preincubated with 0.0 µl (*lane 1*); 0.5 µl (*lane 2*); 2 µl (*lane 3*); 8 µl (*lane 4*); 16 µl (*lane 5*), of purified material after which P68 and eIF-2 α phosphorylation were measured as described (Lee et al. 1990). *Lane 6* shows the result of an assay in which the purified P68 kinase was omitted but which contained 8 µl of inhibitor. The concentration of the purified inhibitor was approximately 0.008 mg/ml

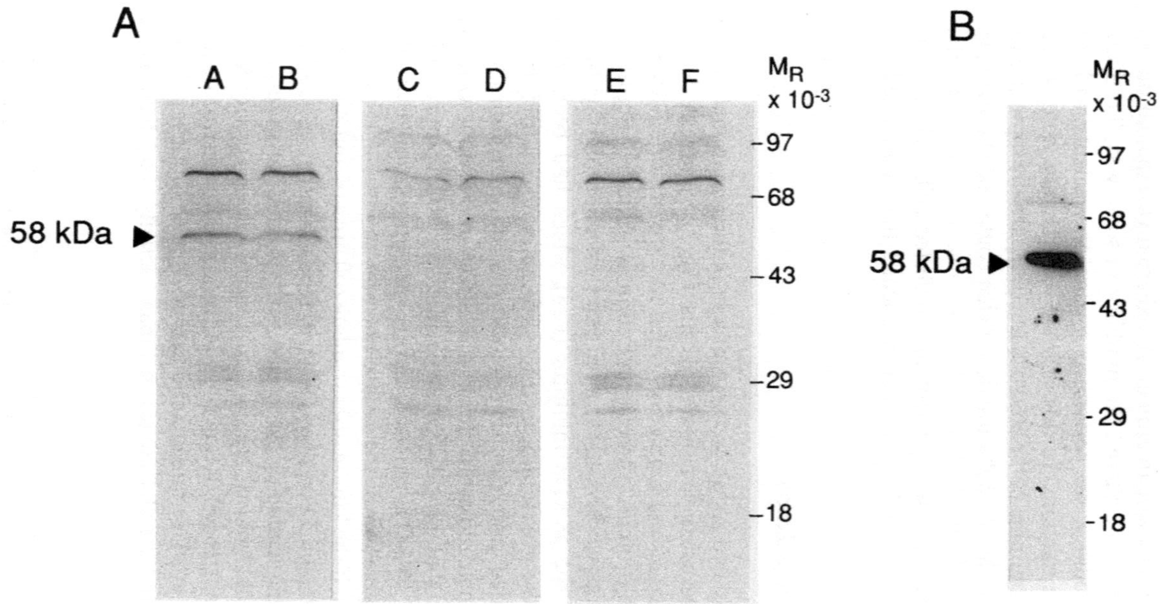

Fig. 2. Identification of the 58-kDa kinase inhibitor in infected and uninfected cells utilizing peptide antibody and Western blot analysis. **A** Cell extracts from uninfected cells (*lanes A, C*, and *E*) or from influenza virus-infected cells (*lanes B, D*, and *F*) were electrophoresed on an SDS/14% PAGE, blotted onto nitrocellulose, and probed with the specific 58-kDa peptide antibody (*lanes A, B*), preimmune antiserum (*lanes C, D*), or with peptide antibody which was prereacted with a 100-fold molar excess of 58-kDa specific peptide (*lanes E, F*). **B** Western blot analysis also was performed on the glycerol gradient purified inhibitor. Proteins were visualized utilizing the Amersham ECL chemiluminescence system. Position of molecular weight standards are shown to the *right of each panel*

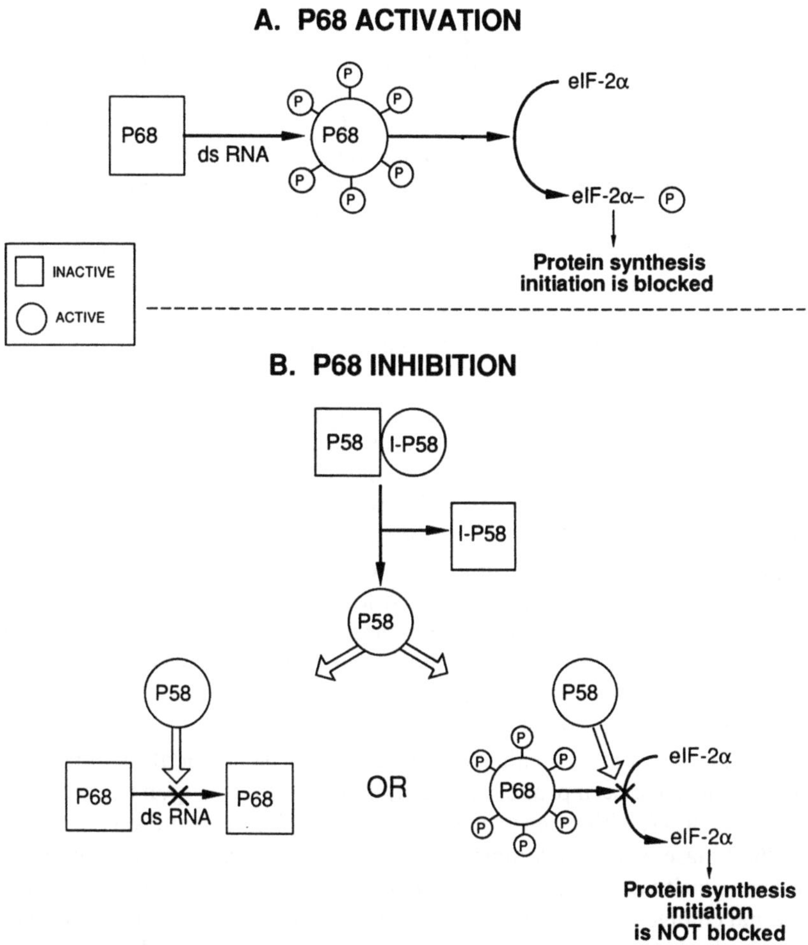

Fig. 3. Model of P68 regulation. **Key:** *P68* interferon-induced protein kinase; *P58* cellular 58-kDa inhibitor of the kinase; *I-P58* the anti-inhibitor, or specific inhibitor of P58; *eIF-2* eukaryotic initiation factor 2 α-subunit; *squares* denote inactive molecules and *circles* active. **A** P68 activation: dsRNAs induce the autophosphorylation of the kinase which then leads to eIF-2 α phosphorylation and protein synthesis inhibition. **B** P68 inhibition: during viral infection (or other environmental signals), the anti-inhibitor (I-P58) becomes dissociated from the kinase inhibitor (P58). The inhibitor, now in an active form, can function either to block the autophosphorylation of the kinase, or if the kinase is already activated, it can block the phosphorylation of eIF-2 α. The end result is that protein synthesis is not compromised

3 Degradation of the P68 Kinase by a Cellular Protease During Poliovirus Infection

Poliovirus has devised a unique mechanism to regulate P68: the kinase was dramatically degraded during infection (Black et al. 1989). Despite P68 degradation, the levels of eIF-2 α phosphorylation increased in poliovirus-infected cells (Black et al. 1989; O'Neill and Racaniello 1989). It remains to be determined why poliovirus encoded a mechanism to degrade P68 and not to block the activation step like many other viruses. One possibility is that enhanced eIF-2 α phosphorylation early after infection may help play a role in the host protein synthesis shutoff (reviewed in Sonenberg 1990). Nevertheless, it is highly likely that without significant P68 proteolysis, eIF-2 α phosphorylation would increase to unacceptable levels in poliovirus-infected cells. Recent work from our laboratory has shown that neither the poliovirus-encoded 2A protease nor 3C protease was directly responsible for the degradation of P68, strongly suggesting that a cellular protease was acting on the kinase. Although the activation of the kinase was not required for P68 degradation, it has been shown that poliovirus RNA replication is necessary for kinase degradation (Black et al. 1993). Data now suggest that the protease was insoluble, required divalent cations to act, and was comprised of both an RNA and protein component since activity was inhibited by the action of both trypsin and single- and double-stranded specific ribonucleases. Mapping of the protease-sensitive sites, using in vitro translated, truncated, and deletion mutants, revealed that the sites required for degradation resided in the amino terminus and colocalized to dsRNA-binding domains (Black et al. 1993). Additional evidence for the role of RNA was from the observation that preincubation of cell-free extracts with dsRNA, poly(I):poly(C), prevented P68 degradation (Black et al. 1993).

Based on these data, two alternative models of P68 degradation in poliovirus-infected cells have been proposed (Fig. 4). In one scenario, upon interaction with dsRNA (or RNA with extensive double-stranded features) and divalent cations, the P68 kinase, which is comprised of an amino terminus containing both the protease-sensitive sites and the dsRNA-binding domains, undergoes a conformational change. This alteration in structure then enables a cellular protease to degrade the kinase. In an alternative model, the dsRNA and protease act in concert in the presence of divalent cations, possibly as a ribonucleoprotein complex, to proteolyze the kinase directly and completely. Both models suggest that the specificity of the reaction or the targeting of P68 results from the kinase interaction with dsRNA or RNA with extensive double-stranded regions. What then is the nature of the RNA participating in the cleavage? A possible candidate is the poliovirus-specific RNA, which has been shown to have a double-stranded structure (Ehrenfeld and Hunt 1971; Celma and Ehrenfeld 1974), and which could activate the P68 kinase as reported (Black et al. 1989). However, we presently cannot rule out that cellular RNAs with extensive double-stranded regions also may be a component of the degradation pathway. It should be stated that poliovirus-specific dsRNAs, when added alone to cell-free extracts prepared from uninfected cells, failed to induce kinase degradation. Thus, we conclude that poliovirus infection must result in the activation of a cellular protease which then acts together with RNA or attacks the kinase after RNA has bound to P68. Further studies on the nature of the cellular protease and

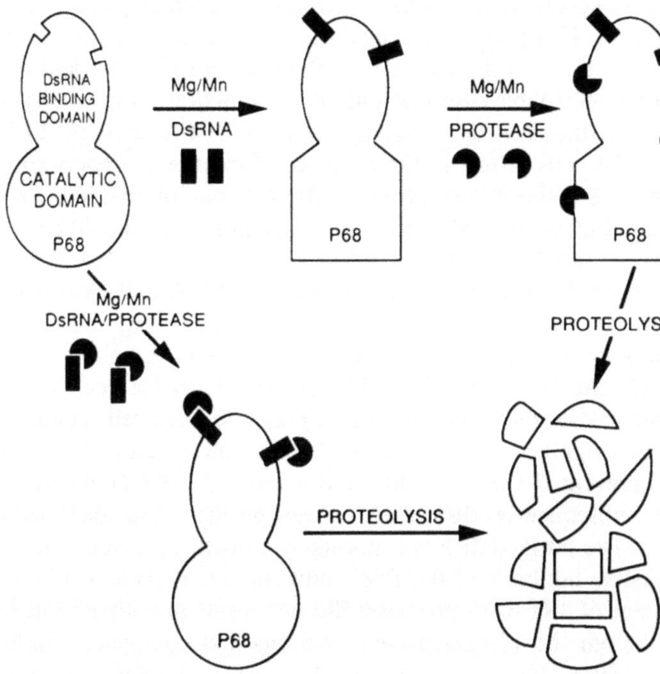

Fig. 4. Hypothetical models depicting the mechanisms of P68 degradation during poliovirus infection. See the text for details

RNA are required before addressing the question whether this degradation pathway also participates in the regulation of the P68 kinase in uninfected cells.

4 A Cellular Inhibitor That Regulates the P68 Kinase in 3T3-F442 Fibroblasts

In the previous sections, we discussed how the P68 kinase was downregulated by a cellular modulator during viral infection. Identification of another cellular inhibitor of the kinase came from studies on cell differentiation. When mouse 3T3-F442A preadipocyte fibroblasts reached confluence in the appropriate culture medium, their growth was arrested, and the cells underwent terminal differentiation to adipocytes (Green and Kehinde 1974). Based on the role of interferon in control of cell differentiation, Petryshyn's group examined the relationship between cell growth and phosphorylation of the P68 kinase. Studies have shown that 3T3-F442A cells produced and secreted interferon spontaneously and exhibited a transient double-stranded RNA-dependent P68 phosphorylation (Petryshyn et al. 1984). The levels of this phosphorylation increased progressively. Furthermore, treatment of 3T3-F442A cells with β-interferon at all stages of growth resulted in an increase in phosphorylation of P68 (Petryshyn et al. 1984). Using antiserum against that kinase, they demonstrated that the phosphorylation of the kinase was observed both in vivo and in vitro and the increased phosphorylation of P68 after induction with interferon was due to de novo synthesis of P68 (Petryshyn et al. 1988). Interestingly, P68 isolated from 3T3-F442A cells grown in the presence of interferon underwent phosphorylation in vitro without the addition of dsRNA, and in vivo in the absence of viral infection (Petryshyn et al. 1988). This phosphorylation could be attributed to the activity of cellular regulatory RNA found in 3T3-F442A (Li and Petryshyn 1991). However, the identity of this cellular RNA and its role in vivo are not known.

It was demonstrated previously that 3T3-F442A cells maintained in 10% domestic cat serum failed to differentiate (Kuri-Harcuch and Green 1978). Further studies showed that extracts prepared from cells cultured in cat serum contained significantly elevated levels of a P68 regulatory factor (termed dRF) which prevented the activation and activity of P68 (Judware and Petryshyn 1991). Subsequently, dRF activity was purified and identified as a 15-kDa protein. The purified dRF was not likely a protein phosphatase or protease but a reversible inhibitor of the P68 (Judware and Petryshyn 1991). Interestingly, treatment of dRF with proteases did not result in a significant loss of activity even though the 15-kDa protein was digested completely (Judware and Petryshyn 1991). It was suggested in this case that dRF activity resided within a small peptide fragment which was resistant to further proteolysis. Recently, it was demonstrated that dRF prevented the binding of ATP to the kinase by blocking the formation of P68-dsRNA complexes. The dRF did not appear to interact with the dsRNA directly. It was proposed that dRF might exert its effect through an interaction with P68 (Judware and Petryshyn 1992). However, the physical interaction between P68 and dRF has not been shown. The dRF appears to be different from the P58 inhibitor purified from influenza virus-infected cells due to its difference in molecular weight and its inhibitory mechanisms.

5 A Cellular Inhibitor of the P68 Kinase in Oncogenic *ras*-Transformed BALB Cells

In addition to its role in the regulation of translation initiation, the P68 kinase has been implicated as a component of signal transduction pathways, specifically those of platelet-derived growth factor (PDGF) and β-interferon. Treatment of BALB/c-3T3 fibroblasts (BALB) with synthetic dsRNA [poly(I) · poly(C)] induced transcription of the PDGF-inducible genes c-*fos* and c-*myc* as well as the PDGF and β-interferon inducible gene *JE* (Zullo et al. 1985; Hall et al. 1989). A nonspecific inhibitor of the P68 kinase, 2-aminopurine (Legon et al. 1974; Farrell et al. 1977), inhibited induction of these genes by serum (Zinn et al. 1988) or by epidermal growth factor (Mahadevan et al. 1990).

Faller's group reported that introduction of an activated *ras* gene (Ki-v-*ras*) into BALB cells blocked signal transduction by PDGF in these cells (Zullo and Faller 1988). They also found that induction of the *JE* gene by β-interferon was blocked in Ki-v-*ras* transformed BALB cells (KBALB cells) (Offermann and Faller 1989). These results suggested the P68 kinase as a potential part of the PDGF and β-interferon signal transduction pathways. Recently, it was reported that dsRNA-mediated activation of the P68 kinase was blocked in KBALB cells in a manner specific to v-*ras* (Mundschau and Faller 1992). The amount of the kinase was similar in the extracts from BALB and KBALB cells, showing that the P68 kinase was present in KBALB in a latent form. By mixing experiments, they demonstrated that the *ras*-induced inhibitor in KBALB lysates functioned in *trans*. In addition, induction of the *ras*-induced inhibitor appeared to be independent of interferon treatment. The *ras*-induced inhibitor was heat-sensitive and was inactivated by phenol-chloroform extraction, suggesting that the inhibitor was a protein. However, it may also be present in a ribonucleoprotein complex. The inhibitor is unlikely to be a P68-specific protease or a dsRNA-binding protein. It was not clear, however, whether the inhibitory activity was associated with phosphatase activity. The *ras*-transformed cells appeared to have higher levels of phosphatases because addition of the phosphatase inhibitor okadaic acid increased the phosphorylation of the P68 kinase in KBALB lysates. Further purification of the inhibitor from crude extracts will help to distinguish the inhibitory activity from other nonspecific phosphatase activities. The inhibitor appeared to be different from the P58 purified from influenza virus-infected cells due to its different affinity for heparin. However, the *ras*-induced inhibitor needs to be purified and characterized further.

6 A Cellular Inhibitor of the P68 Kinase in Human Amnion FL Cells

Treatment of human amnion FL cells with γ-interferon rapidly induced an antiviral state against sindbis virus, possibly by the activation of the dsRNA-dependent protein kinase (Saito 1989). Further biochemical and immunological analysis showed that this protein kinase was indeed the P68 kinase (Saito 1990). Because cultured cells responded to some viral infections with stress responses, such as synthesis of heat-shock proteins, this group examined the dsRNA-dependent phosphorylation of the P68 kinase under stress. When dsRNA was added, phosphorylation of the P68 kinase increased in the ribosomal salt wash fraction obtained from the heat-shocked as well as arsenite-treated cells. However, the physical amounts of the P68 kinase did not increase appreciably after sindbis virus infection or heat shock, implying the conversion of the P68 kinase from a dsRNA-unresponsive form to a responsive form (Saito 1990). In contrast to these results, Dubois et al. (1989) reported that P68 kinase activity was greatly reduced in extracts of heat-shocked HeLa cells, and that the loss of activity was due to a decreased solubility of the enzyme. The insoluble P68 kinase was localized around the nucleus as a thick aggregate (Dubois et al. 1991). Saito's group mentioned that they also had found a decrease in P68 kinase activity in confluent FL cells. Because they isolated the P68 kinase from the ribosomal salt wash fraction of subconfluent FL cells, it was possible that the P68 kinase responded to heat shock in a different manner, depending on the cell confluency. However, further research is necessary in order to explain this discrepancy. Recently, it was reported that the P68 kinase in FL cells was present in a form complexed with a cellular inhibitor (Saito and Kawakita 1991). The inhibitor was dissociated from the P68 kinase by DEAE-cellulose chromatography or by gel filtration, yielding a dsRNA-responsive P68 kinase. Activity of the inhibitor decreased after interferon treatment or after Sindbis virus infection with a concomitant increase in the amount of the dsRNA-responsive P68 kinase. The apparent molecular weight of the inhibitor as estimated by gel filtration was more than 160 kDa. It has not been determined whether the inhibitor is a single macromolecule or a multisubunit complex. Once the inhibitor is purified, it will be of interest to compare it with the other known cellular inhibitors of the P68 kinase.

7 Conclusions and Future Directions

In this review we have discussed different cellular inhibitors of the interferon-induced, dsRNA-activated protein kinase. The P68 kinase has been studied intensively due to its possible role in the antiviral action mediated by interferon. As described in the introduction, many eukaryotic viruses downregulate the kinase activity during infection. Adenovirus synthesizes virus-associated RNA 1 (VAI RNA) which blocks the dsRNA-induced activation of the kinase by forming a stable complex with the kinase (reviewed in Mathews and Shenk 1991). Similar mechanisms are thought to occur during Epstein-Barr virus infection (Clarke et al. 1991). Another interesting viral inhibitor of the kinase is the K3L protein which is synthesized during vaccinia virus infection (Beattie et al. 1991; Davies et al. 1992). The K3L protein has 28% identity to the N-terminal region of the eIF-2 α which contains the serine phosphorylated by the kinase. It is possible that the P68 kinase could be regulated by a cellular inhibitor which has a pseudosubstrate motif (House and Kemp 1990). Once the cDNAs encoding the cellular inhibitors described above are cloned, it will be of interest to check whether the inhibitors show any homology with eIF-2 α.

In order to understand the function of the cellular inhibitor of the P68 kinase, we need more information about the physiological role of the kinase within the cell. Recently, Koromilas et al. (1992) reported that expression of a functionally defective mutant of the human P68 kinase in NIH 3T3 cells resulted in malignant transformation, suggesting that the P68 kinase may function as a tumor suppressor gene. They have proposed that the transdominant negative action of the mutant P68 might be due either to the formation of an inactive hybrid (wild type/mutant) P68 dimer, or to the sequestration of limiting dsRNA activators by the mutant P68. In the case of viruses that do not kill the cells but instead produce a mitogenic response or transform the cells to a tumorigenic phenotype, part of the growth-promoting effect might involve the inactivation of the P68 kinase by viral gene products. Indeed, it has been reported that expression of wild-type P68 has an inhibitory effect on cell growth in yeast (Chong et al. 1992) and in the insect cell line SF9 (Barber et al. 1992), as does expression of the tumor suppressor gene p53 (Fields and Jang 1990).

These results, together with the known function of the kinase in translational control, suggest that the P68 kinase has several biological functions within the cell. Recently, it has been reported that two biochemically distinguishable forms of the kinase can be isolated from the cytosolic fraction as a dimer of partially phosphorylated subunits, and from the ribosome salt wash as an unphosphorylated monomer (Langland and Jacobs 1992). The cellular inhibitors described in this review may participate in regulating the different activities of the kinase in vivo. However, we do not know at present whether these inhibitors are related. Further biochemical studies of the inhibitors and cloning of the genes are needed. Another regulator of the P68 kinase might be the dsRNA activator itself. It has been reported that dsRNA induces β-interferon expression through activation of the transcription factor NF-κB Lenardo et al. 1989; Visvanathan and Goodbourn 1989; Nielsch et al. 1991; Reis et al. 1992). Li and Petryshyn (1991) reported that cellular regulatory RNA isolated from 3T3-F442A cells could phosphorylate the P68 kinase in vitro. However, the identity of this RNA and its role in vivo are unknown. The P68 kinase also might be regulated via a change in its subcellular localization, as suggested from the studies on the heat-shock response (Dubois et al. 1989, 1991). Finally, dephosphorylation of the kinase by a specific phosphatase might also play a role in the regulation of the kinase (Szyszka et al. 1989). Further studies of this fascinating protein will hopefully provide detailed answers about its biological role in the eukaryotic cell.

Acknowledgements. We thank Michele Garfinkel and Ted Strom for helpful comments. The work presented from the author's laboratory was supported by NIH grants AI 22646 and RR 00166. T.G.L. is a recipient of the Dr. Helen Riaboff Whiteley graduate fellowship in microbiology from the University of Washington.

References

Akkaraju GR, Whitaker-Dowling P, Youngner JSY, Jagus R (1989) Specific kinase inhibitory factor of vaccinia virus prevents translational inhibition in rabbit reticulocytes. J Biol Chem 264:10321–10325

Barber GN, Tomita J, Garfinkel MS, Meurs E, Hovanessian A, Katze MG (1992) Detection of protein kinase homologues and viral RNA binding domains utilizing polyclonal antiserum prepared against a baculovirus expressed dsRNA activated 68,000 dalton protein kinase. Virology 191:670–679

Beattie E, Tartaglia J, Paoletti E (1991) Vaccinia virus-encoded elF-2 alpha homolog abrogates the antiviral effect of interferon. Virology 183:419–422

Bischoff JR, Samuel CE (1989) Mechanism of interferon action: activation of the human PI/eIF-2 protein kinase by individual reovirus-class mRNAs: S1 mRNA is a potent activator to S4 mRNA. Virology 172:106–115

Black T, Safer B, Hovanessian A, Katze MG (1989) The cellular 68,000 Mr protein kinase is highly autophosphorylated and activated yet significantly degraded during poliovirus infection: implications for translational regulation. J Virol 63:2244–2252

Black TL, Barber GN, Katze MG (1993) Degradation of the interferon induced 68,000 Mr protein kinase by poliovirus requires RNA. J Virol 67:791–800

Celma ML, Ehrenfeld E (1974) Effect of poliovirus double-stranded RNA on viral and host cell protein synthesis. Proc Natl Acad Sci USA 71:2440–2444

Chang HW, Watson JC, Jacobs BL (1992) The E3L gene of vaccinia virus encodes and inhibition of the interferon induced, dsRNA dependent protein kinase. Proc Natl Acad Sci USA 89:4825–4829

Chen J, Pal J, Petryshn R, Kvo R, Yang J, Throop M, Gehrke L, London I (1991) Amino acid microsequencing of tryptic peptides of heme regulated eIF-2 kinase. Proc Natl Acad Sci USA 88:315–319

Chong KL, Schappert K, Meurs E, Feng F, Donahue TF, Friesen JD, Hovanessian AG, Williams BRG (1992) Human P68 kinase exhibits growth suppression in yeast and homology to the translational regulator GCN2. EMBO J 11:1553–1562

Clarke PA, Schwemmle M, Schickinger J, Hilse K, Clemens MJ (1991) Binding of Epstein-Barr virus small RNA EBER-1 to the double-stranded RNA-induced inhibition of protein synthesis in reticulocyte lysates. Eur J Biochem 193:635–641

Clemens M (1992) Suppression with a difference. Nature 360:210–211

Davies MV, Elroy-Stein O, Jagus R, Moss B, Kaufman RJ (1992) The vaccinia virus K3L gene product potentiates translation by inhibiting double-stranded-RNA-activated protein kinase and phosphorylation of the alpha subunit of eukaryotic initiation factor 2. J Virol 66:1943–1950

Dever TE, Feng L, Wek RC, Cigan AM, Donahue TF, Hinnebusch AG (1992) Phosphorylation of initiation factor 2 alpha by protein kinase GCN2 mediates gene specific translational control of GCN4 in yeast. Cell 68:585–596

Dubois MF, Galabru J, Lebon P, Safer B, Hovanessian AG (1989) Reduced activity of the interferon-induced double-stranded RNA-dependent protein kinase during a heat shock stress. J Biol Chem 264:12165–12171

Dubois MF, Hovanessian AG, Bensaude O (1991) Heat-shock-induced denaturation of proteins characterization of the insolubilization of the interferon-induced p68 kinase. J Biol Chem 266:9707–9711

Edery I, Petryshyn R, Sonenberg N (1989) Activation of double-stranded RNA-dependent kinase (dsl) by the TAR region of HIV-1 mRNA: a novel translation control mechanism. Cell 56:303–312

Ehrenfeld E, Hunt T (1971) Double-stranded poliovirus RNA inhibits initiation of protein synthesis by reticulocyte lysates. Proc Natl Acad Sci USA 68:1075–1078

Farrell PJ, Balkow K, Hunt T, Jackson RJ, Trachsel H (1977) Phosphorylation of initiation factor eIF2 and the control of reticulocyte protein synthesis. Cell 11:187–200

Feng GS, Chong K, Kumara A, Williams BRG (1992) Identification of dsRNA binding domains in the interferon-induced dsRNA-activated P68 kinase. Proc Natl Acad Sci USA 89:5447–5451

Fields S, Jang SK (1990) Presence of a potent transcription sequence in the p53 protein. Science 249:1046–1049

Galabru J, Hovanessian AG (1987) Autophosphorylation of the protein kinase dependent on double-stranded RNA. J Biol Chem 262:15538–15544

Galabru J, Katze MG, Robert N, Hovanessian AG (1989) The binding of double-stranded RNA and adenovirus VAI RNA to the interferon-induced protein kinase. Eur J Biochem 178:581–589

Green H, Kehinde O (1974) Sublines of mouse 3T3 cells that accumulate lipid. Cell 1:113–116

Gunnery SA, Rice P, Robertson HD, Mathews MB (1990) Tat-responsive region RNA of human immunodeficiency virus 1 can prevent activation of the double-stranded RNA-activated protein kinase. Proc Natl Acad Sci USA 87:8687–8691

Hall DJ, Jones SD, Kaplan DR, Whitman M, Rollins BJ, Stiles CD (1989) Evidence for a novel signal transduction pathway activated by platelet-derived growth factor and by double-stranded RNA. Mol Cell Biol 9:1705–1713

Hershey JWD (1991) Translational control in mammalian cells. Annu Rev Biochem 60:717–755

House C, Kemp BE (1990) Protein kinase C pseudosubstrate prototope: structure-function relationships. Cell Signal 2:187–190

Hovanessian AG (1989) The double-stranded RNA-activated protein kinase induced by interferon: dsRNA-PK. J Interferon Res 9:641–647

Hovanessian AG, Galabru J (1987) The dsRNA dependent protein kinase is also activated by heparin. Eur J Biochem 167:467–473

Icely PL, Gross P, Bergeron JJM, Devault A, Afar DEH, Bell JC (1991) TIK, a novel serine/threonine kinase, is recognized by antibodies directed against phosphotyrosine. J Biol Chem 26:16073–16077

Imani F, Jacobs BL (1988) Inhibitory activity for the interferon-induced protein kinase is associated with the reovirus serotype 1 sigma 3 protein. Proc Natl Acad Sci USA 85:7887–7891

Jagus R, Anderson W, Safer B (1981) The regulation of initiation of mammalian protein synthesis. Prog Nucleic Acid Res 25:127–185

Judware R, Petryshyn R (1991) Partial characterization of a cellular factor that regulates the double-stranded RNA-dependent eIF-2 alpha kinase in 3T3-F442A fibroblasts. Mol Cell Biol 11:3259–3267

Judware R, Petryshyn R (1992) Mechanism of action of a cellular inhibitor of the dsRNA-dependent protein kinase from 3T3-F442A cells. J Biol Chem 267:21685–21690

Katze MG (1992) The war against the interferon-induced dsRNA activated protein kinase: can viruses win? J Interferon Res 12:241–248

Katze MG (1993) Games viruses play: a strategic initiative against the interferon-induced dsRNA activated 68,000 M_r protein kinase. Semin Viro 4:259–268

Katze MG, Detjen BM, Safer B, Krug RM (1986) Translational control by influenza virus: suppression of the kinase that phosphorylates the alpha subunit of initiation factor eIF-2 and selective translation of influenza viral mRNAs. Mol Cell Biol 6:1741–1750

Katze MG, DeCorato D, Safer B, Galabru J, Hovanessian AG (1987) Adenovirus VAI RNA complexes with the 68,000 M_r protein kinase to regulate its autophosphorylation and activity. EMBO J 6:689–697

Katze MG, Tomita J, Black T, Krug RM, Safer B, Hovanessian AG (1988) Influenza virus regulates protein synthesis during infection by repressing the autophosphorylation and activity of the cellular 68,000 M_r protein kinase. J Virol 62:3710–3717

Katze MG, Wambach M, Wong M-L, Garfinkel M, Meurs E, Chong K, Williams BRG, Hovanessian AG, Barber GN (1991) Functional expression of interferon-induced, dsRNA activated 68,000 M_r protein kinase in a cell-free system. Mol Cell Biol 11:5497–5505

Koromilas AE, Roy S, Barber GN, Katze MG, Sonneberg N (1992) Malignant transformation by a mutant of the IFN-inducible dsRNA dependent protein kinase. Science 257:1685–1689

Kuri-Harcuch W, Green H (1978) Adipose conversion of 3T3 cells depends on a serum factor. Proc Natl Acad Sci USA 75:6107–6212

Langland JO, Jacobs BL (1992) Cytosolic double-stranded RNA-dependent protein kinase is likely a dimer of partially phosphorylated $M_r = 66,000$ subunits. J Biol Chem 267:10729–10736

Laurent AG, Krust B, Galabru J, Svab J, Hovanessian AG (1985) Monoclonal antibodies to interferon induced 68,000 M_r protein and their use for the detection of double-stranded RNA dependent protein kinase in human cells. Proc Natl Acad Sci USA 82:4341–4345

Lee TG, Tomita J, Hovanessian AG, Katze MG (1990) Purification and partial characterization of a cellular inhibitor of the interferon-induced 68,000 M_r protein kinase from influenza virus-infected cells. Proc Natl Acad Sci USA 87:6208–6212

Lee TG, Tomita J, Hovanessian AG, Katze MG (1992) Characterization and regulation of the 58,000 dalton cellular inhibitor of the interferon-induced, dsRNA activated protein kinase. J Biol Chem 267:14238–14243

Legon S, Brayley A, Hunt T, Jackson RJ (1974) The effect of cyclic AMP and related compounds on the control of protein synthesis in reticulocyte lysates. Biochem Biophys Res Commun 56:745–752

Lenardo MJ, Fan CM, Maniatis T, Baltimore D (1989) The involvement of NF-κB in β-interferon gene regulation reveals its role as a widely inducible mediator of signal transduction. Cell 57:287–294

Lengyel P (1982) Biochemistry of interferons and their actions. Annu Rev Biochem 51:251–282

Li J, Petryshyn RA (1991) Activation of the double-stranded RNA-dependent elF-2α kinase by cellular RNA from 3T3-F442A cells. Eur J Biochem 195:41–48

Lloyd RM, Shatkin AJ (1992) Translational stimulation by reovirus polypeptide s3: substitution for VAI RNA and inhibition of phosphorylation of the α subunit of eukaryotic initiation factor 2. J Virol 66:6878–6884

Mahadevan LC, Wills AJ, Hirst EA, Rathjen PD, Heath JK (1990) 2-Aminopurine abolishes EGF- and TPA-stimulated pp33 phosphorylation and c-*fos* induction without affecting the activation of protein kinase C. Oncogene 5:327–335

Maran A, Mathews MB (1988) Characterization of the dsRNA implicated in the inhibition of protein synthesis in cells infected with a mutant adenovirus defective for VA RNA 1. Virology 164:106–113

Mathews MB, Shenk T (1991) Adenovirus virus-associated RNA and translation control. J Virol 65:5657–5662

Mellits KH, Kostura M, Mathews MB (1990) Interaction of adenovirus VA RNA1 with the protein kinase DAI: nonequivalence of binding and function. Cell 61:843–852

Merrick WC (1992) Mechanism and regulation of eukaryotic protein synthesis. Microbiol Rev 56:291–315

Meurs E, Chong K, Galabru J, Thomas N, Kerr I, Williams BRG, Hovanessian AG (1990) Molecular cloning and characterization of the human double-stranded RNA-activated protein kinase induced by interferon. Cell 62:379–390

Meurs E, Galabru J, Barber GN, Katze MG, Hovanessian AG (1993) Tumour suppressor function of the interferon-induced double-stranded RNA-activated 68,000-M_r protein kinase. Proc Natl Acad Sci USA (in press) 90:232–236

Mundschau LJ, Faller DV (1992) Oncogenic *ras* induces an inhibitor of double-stranded RNA-dependent eukaryotic initiation factor 2α-kinase activation. J Biol Chem 267:23092–23098

Nielsch U, Zimmer SG, Babiss LE (1991) Changes in NF-kappa B and ISGF3 DNA binding activities are responsible for differences in MHC and beta-IFN gene expression in Ad5-versus Ad12-transformed cells. EMBO J 10:4169–4175

Offermann MK, Faller DV (1989) Autocrine induction of major histocompatibility complex class I antigen expression results from induction of beta interferon in oncogene-transformed BALB/c-3T3 cells. Mol Cell Biol 9:1969–1977

O'Neill RE, Racaniello VR (1989) Inhibition of translation in cells infected with a poliovirus 2A-pro mutant correlates with phosphorylation of the alpha subunit of eucaryotic initiation factor 2. J Virol 63:5069–5075

Petryshyn R, Chen J-J, London IM (1984) Growth-related expression of a double-stranded RNA-dependent protein kinase in 3T3 cells. J Biol Chem 259:14735–14742

Petryshyn R, Chen J-J, London IM (1988) Detection of activiated double-stranded RNA-dependent protein kinase in 3T3-F442A cells. Proc Natl Acad Sci USA 85:1427–1431

Petska S, Langer JA, Zoon K, Samuel CE (1987) Interferons and their actions. Annu Rev Biochem 56:727–777

Ramirez M, Wek RC, Hinnebusch AG (1991) Ribosome association of GCN2 protein kinase, a translational activator of the GCN4 gene of *Saccharomyces cerevisiae*. Mol Cell Biol 11:3027–3036

Reis LFL, Harada H, Wolchok JD, Taniguchi T, Vilcek J (1992) Critical role of a common transcriptional factor, IRF-1, in the regulation of IFN-β and IFN inducible genes. EMBO J 11:185–193

Roy S, Katze MG, Parkin NT, Edery I, Hovanessian AG, Sonenberg N (1990) Control of the interferon-induced 68-kilodalton protein kinase by HIV-1 *tat* gene product. Science 247:1216–1219

Roy S, Agy MB, Hovanessian AG, Sonenberg N, Katze MG (1991) The integrity of the stem structure of human immunodeficiency virus type 1 *Tat*-responsive sequence RNA is required for interaction with the interferon-induced 68,000-M_r protein kinase. J Virol 65:632–640

Saito S (1989) Possible involvement of virus-induced protein kinase in the antiviral state induced with interferon-γ against sindbis virus. J Interferon Res 9:23–34

Saito S (1990) Enhancement of the interferon-induced double-stranded protein kinase activity by sindbis virus infection and heat-shock stress. Microbiol Immunol 34:859–870

Saito S, Kawakita M (1991) Inhibitor of Interferon-induced double stranded RNA-dependent protein kinase and its relevance to alteration of cellular protein kinase activity level in response to external stimuli. Microbiol Immunol 35:1105–1114

Samuel CE (1991) Antiviral actions of interferon-regulated cellular proteins and their surprisingly selective antiviral activities. Virology 183:1–11

Sen GC, Lengyel P (1992) The interferon system – a bird's eye view of its biochemistry. J Biol Chem 267:5017–5020

SenGupta DN, Silverman RH (1989) Activation of interferon-regulated, dsRNA-dependent enzymes by human immunodeficiency virus-1 leader RNA. Nucleic Acids Res 17:969–978

Sonenberg N (1990) Measures and countermeasures in the modulation of initiation factor activities by viruses. New Biol 2:4022–4029

Szyszka R, Kudlicki W, Kramer G, Hardesty B, Galabru J, Hovanessian A (1989) A type 1 phosphoprotein phosphatase active with phosphorylated $M_r = 68\,000$ initiation factor 2 kinase. J Biol Chem 264:3827–3831

Thimmappaya B, Weinberger C, Schneider RJ, Shenk T (1982) Adenovirus VA1 RNA is required for efficient translation of viral mRNAs at late times after infection. Cell 31:543–551

Tiwari RK, Kusari J, Sen GC (1988) Gene induction by interferons and double-stranded RNA: selective inhibition by 2-aminopurine. Mol Cell Biol 8:4289–4294

Visvanathan KV, Goodbourn SEY (1989) Double-stranded RNA activates binding of NF-κB to an inducible element in human β-interferon promoter. EMBO J 8:1129–1138

Wathelet MG, Clauss IM, Paillard FC, Huez GA (1989) 2-Aminopurine selectively blocks the transcriptional activation of cellular genes by virus, double-stranded RNA and interferons in human cells. Eur J Biochem 184:503–509

Whitaker-Dowling PA, Youngner JS (1984) Characterization of a specific kinase inhibitory factor produced by vaccinia virus which inhibits the interferon-induced protein kinase activity. Virology 137:171–181

Zinn K, Keller A, Wittlemore L-A, Maniatis T (1988) 2-Aminopurine selectively inhibits the induction of β-interferon, c-*fos*, and c-*myc* gene expression. Science 240:210–213

Zullo JN, Faller DV (1988) P21 v-*ras* inhibits induction of c-*myc* and c-*fos* expression by platelet-derived growth factor. Mol Cell Biol 8:5080–5085

Zullo JN, Cochran BH, Huang AS, Stiles CD (1985) Platelet-derived growth factor and double-stranded ribonucleic acids stimulate expression of the same genes in 3T3 cells. Cell 43:793–800

Mechanism of the Antiretroviral Effect of dsRNA

W.E.G. MÜLLER[1], H. USHIJIMA[2], and H.C. SCHRÖDER[1]

1 Introduction

The development of AIDS seems to be linked to an impairment of processes which are induced or activated by double-stranded RNA (dsRNA), such as the biosynthesis of interferon (IFN), production of 2′,5′-oligoadenylate (2-5A), ribonuclease L (RNase L) activity and different cell-mediated immune functions. A restriction of available bioactive dsRNA (or of dsRNA-dependent enzymes) may play an important role in the disease progression. The results summarized in this review show that defects in dsRNA-dependent pathways exhibited by AIDS patients can be reversed, at least in part, by exogenously supplied dsRNA.

2 Intracellular Antiviral Defence Mechanisms: 2-5A/RNase L and p68 Kinase Pathways

The intracellular, antiviral effects of IFNs are mediated through two groups of enzymes which are activated by dsRNA (Hovanessian 1991): the 2-5A synthetase (2-5OAS) isoenzymes (Chebath et al. 1987), acting in concert with a 2-5A-dependent RNase L (Silverman et al. 1988), and a dsRNA-dependent protein kinase [also called DAI (dsRNA-activated inhibitor) or p68 kinase] (Meurs et al. 1990). The 2-5OAS/RNase L pathway may play an additional role in the regulation of cell growth and differentiation (Etienne-Smekens et al. 1983; Jacobsen et al. 1983; Ferbus et al. 1985; Krause et al. 1985 a,b; Cohrs et al. 1988). The 2-5OAS isoenzymes which form 2-5A (consisting of two to five, or more, 2′,5′-linked adenylate residues with a 5′-triphosphate) are induced by IFN but lower levels of these enzymes can also be detected in untreated cells (Nilsen et al. 1982b). Among the dsRNAs activating 2-5OAS is the "*trans*-acting response element" (TAR) RNA sequence of the human immunodeficiency virus-1 (HIV-1)

[1] Institut für Physiologische Chemie, Abteilung Angewandte Molekularbiologie, Johannes Gutenberg-Universität, Duesbergweg 6, 55099 Mainz, Germany
[2] Division of AIDS Virus, AIDS Research Center, National Institute of Health, Musashimurayama, Tokyo 208, Japan

Progress in Molecular and Subcellular Biology, Vol. 14
W.E.G. Müller/H.C. Schröder (Eds.)
© Springer-Verlag Berlin Heidelberg 1994

which assumes a stable stem-loop structure (Schröder et al. 1990b). 2-5A formed by 2-5OAS in turn binds and thereby activates the RNase L (Silverman et al. 1988) which degrades viral and cellular RNAs (Schröder et al. 1989). The preferential cleavage of viral RNA in infected cells may be due to a localized activation of the enzyme (Schröder et al. 1990b). 2-5A oligomers are rapidly degraded by nuclease (2′,3′-exoribonuclease and 2′-phosphodiesterase) and phosphatase activities (Schröder et al. 1980, 1988; Johnston and Hearl 1987). The 2-5A metabolic enzymes are present in both the nucleus and the cytoplasm (Nilsen et al. 1982b; Laurent et al, 1983; Schröder et al. 1988, 1989).

The intracellular level of 2-5A and the activities of 2-5OAS and RNase L have been shown to change in response to various physiological conditions, e.g. in the course of the cell cycle (Wells and Mallucci 1985), after treatment of cells with hormones (Cohrs et al. 1988), during aging (Floyd-Smith and Denton 1988; Pfeifer et al. 1993) and in the course of virus infections (see below).

The second pathway involved in the intracellular mechanism of IFN consists of the dsRNA-dependent p68 kinase. The dsRNA-dependent kinase is activated by autophosphorylation. The phosphorylated kinase then catalyzes the phosphorylation of the α-subunit of the protein synthesis initiation factor-2. After phosphorylation initiation factor-2 is incapable of recycling and consequently initiation of mRNA translation is impaired (Farrell et al. 1977; Katze 1992).

3 Alterations in the Level of 2-5A

3.1 Cultured Cells

Infection of cells in vitro with HIV-1 causes a transitory increase in 2-5OAS and RNAse L activity. In the human T-cell line H9, maximum activities were found at days 2 to 3 after infection (Schröder et al. 1988, 1989; see also Fig. 1). Subsequently, a rapid decrease in both enzyme activities occurs, simultaneously with the rise in viral protein synthesis. These results show that the intracellular level of 2-5A is inversely correlated with virus production. Evidence has been presented that the release of HIV-1 can be retarded by compounds which enhance the production of 2-5A, e.g. IFN (Ho et al. 1985), certain lectins (Schröder et al. 1990a), or by 3′-azido-3′-deoxythymidine (AZT) (Schröder et al. 1989). Results showing that 2-5A analogues display anti-HIV activity in vitro have been reported (Doetsch et al. 1981; Montefiori et al. 1989c; Suhadolnik et al. 1989; Müller et al. 1991).

3.2 HIV Patients

Many acute viral infections are associated with elevated levels of 2-5OAS (Schattner et al. 1981; Williams et al. 1982). Elevated levels of 2-5OAS were found in extracts of peripheral blood mononuclear cells (PBMC) from individuals with AIDS and AIDS-related complex (ARC) (Preble et al. 1985; Read et al. 1985a,b; Witt et al. 1991). Nevertheless, the levels of 2-5A and RNase L

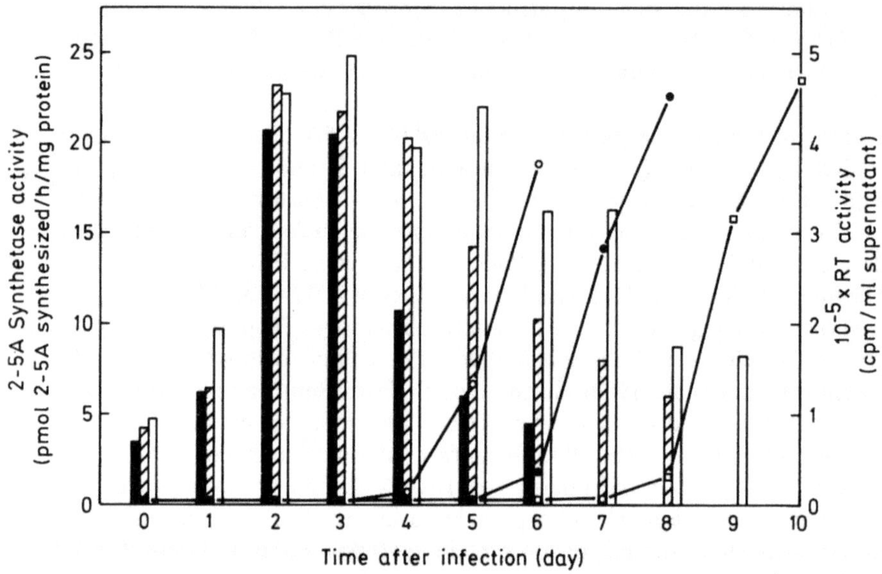

Fig 1. Influence of poly(I)·poly(C_{12}U) on 2-5OAS activity and on release of virus from HIV-infected H9 cells in vitro. Poly(I)·poly(C_{12}U) (0 μg/ml: o and *filled bars*; 40 μg/ml: ● and *hatched bars*; 80 μg/ml: □ and *open bars*) was added 24 h prior to virus infection and was present throughout the incubation period. In the experiment shown the cells were simultaneously treated with IFN (150 IU/ml each of recombinant IFN-α, IFN-β, and IFN-γ). The reverse transcriptase (*RT*) activity in the supernatants (○, ●, □) and the nuclear matrix-bound 2-5OAS activity (*filled*, *hatched* and *open bars*) were determined 0, 1, 2, 3, 4, 5, 6, 7, and 8 days post-infection. The means of five parallel experiments are given. The SD does not exceed 15% (RT activity) and 20% (2-5OAS activity), respectively.

activity in these blood cells were diminished (Carter et al. 1987). A low 2-5A-dependent binding protein activity in PBMC from AIDS patients has also been reported by Wu et al. (1986). There are also some hints that the RNase L is inactive in AIDS patients (Carter et al. 1987), despite a high level of its activator 2-5A, indicating the presence of an inhibitor in HIV-infected cells which interferes with the proper functioning of this antiviral pathway.

4 Activation of the 2-5A System and p68 Kinase by dsRNA

4.1 Activation of 2-5OAS by hnRNA

A large portion of total cellular 2-5A metabolic enzymes is present in the nucleus where they are associated with the nuclear matrix (Schröder et al. 1988, 1989). The nuclear 2-5OAS may be activated by the double-stranded parts of heterogeneous nuclear RNA (hnRNA) (Nilsen et al. 1982a). It has been calculated that the amount of 2-5A synthesized by the uninduced enzyme is high enough to activate the RNase L (Nilsen et al, 1982b). The association of 2-5OAS with the nuclear matrix might be important because the nuclear matrix is thought to be the site of mRNA maturation in the cell (Zeitlin et al. 1987). In some virus-infected cells, e.g. HIV-1-infected H9 cells (Müller et al. 1989, 1990) or adenovirus-infected HeLa cells (Mariman et al. 1982), a preferential association of the viral RNA with the nuclear matrix has been found.

4.2 Activation of 2-5OAS by the TAR Sequence of HIV-1-LTR

HIV-1 contains several regulatory genes, among them also *tat* which encodes the *trans*-activator protein, Tat (Pavlakis and Felber 1990). The target sequence for Tat *trans*-activation has been localized within the long terminal repeat, the *trans*-activating region TAR, immediately downstream of the messenger RNA start site (Karn 1991). The TAR RNA sequence, which is present in the 5′ leader of all HIV-1 mRNAs, is thought to form a stable stem-loop structure and is the binding site of the HIV-1 Tat protein, which is formed soon after infection of cells with HIV-1; the binding of Tat to TAR results in an activation of virus-specific gene expression (Karn 1991). In vitro studies revealed that the TAR sequence is also able to bind and activate purified 2-5OAS (SenGupta and Silverman 1989; Schröder et al. 1990b). Interestingly, the effect of TAR on the 2-5OAS is abolished by the presence of Tat protein (Schröder et al. 1990b).

4.3 Activation of p68 Kinase by the TAR Sequence of HIV-1-LTR

The TAR sequence has also been shown to be able to activate the dsRNA-dependent protein kinase (Edery et al. 1989; SenGupta and Silverman 1989; Ushijima et al. 1993b). However, also contradictory results have been reported (Gunnery et al. 1990). Again, binding and activation of p68 kinase by immobilized TAR RNA were found to be inhibited by Tat protein (Ushijima et al. 1993b). This result suggests that Tat protein can interfere with the intracellular antiviral defence mechanism at both the level of RNA stability (2-5OAS/RNase L pathway) and translation (p68 kinase/initiation factor-2 pathway) (see above).

5 Modulation of Intracellular Antiviral Mechanisms by dsRNA Analogues

Poly(I)·poly(C) is the best-known inducer of IFN (Carter and DeClercq 1974). The applicability of this dsRNA, however, is limited due to its cytotoxicity (Krown et al. 1983). The toxicity of this dsRNA has been reduced by introduction of non-hydrogen-bonded uracil residues into the poly(C) strand at periodic distances (Carter et al. 1987; Montefiori and Mitchell 1987). The resulting mismatched dsRNA, poly(I)·poly($C_{12}U$) or ampligen, is still biologically active; it exhibits immunomodulatory, antitumor (Strayer et al. 1986), and antiviral activities (see below).

5.1 Poly(I)·Poly($C_{12}U$) (Ampligen): Chemistry and Physical Properties

Ampligen has the molecular formula, $(rI_{13}:rC_{12}U_1)_n$, where: n = 20 to 30. The uridylic substitutions in the poly(C) strand occur on average every 12 to 13 nucleotides (Carter et al. 1972). The double-stranded polymer is heterogeneous with respect to chain length, periodicity of mismatched base pairs and molecular configuration (adjacent versus isolated mismatched base pairs).

The mismatched regions consisting of unpaired bases accelerate the rate of hydrolysis of the poly(I)·poly($C_{12}U$); one consequence is a lower toxicity of the molecule (Carter et al. 1985; Brodsky and Strayer 1987). The uninterrupted, base-paired regions serve as trigger regions for modulation of cytokines. Mismatched dsRNA is able to induce the synthesis of IFN as long as the frequency of random insertion is not greater than 0.5 to 1.0 helical turn of perfectly base-paired dsRNA (Greene et al. 1978).

5.2 Modulation of Cytokine Action and Natural Killer Cell Activity

Poly(I)·poly($C_{12}U$) and other dsRNAs have been shown to induce the synthesis of tumor necrosis factor and to regulate the expression or modulate the action of various other cytokines including IFN, interleukin-6 and platelet-derived growth factor (Zullo et al. 1985). In addition, poly(I)·poly($C_{12}U$) augments natural killer cell activity (Zarling et al. 1980) and enhances phagocytosis and tumoricidal activity of macrophages (Burleson et al. 1987; Pinto et al. 1989). Poly(I)·poly($C_{12}U$) exhibits significant antitumor effects in nude mouse model systems, which are accompanied by marked increases in natural killer cell activity (Hubbell et al. 1985, 1987, 1990). Ampligen also enhances B-cell immunity (Diamantstein and Blitstein-Willinger 1975).

5.3 Anti-HIV Activity

DsRNAs such as poly(I)·poly($C_{12}U$) (Laurence et al. 1987; Montefiori and Mitchell 1987; Montefiori et al. 1988; Carter et al. 1991; Ushijima et al. 1993b) or poly(A)·poly(U) (Laurent-Crawford et al. 1992) have been demonstrated to exhibit a strong antiviral and cytoprotective effect on HIV-infected cells in vitro.

Preincubation of C3 or CEM cells with 10–50 μg/ml poly(I)·poly($C_{12}U$) (ampligen) for 24 h affords significant protection against HIV-1 infection (Montefiori and Mitchell 1987). The antiviral effect of poly(I)·poly($C_{12}U$) was even higher if it was added to the medium also after the infection. A one-time exposure to poly(I)·poly($C_{12}U$) (50 μg/ml) had a more pronounced effect than either a one-time exposure to IFN-α (250 IU/ml), IFN-β (250 IU/ml) or IFN-γ (50 IU/ml) (HIV-infected CEM cells), or a one-time exposure to a combination of all three IFNs (150 IU/ml each) (HIV-infected C3 cells). At a concentration of 50 μg/ml, poly(I)·poly($C_{12}U$) displayed no influence on cell division or RNA and protein synthesis in all T-cell lines studied (Montefiori and Mitchell 1987).

Treatment of H9 cells with poly(I)·poly($C_{12}U$) also caused a significant retardation of the onset of virus production after infection of the cells with HIV-1 (Ushijima et al. 1993b). Untreated H9 cells started to release the virus in the supernatant 3 days post-infection, while in the presence of 80 μg/ml poly(I)·poly($C_{12}U$) release of HIV-1 was not observed before day 6. The increase in reverse transcriptase (RT) activity paralleled the increase in p24-positive cells [beginning at day 3, untreated cells; and day 6, presence of 80 μg/ml poly(I)·poly($C_{12}U$)]. An even more pronounced effect was observed in cells treated simultaneously with IFN and poly(I)·poly($C_{12}U$) (Fig. 1): the onset of HIV-1 production and release in H9 cells at continuous presence of these additives during preincubation and whole incubation period was at day 8 post-infection [80 μg/ml poly(I)·poly($C_{12}U$)], whereas cells treated with IFN only, started to release the virus in the supernatant at day 4 (Ushijima et al. 1993b).

More recently, Laurent-Crawford et al. (1992) reported that poly(A)·poly(U) also displays antiviral activity against both HIV-1 and HIV-2. Poly(A)·poly(U) seems to inhibit HIV infection at the level of viral entry into cells. At present, it is unclear whether poly(I)·poly($C_{12}U$) also interferes with this step of the viral replication cycle or whether poly(A)·poly(U) shares some of the mechanisms described below with the mismatched dsRNA.

5.4 Activation of 2-5OAS

Both poly(I)·poly(C) and its mismatched derivative, poly(I)·poly($C_{12}U$) (ampligen), are strong allosteric activators of 2-5OAS (Ilson et al. 1986; Pestka et al. 1987; Li et al. 1990; Ushijima et al. 1993b). Poly(I)·poly($C_{12}U$) exhibits essentially the same capacity to activate the 2-5OAS in IFN-treated HeLa cell extracts as poly(I)·poly(C) (Ushijima et al. 1993b). As shown in Table 1, the 2-5OAS from extracts of IFN-treated HeLa cells was activated dose-dependently by both poly(I)·poly(C) and poly(I)·poly($C_{12}U$) in the range from 1×10^{-6} to 1×10^{-4} g/ml. Purified 100-kDa 2-5OAS, on the other hand, was activated less dramatically and displayed even some activity in the absence of dsRNAs. Similar results were reported by Hovanessian et al. (1988) for the 69- and 100-kDa 2-5OAS. Li et al. (1990) demonstrated by photoaffinity labelling using poly([^{32}P]I,8-azidoI)·poly($C_{12}U$) that ampligen also binds to 2-5OAS. It is well documented that dsRNAs can activate enzymes which are involved in the maintenance of the antiviral state of a cell also in the absence of IFN (Pestka et al. 1987; Schröder et al. 1992).

Table 1. Effect of poly(I)·poly(C) and poly(I)·poly($C_{12}U$) on 2-5OAS in HeLa cell extract

dsRNA	Concentration (g/ml)	2-5OAS activity[a] (nmol ATP converted to 2-5A)
Poly(I)·poly(C)	10^{-6}	1.7 ± 0.3
	10^{-5}	6.0 ± 1.2
	10^{-4}	14.9 ± 2.0
Poly(I)·poly($C_{12}U$)	10^{-6}	1.4 ± 0.2
	10^{-5}	5.6 ± 1.3
	10^{-4}	13.0 ± 2.7

[a]Each value represents the mean (± SD) from 3 independent experiments.

5.5 Modulation of p68 Kinase Activity

Poly(I)·poly($C_{12}U$) also activates the second dsRNA-dependent enzyme present in IFN-treated HeLa cell extracts, the dsRNA-dependent protein kinase (p68 kinase); the consequence is autophosphorylation of the enzyme. In contrast to the 2-5OAS, activation of the kinase in extracts from IFN-treated HeLa cells by poly(I)·poly($C_{12}U$) – as well as poly(I)·poly(C) (Hovanessian and Kerr 1979) – occurs within a limited concentration range (10^{-7} to 10^{-6} g/ml) (Ushijima et al. 1993b). Higher concentrations of dsRNA ($\geq 1 \times 10^{-5}$ g/ml) inhibit the activation of the kinase.

Ushijima et al. (1993b) found that addition of poly(I)·poly($C_{12}U$) strongly suppressed the binding of the kinase to immobilized TAR. As mentioned above, binding (and activation) of dsRNA-dependent kinase by the TAR segment is abolished in the presence of Tat protein. However, Tat displayed no effect on binding and activation of the kinase by poly(I)·poly($C_{12}U$). Tat also suppresses the binding of 2-5OAS to TAR RNA-cellulose (Schröder et al. 1990b). Thus, both effects of Tat (blockage of both the 2-5OAS/RNase L- and the dsRNA-dependent protein kinase pathway) can be antagonized by poly(I)·poly($C_{12}U$).

5.6 Inhibition of DNA Topoisomerase I

Recently, it has been reported that longer 2-5A oligomers consisting of more than four adenylate residues are efficient inhibitors of cellular DNA topoisomerase I activity (Castora et al. 1991; Schröder et al. 1993a). We found that treatment of H9 cells with 10^{-4} g/ml poly(I)·poly($C_{12}U$) results in a strong inhibition of DNA topoisomerase I activity (Schröder et al. 1993a). There is evidence that the decrease in topoisomerase I activity after treatment of cells with dsRNA is due to the production of longer 2-5A oligomers (see below). Poly(I)·poly($C_{12}U$) alone does not inhibit purified topoisomerase I.

The inhibition of cellular DNA topoisomerase I activity may be involved in the antiproliferative mechanism of poly(I)·poly($C_{12}U$). The finding that 2-5A

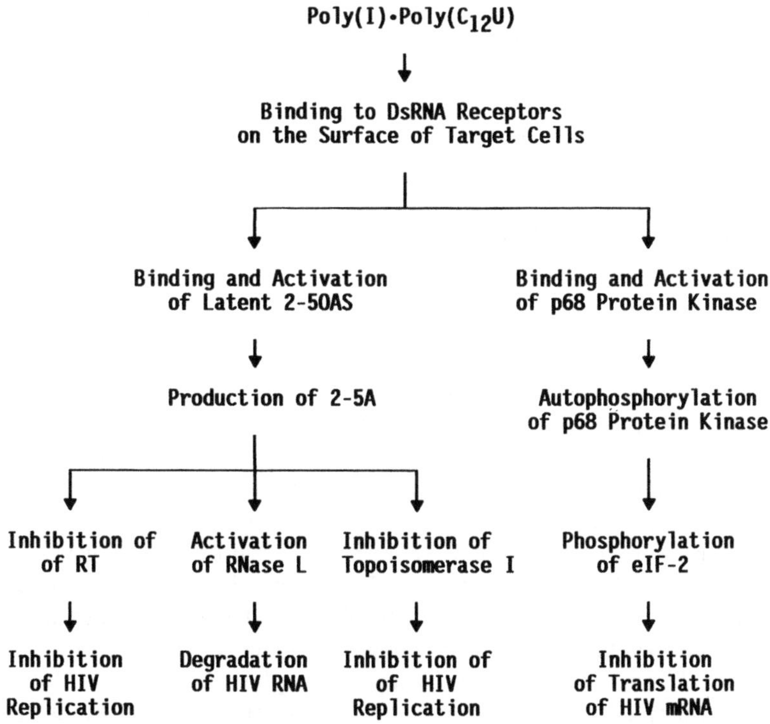

Fig. 2. Antiviral mechanism of poly(I)·poly($C_{12}U$)

inhibits DNA topoisomerase I activity may also have implications for the understanding of the antiviral mechanism of poly(I)·poly($C_{12}U$). From our results (Schröder et al. 1993a) poly(I)·poly($C_{12}U$), via production of bioactive 2-5A, is also capable of inhibiting topoisomerase I activity associated with isolated HIV-1 particles (Priel et al. 1990). Thus, poly(I)·poly($C_{12}U$) may act by more than one intracellular mechanism (Fig. 2): First, it activates the 2-5OAS/RNase L pathway, resulting in the production of bioactive 2-5A, which, in turn, activates RNase L and, hence, destroys HIV RNA. Second, poly(I)·poly($C_{12}U$), through the production of bioactive 2-5A, inhibits HIV RT (see Schröder et al., this Vol.). Third, poly(I)·poly($C_{12}U$), again through the production of bioactive 2-5A, inhibits HIV-associated topoisomerase I activity. Furthermore, poly(I)·poly($C_{12}U$) induces the production of cytokines, such as IFN and tumor necrosis factor, which are themselves antiviral, and causes an activation of the p68 protein kinase pathway (see above).

5.7 Degradation by dsRNase

DsRNAs are cleaved by a specific dsRNase described by Meegan and Marcus (1989). Future studies have to clarify whether poly(I)·poly($C_{12}U$) is also hydrolyzed by this enzyme or able to block the dsRNase, and hence to modulate the level of other dsRNAs.

6 Mechanism of the Antiviral Effect

At the molecular level the antiviral mechanism of dsRNA and mismatched dsRNA is poorly understood. DsRNA activates a group of cellular genes involved in both growth regulation and antiviral response (Zullo et al. 1985; Wathelet et al. 1987; Ray et al. 1988; Goldfeld and Maniatis 1989). Among the dsRNA-inducible genes are the type-I IFNs (IFN-α and IFN-β) (Pestka et al. 1987). Increasing evidence indicates that the production of IFN is not required for dsRNA-caused gene induction (Zarling et al. 1980). In addition, dsRNA can generate an antiviral state without inducing IFN by directly activating dsRNA-dependent enzymes (see Sect. 5.4 and 5.5, and below). Induction of genes by dsRNA may involve a transcriptional event (Memet et al. 1991). The 5′-genomic sequences of the IFN-regulated 2-5OAS gene display a striking similarity to the dsRNA-responsive regions of IFN genes (Wathelet et al. 1987). However, the human 2-5OAS gene does not seem to be directly induced by dsRNA (Raj et al. 1989). Because dsRNA activates the dsRNA-dependent p68 kinase, this kinase might be a receptor involved in transducing the signal from dsRNA via a subsequent phosphorylation event (Marcus 1983). This assumption is corroborated by the fact that the p68 kinase inhibitor, 2-aminopurine, inhibits the poly(I)·poly(C)-caused induction of the IFN-β gene and other dsRNA-responsive genes (Wathelet et al. 1987; Zinn et al. 1988).

It has been shown that the signal transduction process triggered by the IFN receptor involves migration of a factor located in the cytoplasm to the nucleus and binding of this factor to an IFN-stimulated response element (ISRE) (for a review, see Williams 1991). DsRNA appears to modulate directly IFN-stimulated gene factor ISGF3α. This step is inhibited by 2-aminopurine. In contrast, the effect of IFN-α is not affected by 2-aminopurine, suggesting that different signal transduction pathways mediate dsRNA and IFN-α signalling of ISGF3. In the case of dsRNA, the pathway may involve the p68 kinase. Interestingly, dsRNA has also been shown to lead to phosphorylation of the inhibitory subunit IkappaB in the transcription factor NF-kappaB/IkappaB complex, resulting in the dissociation of the complex and translocation of NF-kappaB to the nucleus (Ghosh and Baltimore 1990). However, it remains to be determined whether p68 kinase is actually part of the signal transduction pathway mediating gene induction by dsRNA.

At the enzymic level, poly(I)·poly(C_{12}U) has been shown to enhance the activity of 2-5OAS and dsRNA-dependent protein kinase in cell extracts (Li et al. 1990). Poly(I)·poly(C_{12}U) may block HIV replication very early, possibly by stimulating (RNase L-dependent) degradation of incoming parental virus RNA (Montefiori et al. 1989a); other anti-HIV drugs such as nucleoside analogues inhibit at later steps (Crumpacker 1989).

Table 2. Effect of treatment of HeLa cells with poly(I)·poly($C_{12}U$) on amount of latent and total 2-5OAS activity in cytoplasmic extracts[a]

Treatment of cells with:	Conc. of dsRNA (g/ml)	2-5OAS activity (pmol h^{-1} g^{-1} of protein) Total (activated plus latent)	Latent (non-activated)
-	-	0.1 ± 0.1	≤ 0.1
Poly(I)·poly($C_{12}U$)	10^{-6}	0.3 ± 0.1	≤ 0.1
	10^{-5}	1.0 ± 0.3	0.3 ± 0.2 (28)
	10^{-4}	2.3 ± 0.5	0.7 ± 0.3 (29)
IFN	-	6.9 ± 1.7	6.2 ± 1.1 (90)
IFN plus	10^{-6}	7.4 ± 1.4	6.3 ± 2.1 (85)
Poly(I)·poly($C_{12}U$)	10^{-5}	8.5 ± 1.9	4.4 ± 1.5 (52)
	10^{-4}	9.8 ± 2.0	4.2 ± 1.3 (43)

[a]2-5OAS assays were performed in total cell extracts. Each assay contained 280 μg protein. Incubation was for 3h. The products formed from [^{14}C]ATP were analyzed by thin-layer chromatography on polyethyleneimine-cellulose. The percentage amounts of latent 2-5OAS are given in parentheses.

Poly(I)·poly($C_{12}U$) markedly extends the duration of the transient increase in 2-5OAS activity, preceding virus production after infection of cells with HIV-1 (Ushijima et al. 1993b). In addition, we determined that the presence of poly(I)·poly($C_{12}U$) protracts the time during which elevated levels of 2-5OAS transcripts are present in the cells. More importantly, poly(I)·poly($C_{12}U$) may cause a shift from latent (non-activated) 2-5OAS to the activated form; poly(I)·poly($C_{12}U$) enhances the level of active 2-5OAS, at least in part, at the expense of the inactive enzyme (Ushijima et al. 1993b; Table 2).

It is known that at low concentrations of dsRNA, the 69- and 100kDa isoenzymes of 2-5OAS mainly synthesize the 2-5A dimer, while at higher concentrations longer oligomers are formed (Hovanessian et al. 1988). We could demonstrate that treatment of cells with poly(I)·poly($C_{12}U$) results in the production of longer 2-5A oligomers (Ushijima et al. 1993b). This is important because the 2-5A trimer and higher 2-5A oligomers are potent activators of 2-5A-dependent RNase L; the 2-5A dimer is unable to activate this enzyme (Baglioni et al. 1978).

Moreover, poly(I)·poly($C_{12}U$) may prevent downregulation of dsRNA-dependent enzymes by the HIV *trans*-activator protein Tat (Ushijima et al. 1993b; see above).

The mode of action of poly(I)·poly($C_{12}U$) can be tentatively outlined as follows (Fig. 2). Step 1: poly(I)·poly($C_{12}U$) binds to dsRNA receptors on the surface of target cells. Step 2: poly(I)·poly($C_{12}U$) then binds to 2-5OAS. Step 3: activation of 2-5OAS by the bound poly(I)·poly($C_{12}U$) results in an increase in the intracellular level of 2-5A. The 2-5A can then inhibit HIV multiplication through three different mechanisms. Step 4a: 2-5A can activate RNase L, which degrades HIV RNA. Step 4b: 2-5A can also block HIV replication by inhibiting binding of HIV RT to the primer binding site (see Schröder et al., this Vol.). Step 4c: 2-5A can additionally inhibit the activity of both cellular and HIV-associated topoisomerase I. Thus, the poly(I)·poly($C_{12}U$)-induced bioactive 2-5A has multiple activities which can directly inhibit HIV. In addition, poly(I)·poly($C_{12}U$) can bind and activate p68 protein kinase, resulting in inhibition of translation of HIV mRNA. In the following, we will present some experimental support or disproof for this working model.

6.1 Binding to Cell Surface Receptors

At present, it is still unclear whether the shift of the non-activated ("latent") 2-5OAS to the activated form is caused by cellular uptake of poly(I)·poly($C_{12}U$) and direct binding of the dsRNA to the enzyme, or mediated by some other mechanism occurring after binding of poly(I)·poly($C_{12}U$) to specific cell surface receptors. There are several hints for the existence of dsRNA receptors, to which poly(I)·poly($C_{12}U$) can bind (Greene et al. 1978; Hubbell et al. 1991). For example, from studies with RK-13 cells and a poly(I)·poly(C)-resistant variant of these cells, PR-RK, Yoshida and Azuma (1985) concluded that specific receptors for poly(I)·poly(C) exist. However, non-specific adsorption of poly(I)·poly(C) on the cell membrane occurs. Later, the putative receptor for poly(I)·poly(C) on these cells was identified (Yoshida and Azuma 1991). Diamantstein and Blitstein-Willinger (1978) also described specific binding of poly(I)·poly(C) on murine B-cells.

Recent studies demonstrated that the antiproliferative action of poly(I)·poly($C_{12}U$) can be mediated by cAMP (Hubbell et al. 1991; see also below). Since adenylate cyclase is rapidly induced in response to dsRNA (Hubbell et al. 1991), it seems to be unlikely that dsRNA becomes first internalized then inducing intracellularly the production of cAMP. Therefore, the authors concluded that poly(I)·poly($C_{12}U$) more likely affects its intracellular targets via binding to a cell surface receptor, resulting in activation of the intracellular cAMP signalling pathway (Hubbell et al. 1991).

6.2 Binding to 2-5OAS and p68 Kinase

There are also reports suggesting that poly(I)·poly(C) acts after cellular internalization (Milhaud et al. 1987; Chapekar et al. 1988). In photoaffinity labelling experiments, Li et al. (1990) demonstrated that poly(I)·poly($C_{12}U$) can directly bind to 2-5OAS. Evidence suggesting that the 40-kDa 2-5OAS binds to viral RNA in intact cells has also been presented by immunoprecipitation experiments in IFN-treated HeLa cells infected with encephalomyocarditis virus (Gribaudo et al. 1991). Evidence has been presented that poly(I)·poly($C_{12}U$) may also bind to p68 kinase (Ushijima et al. 1993b).

6.3 Activation of 2-5OAS

Poly(I)·poly(C_{12}U) also activates the 2-5OAS (Li et al. 1990); it exhibits essentially the same capacity to activate the 2-5OAS in IFN-treated HeLa cell extracts as poly(I)·poly(C) (Ushijima et al. 1993b). The purified 100-kDa 2-5OAS is activated by dsRNA less dramatically than the 69-kDa enzyme and displays some activity also in the absence of dsRNA (Hovanessian et al. 1988; Ushijima et al. 1993b). In addition, poly(I)·poly(C_{12}U) is also able to activate the dsRNA-dependent p68 protein kinase (Li et al. 1990; Ushijima et al. 1993b). Minks et al. (1979) have reported that double-stranded stretches of 50–100 base pairs are required to activate dsRNA-dependent enzymes. Obviously, the chain lengths of the base-paired regions (or at least of some of these regions) within the poly(I)·poly(C_{12}U) are sufficient for yielding the biological effect.

7 Antiproliferative Activity of dsRNA

In addition to its antiviral effect, 2-5A exhibits antiproliferative activity (Etienne-Smekens et al. 1983). 2-5OAS has been shown to be present at low levels (Kimchi et al. 1981; Jacobson et al. 1983) and 2-5A nuclease at high levels (Kimchi et al. 1981) in growing cells. RNase L also correlates inversely with the rate of cell growth (Krause et al. 1985a,b). Therefore dsRNA and dsRNA analogues such as poly(I)·poly(C_{12}U) have potential antineoplastic activity (Hubbell et al. 1984; Hubbell 1986; Strayer et al. 1986; Chapekar et al. 1988). DsRNAs have indeed been shown to decrease clonogenicity of fresh human tumors (Strayer et al. 1986) and, in tissue culture, to inhibit the growth of tumor cell lines (Hubbell et al. 1984). Evidence has been presented that dsRNA can inhibit tumor cell growth by an IFN-independent mechanism (Chapekar et al. 1988). Future studies are directed towards determining whether poly(I)·poly(C_{12}U) can also be used in therapy of AIDS-associated tumors such as AIDS-Kaposi sarcoma.

8 Mechanism of the Antiproliferative Effect of Poly(I)·Poly(C_{12}U)

The antiproliferative action of dsRNA may also be associated with the induction of dsRNA-dependent 2-5OAS and subsequent activation of 2-5A-dependent RNase L which degrades cellular RNA (Cohrs at al. 1988). The potential role of the cAMP system in the mediation of the antiproliferative effect of poly(I)·poly(C_{12}U) has been studied recently (Hubbell et al. 1991). It was found that (1) inhibition of cAMP-dependent protein kinase activity in the human glioma cell line A1235, which does not produce detectable levels of IFN, diminished the dsRNA-induced antiproliferative effect, and (2) the activity of adenylate cyclase increased dose-dependently at antiproliferative dsRNA concentrations. Hubbell et al. (1991) concluded that the antiproliferative effects of IFN and poly(I)·poly(C_{12}U) can occur by different mechanisms of action.

DsRNA-induced changes in cAMP level may affect the 2-5A system. Itkes et al. (1984a) have shown that cAMP can induce 2-5OAS and inhibit 2′-phosphodiesterase. Elevation of the intracellular cAMP levels by treatment of cells with epinephrine and theophylline yields an increase in 2-5OAS activity and an inhibition of 2-5A-degrading 2′-phosphodiesterase (Itkes et al. 1984b). These results suggest that the inhibition of tumor cell growth (or virus multiplication) by dsRNA may be caused by activation of the 2-5OAS/RNase L system through the cAMP signal transduction mechanism. It is interesting to note that changes in the intracellular levels of cAMP (and cGMP) have also been found in HIV-infected cells (Nokta and Pollard 1991).

The activation of genes responsive to cAMP has been shown to involve the phosphorylation of specific proteins binding to cAMP-responsive elements. The cAMP regulatory elements of those genes which rapidly respond to cAMP are located within the 5′-flanking region of the gene (Roesler et al. 1988). These genes may additionally possess an inducible promoter site, called the activation protein 2 binding site, which is rapidly induced by either cAMP or activation of protein kinase C (P.J. Mitchell et al. 1987). The transcription of the slow cAMP-responding genes (Jungmann et al. 1983) requires protein synthesis (Andersen et al. 1988).

9 Clinical Experience

9.1 AIDS

Ampligen has already been used in the treatment of AIDS patients. A pilot study with ten patients with AIDS, ARC or lymphadenopathy syndrome (LAS) and treated with mismatched dsRNA (ampligen, 200–250 mg IV twice weekly) revealed a decline in HIV load (as determined by cocultivation and direct molecular hybridization of HIV RNA) and an enhancement of both T- and B-cell immunity, which was statistically significant (Carter et al. 1987; Strayer et al. 1991). The biological activity and function of the 2-5OAS/RNase L system were monitored in PBMC extracts from patients. Although the 2-5OAS activity in the extracts before therapy was enhanced up to 40-fold (Preble et al. 1985; Read et al. 1985a,b), little or no bioactive 2-5A and no measurable RNase L activity could be detected in all patients studied (Carter et al. 1987). These biochemically detectable defects in the intracellular antiviral defence mechanism were corrected towards normal during ampligen therapy (Carter et al., 1987). The lack of RNase L activity in PBMC from HIV patients before therapy cannot be explained by an accumulation of a competitive inhibitor of 2-5A, because RNase L activity in the studied patient samples could not be restored by adding authentic 2-5A. Ampligen might also be useful in treatment of impaired brain cells by HIV because it has been shown to be able to cross the blood-brain barrier (Hearl and Johnston 1986).

9.2 Chronic Fatigue Syndrome

Chronic fatigue syndrome is a symptom complex characterized by chronic or recurrent disabling fatigue (Holmes et al. 1987, 1988; Straus 1988; Shafran 1991). This illness is associated with a general viral reactivation; it was suggested that human herpes virus-6 (HHV-6) may play a role in the pathogenesis of the disease (Ablashi et al. 1988). HHV-6 can upregulate expression of HIV genes in co-infected CD4 lymphocytes (Lusso et al. 1989) and can also induce expression of the CD4 cell surface molecule on $CD4^-CD8^+$ cells, conferring susceptibility to HIV infection on these cells (Lusso et al. 1991). Chronic fatigue syndrome is associated with changes in serum levels of various cytokines, such as interleukin-2 and IFN (Straus 1988). Suhadolnik et al. (1993) reported additional defects in the 2-5OAS/RNase L pathway in PBMC extracts from patients with chronic fatigue syndrome: abnormally low levels of latent 2-5OAS (= 2-5OAS not bound to its dsRNA activator), elevated concentrations of bioactive 2-5A and elevated RNase L activity. All these defects were found to be corrected during therapy with ampligen (Suhadolnik et al. 1990).

9.3 Cancer

Poly(I)·poly(C_{12}U) has also been used in long-term phase I–II clinical trials with patients suffering from cancer; beneficial therapeutic effects without cumulative toxicity or significant side effects have been reported (Brodsky et al. 1985; Strayer et al. 1990).

10 Drug-Resistant HIV

One major problem in HIV therapy with nucleoside analogues is the development of drug resistance (Larder and Kemp 1989; Larder et al. 1989). HIV-1 is subject to rapid mutational changes when exposed to nucleoside analogues which directly interfere with virus replication. Therefore, only a temporary beneficial effect is obtained during treatment with these compounds. By mutational changes HIV will also become resistant to antisense oligonucleotides and targeted ribozymes. Therefore, current efforts are directed toward finding ways by which the development of resistance of the virus (and the cell) can be avoided.

The advantage of the strategy to develop compounds which activate the intracellular antiviral 2-5OAS/RNase L system is to circumvent in this way the problem of developing drug resistance. Development of ampligen resistance was observed in any of the treated patients (Carter et al. 1991). On the contrary, ampligen seems to prevent the appearance of drug-resistant HIV. Given in combination with AZT, ampligen exhibits a significant synergistic effect in inhibition of HIV in vitro, including AZT-resistant HIV (see below).

11 Combination with Other Anti-HIV Compounds

11.1 AZT

Both additive and synergistic effects can be observed when ampligen is given in combination with other drugs. Ampligen has been shown to act synergistically with AZT in inhibiting HIV replication; it displays also synergistic effects in combined treatment of AZT-resistant HIV (W. Mitchell et al. 1987; Montefiori et al. 1989b).

11.2 dsRNA Intercalating Agents

Ampligen has also been used in combination with a drug interfering with Rev-RRE (Rev-responsive element) interaction.

The Rev protein of HIV-1 acts through binding to a *cis*-acting RNA sequence, called the RRE (Cullen and Malim 1991; Rosen 1991). The RRE which is present in the 9- and 4-kb viral RNA is assumed to form a complex secondary structure (Heaphy et al. 1991). Rev promotes the nuclear export of the RRE-containing structural mRNAs either by direct interaction with the nucleocytoplasmic mRNA transport apparatus (Pfeifer et al. 1991) or by blocking spliceosome assembly of HIV transcripts, thus permitting transport of the unspliced or incompletely spliced HIV mRNA to the cytoplasm (Chang and Sharp 1989).

The Rev-RRE complex appears to be a promising target for therapeutic intervention of HIV infection. Schröder et al. (1993b) showed that pyronin Y, an intercalating cationic dye that exhibits high selectivity towards dsRNA (Darzynkiewicz et al. 1986), is able to inhibit the formation of the Rev-RRE complex. In addition, this group demonstrated that pyronin Y suppresses the RRE-nucleated assembly of Rev protein along RRE-containing RNA, which was recently observed in electron microscopy (Heaphy et al. 1991). Although at present a direct application of pyronin Y in anti-HIV therapy cannot be envisaged due to its high cytotoxicity; this drug might be usable in combination therapy, e.g. together with immunomodulatory and antiviral dsRNAs. Ushijima et al. (1993a) demonstrated that poly(I)·poly(C_{12}U) interacts synergistically with pyronin Y. At present, the molecular mechanism through which pyronin Y potentiates the antiviral effect of poly(I)·poly(C_{12}U) is not known. Pyronin Y might intercalate into poly(I)·poly(C_{12}U) and thereby stabilize the dsRNA. Using a similar approach, Jamison et al. (1990) demonstrated that a combination of poly(A)·poly(U) with intercalating agents such as ethidium bromide or adriamycin resulted in a strong antiviral activity in the vesicular stomatitis virus system without affecting the IFN production. Under the experimental conditions, neither poly(A)·poly(U) alone nor the intercalating agents were efficient antiviral agents. The authors suggested that the dsRNA together with the intercalating dyes directly activate the IFN response genes, which are known to maintain the antiviral state (Lammers et al. 1988).

12 Perspectives

An activation (or restoration) of the natural 2-5OAS/RNase L and dsRNA-dependent protein kinase pathways may help to overcome the depressed immune status of AIDS patients. Progression of the disease is accompanied by various immunologic dysfunctions. These include disorders in the production of cytokines; e.g. the IFN production has been found to be diminished or totally blocked at later stages of HIV infection (Murray et al. 1984; Rossol et al. 1989). Application of exogenous dsRNA such as poly(I)·poly(C_{12}U) (ampligen) may be a way to reverse the defects in dsRNA-dependent pathways exhibited by the AIDS-related complex and AIDS patients.

Acknowledgements. This work was supported by grants from the Bundesgesundheitsamt (FVP 5/88; A2 and A3).

References

Ablashi DV, Josephs SF, Buchbinder A, Hellman K, Nakamura S, Llana T, Lusso P, Kaplan M, Dahlberg J, Memon S, Imam F, Ablashi KL, Markham PD, Kramarsky B, Krueger GRF, Biberfeld P, Wong-Staal F, Salahuddin SZ, Gallo RC (1988) Human B-lymphotropic virus (human herpesvirus-6). J Virol Methods 21:29–48

Andersen B, Milsted A, Kennedy G, Nilson JH (1988) Cyclic AMP and phorbol esters interact synergistically to regulate expression of the chorionic gonadotropin genes. J Biol Chem 263:15578–15583

Baba M, Pauwels R, Balzarini J, Schols D, De Clercq E (1989) Coumermycin A1 is a potent inhibitor of human immunodeficiency virus (HIV) replication in vitro. Int J Exp Clin Chemother 2:15–20

Baglioni C, Minks MA, Maroney PA (1978) Interferon action may be mediated by activation of a nuclease by pppA2'p5'A2'p5'A. Nature 273:684–687

Brodsky I, Strayer DR (1987) Therapeutic potential of ampligen. Am Fam Physician 36:253–256

Brodsky I, Strayer DR, Krueger LJ, Carter WA (1985) Clinical studies with ampligen (mismatched double-stranded RNA). J Biol Response Mod 4:669–675

Burleson GR, Fuller LB, Menache MG, Graham JA (1987) Poly(I):poly(C)-enhanced alveolar and peritoneal macrophage phagocytosis: quantification by a new method utilizing fluorescent beads. Proc Soc Exp Biol Med 184:468–476

Carter WA, DeClercq E (1974) Viral infection-host defense: modulatory role of double-stranded RNA. Science 186:1172–1178

Carter WA, Pitha PM, Marshall LW, Tazawa I, Tazawa S, Ts'o POP (1972) Structural requirements of the rI_n:rC_n complex for induction of human interferon. J Mol Biol 70:567–587

Carter WA, Hubbell HR, Krueger LJ, Strayer DR (1985) Comparative studies of ampligen (mismatched double-stranded RNA) and interferons. J Biol Response Mod 4:613–620

Carter WA, Strayer DR, Brodsky I, Lewin M, Pellegrino MG, Einck L, Henriques HF, Simon GL, Parenti DM, Scheib RG, Schulof RS, Montefiori DC, Robinson WE, Mitchell WM, Volsky DJ, Paul D, Paxton H, Meyer III WA, Karikó K, Reichenbach N, Suhadolnik RJ, Gillespie DH (1987) Clinical, immunological, and virological effects of ampligen, a mismatched double-stranded RNA, in patients with AIDS or AIDS-related complex. Lancet i:1286–1292

Carter WA, Ventura D, Shapiro DE, Strayer DR, Gillespie DH, Hubbell HR (1991) Mismatched double-stranded RNA, ampligen (poly(I)·poly(C_{12}U)), demonstrates antiviral and immunostimulatory activities in HIV disease. Int J Immunopharmacol 13 (Suppl 1):69–76

Castora FJ, Erickson CE, Kovacs T, Lesiak K, Torrence PF (1991) 2′,5′-Oligoadenylate inhibits relaxation of supercoiled DNA by calf thymus DNA topoisomerase I. J Interferon Res 11:143–149

Chang DD, Sharp PA (1989) Regulation by HIV Rev depends upon recognition of splice sites. Cell 59:789–795

Chapekar MS, Knode MC, Glazer RI (1988) The epidermal growth factor- and interferon-independent effects of double-stranded RNA in A431 cells. Mol Pharmacol 34:461–466

Chebath J, Benech P, Hovanessian A, Galabru J, Revel M (1987) Four different forms of interferon-induced 2′,5′-oligo(A) synthetase identified by immunoblotting in human cells. J Biol Chem 262:3852–3857

Cohrs RJ, Goswami BB, Sharma OK (1988) Occurrence of 2-5A and RNA degradation in the chick oviduct during rapid estrogen withdrawal. Biochemistry 27:3246–3252

Crumpacker CS (1989) Molecular targets for antiviral therapy. N Engl J Med 321:163–172

Cullen BR, Malim MH (1991) The HIV-1 Rev protein: prototype of a novel class of eukaryotic post-transcriptional regulators. TIBS 16:346–350

Darzynkiewicz Z, Kapuscinski J, Carter SP, Schmid FA, Melamed MR (1986) Cytostatic and cytotoxic properties of pyronin Y: relation to mitochondrial localization of the dye and its interaction with RNA. Cancer Res 46:5760–5766

Diamantstein T, Blitstein-Willinger E (1975) Relationship between biological activities of biopolymers. Immunology 29:1087–1092

Diamantstein T, Blitstein-Willinger E (1978) Specific binding of poly(I)·poly(C) to the membrane of murine B lymphocyte subsets. Eur J Immunol 8:896–899

Doetsch PW, Suhadolnik RJ, Sawada Y, Mosca JD, Flick MB, Reichenbach NJ, Dang AQ, Wu JM, Charubala R, Pfleiderer W, Henderson EE (1981) Core (2′-5′)oligoadenylate and the cordycepin analog: inhibitors of Epstein-Barr virus-induced transformation of human lymphocytes in the absence of interferon. Proc Natl Acad Sci USA 78:6699–6703

Edery I, Petryshyn R, Sonenberg N (1989) Activation of double-stranded RNA-dependent kinase (dsI) by the TAR region of HIV-1 mRNA: a novel translational control mechanism. Cell 56:303–312

Etienne-Smekens M, Vandenbussche P, Content J, Dumont JE (1983) (2′-5′)Oligoadenylate in rat liver: modulation after partial hepatectomy. Proc Natl Acad Sci USA 80:4609–4613

Farrell PJ, Balkow K, Hunt T, Jackson RJ, Trachsel H (1977) Phosphorylation of initiation factor eIF-2 and the control of reticulocyte protein synthesis. Cell 11:187–200

Ferbus D, Testa V, Titeux M, Louache F, Thang MN (1985) Induction of (2′-5′) oligoadenylate synthetase activity during granulocyte and monocyte differentiation. Mol Cell Biochem 67:125–133

Floyd-Smith G, Denton JS (1988) Age-dependent changes are observed in the levels of an enzyme mediator of interferon action: a (2′-5′)A_n-dependent endoribonuclease. Proc Soc Exp Biol Med 189:329–337

Ghosh S, Baltimore B (1990) Activation in vitro of NF-kappaB by phosphorylation of its inhibitor IkappaB. Nature 344:678–682

Goldfeld AE, Maniatis T (1989) Coordinate viral induction of tumor necrosis factor α and interferon β in human B cells and monocytes. Proc Natl Acad Sci USA 86:1490–1494

Greene JJ, Alderfer JL, Tazawa I, Tazawa S, Ts'o POP, O'Malley JA, Carter WA (1978) Interferon induction and its dependence on the primary and secondary structure of poly(inosinic acid)·poly(cytidylic acid). Biochemistry 17:4214–4220

Gribaudo G, Lembo D, Cavallo G, Landolfo S, Lengyel P (1991) Interferon action: binding of viral RNA to the 40-kilodalton 2′-5′ -oligoadenylate synthetase in interferon-treated HeLa cells infected with encephalomyocarditis virus. J Virol 65:1748–1757

Gunnery S, Rice AP, Robertson HD, Mathews MB (1990) Tat-responsive region RNA of human immunodeficiency virus 1 can prevent activation of the double-stranded-RNA-activated protein kinase. Proc Natl Acad Sci USA 87:8687–8691

Heaphy S, Finch JT, Gait MJ, Karn J, Singh M (1991) Human immunodeficiency virus type 1 regulator of virion expression, rev, forms nucleoprotein filaments after binding to a purine-rich "bubble" located within the rev-responsive region of viral mRNAs. Proc Natl Acad Sci USA 88:7366–7370

Hearl WG, Johnston MI (1986) A misaligned double-stranded RNA, poly(I)·poly($C_{12}U$), induces accumulation of 2′,5′-oligoadenylates in mouse tissues. Biochem Biophys Res Commun 138:40–46

Ho DD, Hartshorn KL, Rota TR, Andrews CA, Kaplan JC, Schooley RT, Hirsch MS (1985) Recombinant human interferon alpha-A suppresses HTLV-III replication in vitro. Lancet i:602–604

Holmes GP, Kaplan JE, Stewart JA, Hunt B, Pinsky PF, Schonberger LB (1987) A cluster of patients with a chronic mononucleosis-like syndrome: is Epstein-Barr virus the cause? JAMA 257:2297–2303

Holmes GP, Kaplan JE, Glantz NM, Komaroff AL, Schonberger LB, Straus SE, Jones JF, Dubois RE, Cunningham-Rundles C, Pahwa S, Tosato G, Zegano LS, Purtilo DT, Brown N, Schooley RT, Bris I (1988) Chronic fatigue syndrome: a working case definition. Ann Intern Med 108:387–389

Hovanessian AG (1991) Interferon-induced and double-stranded RNA-activated enzymes: a specific protein kinase and 2′,5′-oligoadenylate synthetases. J Interferon Res 11:199–205

Hovanessian AG, Kerr IM (1979) The (2′-5′) oligoadenylate (pppA2′p5′A2′p5′A) synthetase and protein kinase(s) from interferon-treated cells. Eur J Biochem 93:515–526

Hovanessian AG, Svab J, Marie I, Robert N, Chamaret S, Laurent AG (1988) Characterization of 69- and 100-kDa forms of 2-5A-synthetase from interferon-treated human cells. J Biol Chem 263:4945-4949

Hubbell HR (1986) Synergistic antiproliferative effect of human interferons in combination with mismatched double-stranded RNA on human tumor cells. Int J Cancer 37:359–365

Hubbell HR, Liu R-S, Maxwell BL (1984) Independent sensitivity of human tumor cell lines to interferon and double-stranded RNA. Cancer Res 44:3252–3257

Hubbell HR, Kvalnes-Krick K, Carter WA, Strayer DR (1985) Antiproliferative and immunomodulatory actions of β-interferon and double-stranded RNA, individually and in combination, on human bladder tumor xenografts in nude mice. Cancer Res 45:2481–2486

Hubbell HR, Pequignot EC, Todd J, Raymond LC, Mayberry SD, Carter WA, Strayer DR (1987) Augmented antitumor effect of combined human natural interferon-alpha and mismatched double-stranded RNA treatment against a human malignant melanoma xenograft. J Biol Response Mod 6:525–536

Hubbell HR, Pequignot EC, Shanabrook KR, Carter WA, Williams RD, Strayer DR (1990) Differential effects of human natural interferon-alpha and mismatched double-stranded RNA against a human renal cell carcinoma xenograft. Anticancer Res 10:795–802

Hubbell HR, Boyer JE, Roane P, Burch RM (1991) Cyclic AMP mediates the direct antiproliferative action of mismatched double-stranded RNA. Proc Natl Acad Sci USA 88:906–910

Ilson DH, Torrence PF, Vilcek J (1986) Two molecular weight forms of human 2′,5′-oligoadenylate synthetase have different activation requirements. J Interferon Res 6:5–12

Itkes AV, Turpaev KT, Kartesheva ON, Kafiani CA, Severin ES (1984a) Cyclic AMP-dependent regulation of activities of synthetase and phosphodiesterase of 2′,5′-oligoadenylate in NIH 3T3 cells. Mol Cell Biochem 58:165–171

Itkes AV, Turpaev KT, Kartasheva ON, Tunitskaya VL, Kafiani CA (1984b) The mechanisms of the cyclic AMP-dependent regulation of the enzymes of the 2′,5′-oligoadenylate system. FEBS Lett 171:101–105

Jacobsen H, Krause D, Friedman RM, Silverman RH (1983) Induction of ppp(A2'p)$_n$A-dependent RNase in murine JLS-V9R cells during growth inhibition. Proc Natl Acad Sci USA 80:4954–4958

Jamison JM, Krabill K, Allen KA, Stuart SH, Tsai CC (1990) RNA-intercalating agent interactions: in vitro antiviral activity studies. Antiviral Chem Chemother 1:333–347

Johnston MI, Hearl WG (1987) Purification and characterization of a 2'-phosphodiesterase from bovine spleen. J Biol Chem 262:8377–8382

Jungmann RA, Kelley DC, Miles MF, Milkowski DM (1983) Cyclic AMP regulation of lactate dehydrogenase. J Biol Chem 258:5312–5318

Karn J (1991) Control of human immunodeficiency virus replication by the *tat*, *rev*, *nef* and protease genes. Curr Opinion Immunol 3:526–536

Katze MG (1992) The war against the interferon-induced dsRNA-activated protein kinase: can viruses win? J Interferon Res 12:241–248

Kimchi A, Shure H, Revel H (1981) Anti-mitogenic function of interferon-induced (2'-5') oligo(adenylate) and growth-related variations in enzymes that synthesize and degrade this oligonucleotide. Eur J Biochem 114:5–10

Krause D, Panet A, Arad G, Dieffenbach CE, Silverman RH (1985a) Independent regulation of ppp(A2'p)$_n$A-dependent RNase in NIH 3T3, clone 1 cells by growth arrest and interferon treatment. J Biol Chem 260:9501–9507

Krause D, Silverman RH, Jacobsen H, Leisy SA, Dieffenbach C, Friedman RM (1985b) Regulation of ppp(A2'p)$_n$A-dependent RNase levels during interferon treatment and cell differentiation. Eur J Biochem 146:611–618

Krown SE, Friden GB, Khansur T, Davies ME, Oettegen HF, Field AK (1983) Phase I trial with the interferon inducer poly I:C/L-lysine (Poly ICL). J Interferon Res 3:281–290

Lammers R, Gross G, Mayr U, Collins J (1988) Alternative mechanisms for gene activation induced by poly(rI)·poly(rC) and Newcastle disease virus. Eur J Biochem 178:93–99

Larder BA, Kemp SD (1989) Multiple mutations in HIV-1 reverse transcriptase confer high-level resistance to zidovudine (AZT). Science 246:1155–1158

Larder BA, Darby G, Richman DD (1989) HIV with reduced sensitivity to zidovudine (AZT) isolated during prolonged therapy. Science 243:1731–1734

Laurence J, Kulkosky J, Friedman SM, Posnett DN, Ts'o POP (1987) poly I·polyC$_{12}$U-mediated inhibition of alloantigen responsiveness and viral replication in human CD4 + T cell clones exposed to human immunodeficiency virus in vitro. J Clin Invest 80:1631–1639

Laurent G St, Yoshie O, Floyd-Smith G, Samanta H, Sehgal PB (1983) Interferon action: two (2'-5')(A)$_n$ synthetases specified by distinct mRNAs in Ehrlich ascites tumor cells treated with interferon. Cell 33:95–102

Laurent-Crawford AG, Krust B, Deschamps de Paillette E, Montagnier L, Hovanessian AG (1992) Antiviral action of polyadenylic-polyuridylic acid against HIV in cell cultures. AIDS Res Hum Retroviruses 8:285–290

Li SW, Moskow JJ, Suhadolnik RJ (1990) 8-Azido double-stranded RNA photoaffinity probes. Enzymatic synthesis, characterization, and biological properties of poly(I,8-azidoI)·poly(C) and poly(I,8-azidoI)·poly(C$_{12}$U) with 2',5'-oligoadenylate synthetase and protein kinase. J Biol Chem 265:5470–5474

Lusso P, Ensoli B, Markham PD, Ablashi DV, Salahuddin SZ, Tschachler E, Wong-Staal F, Gallo RC (1989) Productive dual infection of human CD4$^+$ lymphocytes by HIV-1 and HHV-6. Nature 337:370–373

Lusso P, DeMaria A, Malnati M, Lori F, DeRocco SE, Baseler M, Gallo RC (1991) Induction of CD4 and susceptibility to HIV-1 infection in human CD8$^+$ T lymphocytes by human herpesvirus 6. Nature 349:533–535

Marcus PI (1983) Interferon induction by viruses: one molecule of dsRNA as the threshold for interferon induction. In: Gresser I (ed) Interferon 5. Academic Press, New York, pp 115–180

Mariman ECM, van Eekelen CAG, Reinders RJ, Berns AJM, van Venrooij WJ (1982) Adenoviral heterogeneous nuclear RNA is associated with the host nuclear matrix during splicing. J Mol Biol 154:103–119

Meegan JM, Marcus PI (1989) Double-stranded ribonuclease coinduced with interferon. Science 244:1089–1091

Memet S, Besancon F, Bourgeade MF, Thang MN (1991) Direct induction of interferon-gamma- and interferon-alpha/beta-inducible genes by double-stranded RNA. J Interferon Res 11:131–141

Meurs E, Chong K, Galabru J, Thomas N, Kerr I, Williams BRG, Hovanessian AG (1990) Molecular cloning and characterization of the human double-stranded RNA-activated protein kinase induced by interferon. Cell 62:379–390

Milhaud PG, Silhol M, Salehzada T, Lebleu B (1987) Requirement for endocytosis of poly(rI)·poly(rC) to generate toxicity on interferon-treated LM cells. J Gen Virol 68:1125–1134

Minks MA, West DK, Benvin S, Baglioni C (1979) Structural requirements of double-stranded RNA for the activation of 2′-5′ oligo(A) polymerase and protein kinase of interferon treated HeLa cells. J Biol Chem 254:10180–10183

Mitchell PJ, Wang C, Tjian R (1987) Positive and negative regulation of transcription in vitro: enhancer-binding protein AP-2 is inhibited by SV40 T antigen. Cell 50:847–861

Mitchell W, Montefiori DC, Robinson WE, Strayer DR, Carter WA (1987) Mismatched double-stranded RNA (ampligen) reduces concentration of zidovudine (azidothymidine) required for in vitro inhibition of human immunodeficiency virus. Lancet i:890–892

Montefiori DC, Mitchell WM (1987) Antiviral activity of mismatched double-stranded RNA against human immunodeficiency virus in vitro. Proc Natl Acad Sci USA 84:2985–2989

Montefiori DC, Robinson WE, Schuffman SS, Mitchell WM (1988) Evaluation of antiviral drugs and neutralizing antibodies against human immunodeficiency virus by a rapid and sensitive microtiter infection assay. J Clin Microbiol 26:231–235

Montefiori DC, Pellegrino MG, Robinson WE, Engle K, Field M, Mitchell WM, Gillespie DH (1989a) Inhibition of HIV-1 proviral DNA synthesis and RNA accumulation by mismatched dsRNA. Biochem Biophys Res Commun 158:943–950

Montefiori DC, Robinson WE, Mitchell WM (1989b) In vitro evaluation of mismatched double-stranded RNA (ampligen) for combination therapy in the treatment of acquired immunodeficiency syndrome. AIDS Res Hum Retroviruses 5:193–203

Montefiori DC, Sobol RW, Li SW, Reichenbach NL, Suhadolnik RJ, Charubala R, Pfleiderer W, Modliszewski A, Robinson WE, Mitchell WM (1989c) Phosphorothioate and cordycepin analogues of 2′,5′-oligoadenylate: inhibition of human immunodeficiency virus type 1 reverse transcriptase and infection in vitro. Proc Natl Acad Sci USA 86:7191–7194

Müller WEG, Wenger R, Reuter P, Renneisen K, Schröder HC (1989) Association of tat protein and viral mRNA with nuclear matrix from HIV-1 infected cells. Biochim Biophys Acta 1008:208–212

Müller WEG, Okamoto T, Reuter P, Ugarkovic D, Schröder HC (1990) Functional characterization of Tat protein from human immunodeficiency virus; evidence that Tat links viral RNAs to nuclear matrix. J Biol Chem 265:3803–3808

Müller WEG, Weiler BE, Charubala R, Pfleiderer W, Leserman L, Sobol RW, Suhadolnik RJ, Schröder HC (1991) Cordycepin analogues of 2′,5′-oligoadenylate inhibit human immunodeficiency virus infection via inhibition of reverse transcriptase. Biochemistry 30:2027–2033

Murray HW, Rutin BY, Masur H, Roberts R (1984) Impaired production of lymphokines and immune (gamma) interferon in the acquired immunodeficiency syndrome. N Engl J Med 310:883–889

Nilsen TW, Maroney PA, Robertson HD, Baglioni C (1982a) Heterogeneous nuclear RNA promotes synthesis of (2′,5′)oligoadenylate and is cleaved by the (2′,5′)oligoadenylate-activated endoribonuclease. Mol Cell Biol 2:154–160

Nilsen TW, Wood DL, Baglioni C (1982b) Presence of 2′,5′-oligo(A) and of enzymes that synthesize, bind, and degrade 2′,5′-oligo(A) in HeLa cell nuclei. J Biol Chem 257:1602–1605

Nokta M, Pollard R (1991) Human immunodeficiency virus infection: association with altered intracellular levels of cAMP and cGMP in MT-4 cells. Virology 181:211–217

Pavlakis GN, Felber BK (1990) Regulation of expression of human immunodeficiency virus. New Biol 2:20–31

Pestka S, Langer JA, Zoon KC, Samuel CE (1987) Interferons and their actions. Annu Rev Biochem 56:727–777

Pfeifer K, Weiler BE, Ugarkovic D, Bachmann M, Schröder HC, Müller WEG (1991) Evidence for a direct interaction of Rev protein with nuclear envelope mRNA-translocation system. Eur J Biochem 199:53–64

Pfeifer K, Ushijima H, Lorenz B, Müller WEG, Schröder HC (1993) Evidence for age-dependent impairment of antiviral 2′,5′-oligoadenylate synthetase/ribonuclease L-system in tissues of rat. Mech Ageing Dev 67:101–114

Pinto AJ, Morahan PS, Brinton MA (1989) Comparative study of various immunomodulators for macrophage and natural killer cell activation and antiviral efficacy against exotic RNA viruses. Int J Immunopharmacol 10:197–209

Preble OT, Rook AH, Steis R, Silverman RH, Krause D, Quinnan GV, Masur H, Jacob J, Longo D, Gelmann EP (1985) Interferon-induced 2′-5′ oligoadenylate synthetase during interferon therapy in homosexual men with Kaposi's sarcoma: marked deficiency in biochemical response to interferon in patients with acquired immunodeficiency syndrome. J Infect Dis 152:457–465

Priel E, Showalter SD, Roberts M, Oroszlan S, Segal S, Aboud M, Blair DG (1990) Topoisomerase I activity associated with human immunodeficiency virus (HIV) particles and equine infectious anemia virus core. EMBO J 9:4167–4172

Raj NBK, Engelhardt J, Au W-C, Levy DE, Pitha PM (1989) Virus infection and interferon can activate gene expression through a single synthetic element, but endogenous genes show distinct regulation. J Biol Chem 264:16658–16666

Ray A, Tatter SB, May LT, Sehgal PB (1988) Activation of the human "β_2-interferon/hepatocyte-stimulating factor/interleukin 6" promoter by cytokines, viruses, and second messenger antagonists. Proc Natl Acad Sci USA 85:6701–6705

Read SE, LeBrocq FJ, Williams BRG (1985a) Persistent elevation of 2-5A synthetase and prognosis in the AIDS-related complex (ARC). In: Williams BRG, Silverman RH (eds) The 2-5A system: molecular and clinical aspects of the interferon-regulated pathway. Liss, New York, pp 405–413

Read SE, Williams BR, Coates RA, Evans WK, Fanning MM, Garvey MB, Shepherd FA (1985b) Elevated levels of interferon-induced 2′-5′ oligoadenylate synthetase in generalized persistent lymphadenopathy and the acquired immunodeficiency syndrome. J Infect Dis 152:466–472

Roesler WJ, Vandenbark G R, Hanson RW (1988) Cyclic AMP and the induction of eukaryotic gene transcription. J Biol Chem 263:9063–9066

Rosen CA (1991) Regulation of HIV gene expression by RNA-protein interactions. Trends Genet 7:9–14

Rossol S, Voth R, Laubenstein HP, Müller WEG, Schröder HC, Meyer zum Büschenfelde KH, Hess G (1989) Interferon production in patients infected with HIV-1. J Infect Dis 159:815–821

Schattner A, Wallach D, Merlin G, Hahn T, Levin S, Revel M (1981) Assay of an interferon-induced enzyme in white blood cells as a diagnostic aid in viral diseases. Lancet i:497–500

Schröder HC, Zahn RK, Dose K, Müller WEG (1980) Purification and characterization of a poly(A)-specific exoribonuclease from calf thymus. J Biol Chem 255:4535–4538

Schröder HC, Wenger R, Rottmann M, Müller WEG (1988) Alteration of nuclear (2′-5′)oligoriboadenylate synthetase and nuclease activities preceding replication of human immunodeficiency virus in H9 cells. Biol Chem Hoppe-Seyler 369:985–995

Schröder HC, Wenger R, Kuchino Y, Müller WEG (1989) Modulation of nuclear matrix-associated (2′-5′)oligoadenylate metabolism and ribonuclease L activity in H9 cells by human immunodeficiency virus. J Biol Chem 264:5669–5673

Schröder HC, Kljajic Z, Weiler BE, Gasic M, Uhlenbruck G, Kurelec B, Müller WEG (1990a) The galactose-specific lectin from the sponge *Chondrilla nucula* displays anti-human immunodeficiency virus activity in vitro via stimulation of the (2′-5′)oligoadenylate metabolism. Antiviral Chem Chemother 1:99–105

Schröder HC, Ugarkovic D, Wenger R, Okamoto T, Müller WEG (1990b) Binding of Tat protein to TAR region of human immunodeficiency virus type 1 blocks TAR-mediated activation of (2′-5′)oligoadenylate synthetase. AIDS Res Hum Retroviruses 6:659–672

Schröder HC, Suhadolnik RJ, Pfleiderer W, Charubala R, Müller WEG (1992) (2'-5')Oligoadenylate and intracellular immunity against retrovirus infection. Int J Biochem 24:55–63
Schröder HC, Kelve M, Schäcke H, Pfleiderer W, Charubala R, Suhadolnik RJ, Müller WEG (1993a) Inhibition of DNA topoisomerase I activity by 2',5'-oligoadenylates and mismatched double-stranded RNA (ampligen) in uninfected and HIV-1-infected H9 cells. Chem-Biol Interactions (in press)
Schröder HC, Ushijima H, Bek A, Merz H, Pfeifer K, Müller WEG (1993b) Inhibition of formation of Rev-RRE complex by pyronin Y. Antiviral Chem Chemother 4:103–111
SenGupta DN, Silverman RH (1989) Activation of interferon-regulated, dsRNA-dependent enzymes by human immunodeficiency virus-1 leader RNA. Nucleic Acids Res 17:969–978
Shafran SD (1991) The chronic fatigue syndrome. Am J Med 90:730–739
Silverman RH, Jung DD, Nolan-Sorden NL, Dieffenbach CW, Kedar VP, SenGupta DN (1988) Purification and analysis of murine 2-5A-dependent RNase. J Biol Chem 263:7336–7341
Straus SE (1988) The chronic mononucleosis syndrome. J Infect Dis 157:405–412
Strayer DR, Watson P, Carter WA, Brodsky I (1986) Antiproliferative effect of mismatched double-stranded RNA on fresh human tumor cells analyzed in a clonogenic assay. J Interferon Res 6:373–379
Strayer DR, Brodsky I, Pequignot EC, Crilley PA, Carter WA, Fenning R, Kariko K, Reichenbach NL, Sobol RW, Li SW, Suhadolnik RJ (1990) The antitumor activity of ampligen, a mismatched double-stranded RNA that modulates the 2-5A synthetase/RNase L pathway in cancer and AIDS. In: Diasio RB, Sommadossi JP (eds) Advances in chemotherapy of AIDS. Pergamon Press, New York, pp 23–31
Strayer DR, Carter WA, Pequignot E, Topolsky D, Brodsky I, Suhadolnik RJ, Reichenbach N, Paul D, Einck L, Hubbell HR, Pinto A, Strauss K, Gillespie D (1991) Activity of Ampligen in HIV disease. Clin Biotechnol 3:169–175
Suhadolnik RJ, Lebleu B, Pfleiderer W, Charubala R, Montefiori DC, Mitchell WM, Sobol RW, Li SW, Kariko K, Reichenbach NL (1989) Phosphorothioate analogs of 2-5A: activation/inhibition of RNase L and inhibition of HIV-1 reverse transcriptase. Nucleosides Nucleotides 8:987–990
Suhadolnik RJ, Reichenbach NL, Sobol RW, Hart R, Peterson DL, Strayer DR, Henry B, Ablashi DV, Gillespie DH, Carter WA (1990) Biochemical defects in the 2-5A synthetase/RNase L pathway associated with chronic fatígue syndrome with encephalopathy. In: Hyde BM, Goldstein J, Levine P (eds) The clinical and scientific basis of myalgic encephalomyelitis/chronic fatique syndrome. The Nightingale Research Foundation, London, pp 613–617
Suhadolnik RJ, Carter WA, Reichenbach NL, Gillespie DH, Strayer DR, Elsasser W, Ventura DL, Henry B, Ablashi DV (1993) RNase L: modulation in HIV disease and chronique fatigue syndrome by ampligen. Can J Infect Dis (in press)
Ushijima H, Daum T, Schröder HC, Matthes E, Engels JW, Mag M, Muth J, Müller WEG (1993a) Synergistic anti-human immunodeficiency viral (HIV-1) effect between the immunomodulator ampligen (mismatched double-stranded RNA) and inhibitors of reverse transcriptase and HIV-1 regulatory proteins. Antiviral Chem Chemother (in press)
Ushijima H, Rytik PG, Schäcke H, Scheffer U, Müller WEG, Schröder HC (1993b) Mode of action of the anti-AIDS compound poly(I)·poly(C_{12}U) (ampligen): activator of 2',5'-oligoadenylate synthetase and double-stranded RNA-dependent kinase. J Interferon Res 13:161–171
Wathelet MG, Clauss IM, Nols CB, Content J, Huez GA (1987) New inducers revealed by the promoter sequence analysis of two interferon-activated human genes. Eur J Biochem 169:313–321
Wathelet MG, Clauss IM, Paillard FC, Huez GA (1989) 2-Aminopurine selectively blocks the transcriptional activation of cellular genes by virus, double-stranded RNA and interferons in human cells. Eur J Biochem 184:503–509
Wells V, Mallucci L (1985) Expression of the 2-5A system during the cell cycle. Exp Cell Res 159:27–36
Williams BRG (1991) Transcriptional regulation of interferon-stimulated genes. Eur J Biochem 200:1–11

Williams BRG, Read SE, Freedman MH, Carver DH, Gelfand EW (1982) The assay of 2-5A synthetase as an indicator of interferon activity and virus infection in vivo. In: Merigan TC, Friedman RM (eds) Interferons. UCLA Symp on Molecular and cellular biology, vol 25. Academic Press, New York, pp 253–267

Witt PL, Spear GT, Lindstrom MJ, Kessler HA, Borden EC, Phair J, Landay AL (1991) 2′,5′-Oligoadenylate synthetase, neopterin and beta 2-microglobulin in asymptomatic HIV-infected individuals. AIDS 5:289–293

Wreschner DH, Nathanel T, Herzberg M (1985) Double stranded RNA and the nuclear matrix – implications for the 2-5A system. In: Williams BRG, Silverman RH (eds) Molecular and clinical aspects of the interferon-regulated pathway. Liss, New York, pp 47–66

Wu JM, Chiao JW, Maayan S (1986) Diagnostic value of the determination of an interferon-induced enzyme activity: decreased 2′,5′-oligoadenylate dependent binding protein activity in AIDS patient lymphocytes. AIDS Res 2:127–131

Yoshida I, Azuma M (1985) Adsorption of poly rI:rC on cell membrane participating and nonparticipating in interferon induction. J Interferon Res 5:1–10

Yoshida I, Azuma M (1991) Analysis of specific receptor for poly(I)·poly(C) on human embryonic lung cell surface. J Interferon Res 11:S209

Zarling JM, Schlais J, Eskra L, Greene JJ, Ts'o POP, Carter WA (1980) Augmentation of human natural killer cell activity by polyinosinic acid-polycytidylic acid and its nontoxic mismatched analogues. J Immunol 124:1852–1857

Zeitlin S, Parent A, Silverstein S, Efstratiadis A (1987) Pre-mRNA splicing and the nuclear matrix. Mol Cell Biol 7:111–120

Zinn K, Keller A, Whittemore LA, Maniatis T (1988) 2-Aminopurine selectively inhibits the induction of β-interferon, c-*fos*, and c-*myc* gene expression. Science 240:210–213

Zullo JN, Cochran BH, Huang AS, Stiles CD (1985) Platelet-derived growth factor and double-stranded ribonucleic acids stimulate expression of the same genes in 3T3 cells. Cell 43:793–800

The Antiviral Activity of RNA-Dye Combinations

J.M. Jamison[1], J. Gilloteaux[2], and J.L. Summers[3]

1 Introduction

The primary challenge of viral chemotherapy is the development of drugs with good pharmacokinetic profiles which are nontoxic, nonmutagenic, noncarcinogenic, and specific for viral functions. Effective viral chemotherapy has been hindered because viruses employ host cell systems to replicate. Recent advances in molecular virology provide a rational basis for the development of antiviral compounds that will interfere with biochemically defined, virus-specific functions, such as the herpes virus-coded thymidine kinase (Elion 1984). One strategy is to inhibit key activities in the target cell which are also required for viral replication, such as the inhibition of cellular protein kinases by tricyclodecan-9-yl-xanthogenate (D609) (Warlo and Rosenthal 1988) and the inhibition of glycosylation by prostaglandin A (Santoro et al. 1989) or 2-deoxyglucose (Blough and Ray 1980). However, if the virus is not killed, such a strategy can lead to the development of a drug-resistant virus (Field and Goldthorpe 1989).

A second strategy is to employ cytokines, such as interferons (IFN), that activate multiple genes whose products generate an activiral state that can discriminate between cellular processes and the processes incumbent to a wide variety of DNA and RNA viruses (Becker 1984). However, interferon antiviral therapy is complicated by its rapid clearance from the blood, its toxicity, the emergence of hyporesponsiveness with continued use, and the existence of virus-mediated, anti-interferon effects (Becker 1984; Sen and Ransohoff 1993). These difficulties have led some investigators to employ interferon inducers instead of interferon. Polyribonucleotides, such as poly (rI)·poly (rC) are among the most effective inducers of interferon and are advantageous because of their extreme potency and of the mosaic nature of the interferons they induce (Nosik et al. 1988). The major disadvantages of polyribonucleotides as interferon inducers are their toxicity and their lability in human serum (Becker 1984).

[1] Department of Microbiology and Immunology, Northeastern Ohio Universities College of Medicine, 4209 State Route 44, P.O. Box 95, Rootstown, Ohio 44272, USA
[2] Department of Anatomy, Northeastern Ohio University College of Medicine, 4209 State Route 44, P.O. Box 95, Rootstown, Ohio 44272, USA
[3] Professor of Urology, 75 Arch Street, Suite 101, Akron City Hospital, Akron, Ohio 44304, USA

Progress in Molecular and Subcellular Biology, Vol. 14
W.E.G. Müller/H.C. Schröder (Eds.)
© Springer-Verlag Berlin Heidelberg 1994

Several strategies have been employed to address these problems including: development of covalently modified polyribonucleotides (ampligen and thiophosphorate derivatives) (Carter et al. 1976; Eckstein and Gish 1989), administration of polyribonucleotides in liposomes (Grigorian et al. 1987), the use of polyribonucleotides with different base sequences, such as poly r(A-U) (De Clercq and Torrence 1977) and co-administration of polyribonucleotides with dyes and other substances (De Clercq et al. 1971; Lodemann et al. 1973; Sehgal et al. 1975; Iliescu et al. 1983).

Studies examining the antiviral activity of dye-RNA combinations can be divided into two categories: pulse-chase studies and co-administration studies. In the pulse-chase studies, the RNA is co-incubated with cell monolayers for several hours and is subsequently removed and replaced with the dye. This type of protocol is exemplified by the work of Sehgal and coworkers (1975) who demonstrated that while neutral red or chloroquine alone is not an effective interferon inducer, treatment of human diploid fibroblasts with these dyes 2.5 to 3.5 h after poly (rI)·poly (rC) treatment potentiated interferon induction 16- to 64-fold and 4- to 16-fold, respectively.

In co-administration studies, the dye and RNA are combined before either component is added to the cell monolayer. De Clercq and his associates (1971) employed this technique to show that poly r(A-U) exhibits a potentiated antiviral activity against vesicular stomatitis virus (VSV) and an enhanced resistance to endonuclease degradation when it is thermally activated in the presence of divalent cations (Ca^{2+}, Mg^{2+}). Lodemann and coworkers (1973) observed that cationic dyes such as toluidine blue O, methylene blue, acridine orange, and tryptaflavine also potentiate the antiviral and interferon-inducing activities of poly (rI)·poly (rC) in L-cells. The basis of this potentiated antiviral activity has not been elucidated, but an interaction of the dyes with the polyribonucleotide and the resulting charge shielding has been suggested. Iliescu and co-investigators (1983) have shown that 2-h treatment of human embryo fibroblasts, chicken embryo fibroblasts, and R_9CA monkey kidney with *E. coli* chromosomal DNA, mouse liver and rat liver total RNA, rat liver rRNA, or rat liver tRNA prior to inoculation with herpes simplex virus type I (VR_3, Rapp) reduced the infectious titer by up to 1.75 log. Maximum reduction in virus titer occurred at a dose of 50 µg nucleic acid/culture tube (250 µg/ml). Complexation of the nucleic acids with the intercalative dyes, ethidium bromide and violamycin BI, enhanced the reduction in virus titer by 0.5 to 1.5 log. Furthermore, the enhanced antiviral activity of these dye/nucleic acid combinations did not appear to be due to the induction of interferon. The results of these studies suggest that dyes may potentiate the antiviral activity of RNA in ways other than increasing interferon induction.

Carter has stated that double-stranded ribonucleic acids masquerade in the human body as "artificial viruses" which results in the amplification of the body's immune response by enhancing the interferon-inducing and natural killer cell activity (Bennett 1985). The objective of our research has been to design nontoxic, virus-like molecules capable of potentiating the immune system. Specifically, intercalating and putative intercalating dyes have been employed to perturb the conformation of A-form polyribonucleotides in an effort to enhance the antiviral activity of the polyribonucleotide. Our studies are co-administration studies in which the dye and RNA are combined and then the combination is added to cell monolayers for 3 h. The test agent is then removed and replaced with culture medium for 17 h at which time viral challenge (VSV, multiplicity of infection = 0.1) occurs (Jamison et al. 1988, 1989, 1990a–g).

2 The Structure of Double-Stranded RNA

Before dye-double-stranded RNA (dsRNA) interactions can be appreciated, the basic structure of dsRNA must be reviewed. The results of structural studies on DNA and dsRNA illustrate that DNA displays a structural polymorphism (right-handed A, B, C, D and E and left-handed Z) induced by changes in humidity and ionic strength as well as oligonucleotide sequence, while dsRNA exhibits a structural conservatism in which the dsRNA is restricted to the right-handed A or A′ forms (Saenger 1984). B-DNA is characterized by a negative base pair tilt, base pairs which are pierced by the helix axis and a broad major groove and a narrow minor groove of nearly equal depth. A-form of dsRNA is characterized by a positive base pair tilt with respect to the helix axis and the displacement of the helix towards the major groove. This displacement produces a macroscopic arrangement in which the polynucleotides are wrapped around the helix axis creating an open cylinder with a deep, narrow major groove on the inner surface of the cylinder and a reduced, shallow, wide minor groove on the outer surface. The major groove lies on the inner surface of the cylinder and is accessible only to water molecules and metal ions. This suggests that both the major and minor grooves are available for intercalation in B-DNA while only the minor groove is readily accessible for intercalation in A-dsRNA. Base overlaps also differ greatly in A- and B-type double helices, with the degree of intra- and interstrand base stacking in A-type double helices being much greater than in B-type double helices because of the positive base pair tilt and the smaller rotation per nucleotide. The degree of overlap in A-type helices is also dependent on the sequence of the bases. Pyrimidine-3′, 5′-purine sequences, such those found in poly r(A-U), are the preferred binding site for the planar chromophores of dyes because their interstrand base overlaps stabilize intercalation and because intercalation into these sequences is energetically favorable (Saenger 1984).

3 Structural Consequences of Intercalation

Dyes interact with double-stranded nucleic acids in either a covalent or a reversible, noncovalent manner (Waring 1981). Covalent binding may occur by a number of mechanisms, including alkylating substituents of the dye (Waring 1981), chemical conjugation of the dye to the nucleic acid via linker arms (Hélène and Thuong 1988) or photo-induced cross-linking of the dye to the nucleic acid

(Content and Cogniaux-LeClercq 1968). Dyes that interact with nucleic acids in a noncovalent fashion fall into three categories: electrostatic binders, groove binders, and intercalators (Waring 1981). Electrostatic binders, like methyl green, are electrostatically bound to the phosphate groups in the backbone of the nucleic acid. Groove binders, such as Netropsin, usually bind in the minor groove of the B-DNA double helix because of their requirements for steric and electronic complementarity. Intercalative binders, like ethidium bromide, bind by inserting their planar, aromatic chromophores between adjacent base pairs of the nucleic acid.

Intercalation is a two-stage, anticooperative binding process (Crothers 1968). Initial binding entails an electrostatic interaction between the phosphate groups in the double-stranded nucleic acids and the dye molecules. Within milliseconds the planar, aromatic chromophore of the dye is inserted or intercalated between adjacent base pairs in the DNA/dsRNA helix. At low dye to nucleoside phosphate ratios (D/P), the predominant mode of binding is generally stronger and involves intercalation. At higher D/P ratios (once the intercalative sites have been saturated), a weaker electrostatic binding occurs between the phosphate groups of the double-stranded nucleic acids and the dye molecules. Intercalation of the planar, aromatic chromophore of the dye perturbs the double helix such that the helix: increases 3.4 Å in length per bound dye molecule, becomes more stable and exhibits an elevated melting temperature, and adopts an altered sugar puckering. The altered sugar puckering diminishes the positive base tilt; produces an angular unwinding of the helix; and results in a residual "kink" or "bend" at the intercalation site such that the helix axes of consecutive base pairs are not co-linear and the helix itself assumes higher-order conformations (Tsai 1978). In the case of dsRNA, the increased melting temperature and decreased positive base tilt are conducive to enhanced interferon induction (Bobst et al. 1976). As a consequence of these perturbations, the geometry of the nucleotides flanking the intercalative dye is altered such that steric constaints prohibit the intercalation of dye at adjacent sites (neighbor exclusion). For dyes without side chains, the limit in intercalative sites is one per 2-2.5 base pairs, whereas for dyes with side chains the limit is one dye per 3 base pairs (or more) depending on the nature of the side chains (Berman and Young 1981). At high binding density and high dye concentration, the interactions between cationic, aromatic intercalating dyes and nucleic acids often result in condensation and precipitation of the product (Kapuscinski and Darzynkiewicz 1984).

4 Antiviral Activity of Intercalative Dyes

The results of single crystal X-ray analysis of intercalative dye-ribonucleotide complexes (reviewed in Tsai 1978) indicate that intercalative dyes may be divided into three classes which are distinguished by their directional entrance into double-stranded nucleic acids: dyes that intercalate exclusively from the minor groove, dyes that intercalate exclusively from the major groove, and dyes

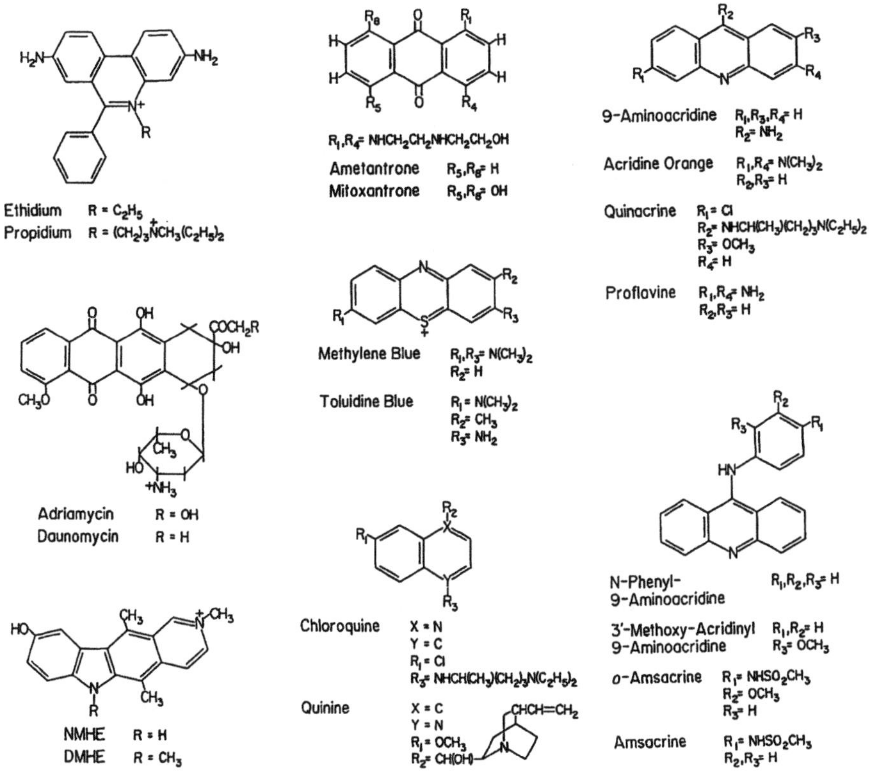

Fig. 1. Classic intercalators and anthraquinones

that may intercalate from either the minor or major groove. Representatives from each of these three classes are employed in our previous study (Jamison et al. 1988, 1989, 1990a-g). In this study, each dye and poly r(A-U) are evaluated individually for concentrations ranging from 0.8 to 400 µM. All dyes are also tested in conjunction with poly r(A-U) with the concentration of the poly r(A-U) being fixed at 200 µM while the dye concentration is varied to produce dye/ribonucleotide ratios of 1/16, 1/8, 1/6, 1/4, 1/2, 1/1 and 2/1.

Four minor groove intercalating dyes are examined in this study: ethidium bromide (EB), propidium iodide (PI), adriamycin (ADR), and daunomycin (DMN) (Fig. 1). The results of previous studies (Tsai 1978; Nelson and Tinoco 1984) indicate that EB and PI bind to dsRNA primarily by intercalation until the intercalative binding sites are saturated at a drug/nucleoside phosphate (D/P) ratio of 1:4 (one EB/PI molecule per four nucleotides). At D/P ratios greater than 1:4 (once the intercalative sites have been saturated), electrostatic binding occurs. If intercalation is important to the enhancement of the antiviral activity of the poly r(A-U), one would expect the antiviral activity of poly r(A-U) to increase as the ethidium bromide is added until the intercalative sites are saturated at a D/P ratio of 1:4. For these two phenanthridines (Table 1), the ratios of the 50% effective doses (ED_{50}) decrease as the EB (or PI)/ribonucleotide ratios increase, reach a minimum at an EB (or PI)/ribonucleotide ratio in the region of 1/4, and increase as the EB (or PI)/ribonucleotide ratios approach 2/1. The ED_{50} for EB alone is 15.4 μM, while the ED_{50} for poly r(A-U) alone is 6.8 μM. When EB is combined with poly r(A-U) to produce an EB/poly r(A-U) ratio of 1:4, the ratio of ED_{50} is 0.10/0.40. These results demonstrate that when EB is combined with poly r(A-U), the ED_{50} of the poly r(A-U) decreases from 6.8 to 0.40 μM (a 17-fold decrease), while the ED_{50} of the EB decrease from 15.4 to 0.10 μM (a 154-fold decrease). For the PI/poly r(A-U) combination the ED_{50} of the poly r(A-U) decreases 24-fold, while the ED_{50} of the PI decrease 299-fold. These results suggest that the enhancement process is affected by the alkyl substituent or the additional positive charge of the propidium iodide. Experiments (Jamison et al. 1990g)

Table 1. Antiviral activity of dyes alone and dye/poly r(A-U) combinations[a]

Dye	Dye alone 50% effective dose (μM)	Dye/ribonucleotide ratio of 50% effective doses (μM)	Enhancement of antiviral activity (n-fold)	Fractional inhibitory concentration[b] (FIC)
Acridine orange*	14.3	2.00/ 8.00	1	1.330
Adriamycin*	13.5	0.25/ 1.00	8	0.166
Alizarin (A.)	25.0	2.50/10.00	1	1.570
A. blue black B	NI	1.60/ 6.30	1	NA
A. complexone	NI	1.60/ 6.30	1	NA
A. sulfonate	NI	2.50/10.00	1	NA
Ametantrone	25.0	0.20/ 0.80	11	0.126
9-Aminoacridine* (9-AA)	13.0	0.60/ 2.40	5	0.399
3′-Methoxy-A-9-AA	NI	1.13/ 4.52	2	0.665
N-phenyl-9-AA	NI	0.45/ 1.80	5	0.265
AMSA	NI	1.13/ 4.52	2	0.665
o-AMSA	NI	1.13/ 4.52	2	0.665
Carminic acid	NI	0.20/ 0.80	12	NA
Chloroquine diphosphate	13.0	0.20/ 0.80	10	0.133
Cibacron blue	NI	1.60/ 6.30	1	NA
Daunomycin	11.1	0.14/ 0.56	15	0.095
DMHE	21.9	0.08/ 0.32	20	0.051
Emodin	25.0	1.00/ 4.00	2	0.628
Eosin B	NI	0.40/ 1.60	6	NA
Ethidium bromide	15.4	0.10/ 0.40	22	0.065
Fluoresceinamine isomer I	25.0	1.00/ 4.00	2	0.628
Fluoresceinamine isomer II	25.0	0.40/ 1.60	6	0.251
Methylene blue	25.0	1.30/ 5.20	2	0.817
Mitoxantrone	25.0	0.20/ 0.80	11	0.126

Table 1. Continued

Dye	Dye alone 50% effective dose (μM)	Dye/ribonucleotide ratio of 50% effective doses (μM)	Enhancement of antiviral activity (n-fold)	Fractional inhibitory concentration[b] (FIC)
NMHE	17.6	0.07/ 0.28	16	0.045
Proflavine	8.5	1.60/ 6.40	1	1.129
Propidium iodide	20.9	0.07/ 0.28	20	0.045
Pyronin B	13.0	0.28/ 1.13	7	0.188
Pyronin Y	13.0	0.40/ 1.60	7	0.266
Quinacrine	3.1	0.80/ 3.20	2	0.729
Quinine hydrochloride	6.3	0.20/ 0.80	9	0.149
Rhein	NI	0.80/ 3.20	2	NA
Rhodamine B	NI	0.20/ 0.80	12	NA
Rhodamine 6G	NI	0.32/ 1.27	8	NA
Rhodamine 123	55.5	0.16/ 0.64	15	0.097
Sulforhodamine B	NI	0.16/ 0.64	15	NA
Sulforhodamine 101	NI	0.36/ 1.43	7	NA
Toluidine blue O	NI	0.53/ 2.10	5	0.309
0.2 mM Poly r(A-U)	6.8	NA	NA	NA

NI = not inhibitory NA = not applicable

Mean titers are the arithmetic averages of six readings, except as noted (* = triplicate readings). Mean inhibitory concentrations are the geometric averages of six readings, except as noted (* = triplicate readings).

[a] The role of dyes in modulating the antiviral activity of poly r(A-U) was examined by experiments in which the concentration of the poly r(A-U) was fixed at 200 μM, while dye concentrations were fixed at 50 μM to produce dye/poly r(A-U) ratios of 1/4. Each of these dye/poly r(A-U) solutions was serially diluted with PBS in 12 twofold dilutions to form test solutions. These test solutions were co-incubated with HSF cells for 3 h at 37°C. Subsequently, the test solutions were removed and the cells were washed twice with 100 μl PBS, overlaid with 50 μl EMEM (5% FCS), incubated for 17 h at 37°C and challenged with VSV (MOI = 0.1). Antiviral activity was evaluated by 50% CPE 17 h postinfection. The enhancement factor was calculated by dividing the equivalent IFN titer of the dye-poly r(A-U) combination by the sum of the equivalent IFN titers of the poly r(A-U) alone and the dye alone.

[b] Fractional Inhibitory Concentration = $ED_{50}^{A\ comb}/ED_{50}^{A\ alone} + ED_{50}^{B\ comb}/ED_{50}^{B\ alone}$, where $ED_{50}^{A\ alone}$ and $ED_{50}^{B\ alone}$ are concentrations of each drug alone producing CPE_{50}; $ED_{50}^{A\ comb}$ is the concentration of drug A in combination with drug B [poly r(A-U)] yielding CPE_{50}; $ED_{50}^{B\ comb}$ is the concentration of drug B in combination with drug A yielding CPE_{50}.

performed with polynucleotides with other base sequences indicate that when EB is combined with poly r(G-C), the ED_{50} of the poly r(G-C) decreases from 5.9 to 0.40 μM (a 15-fold decrease), while the ED_{50} of the EB decreases from 15.4 to 0.10 μM (154-fold decrease). However, when EB is combined with poly (rI)·poly (rC), the ED_{50} of the poly (rI)·poly (rC) increases from 0.71 to 1.28 μM (a 2-fold increase), while the ED_{50} of the EB decreases from 15.4 to 0.32 μM (a 48-fold decrease). When poly d(A-T) is employed, neither the d(A-T) alone nor the EB/poly d(A-T) combination exhibits antiviral activity. These results suggest that potentiated antiviral activity occurs only when EB is added to poly r(A-U) or poly r(G-C).

To insure that the enhancement of antiviral activity is truly related to the maximum amount of intercalated dye, additional experiments have been designed using dyes whose ratio of maximum intercalation is 1:6 not 1:4 (Wang et al. 1987). When ADR or DMN is combined with poly r(A-U), the ED_{50} values decrease as the ADR (or DMN)/ribonucleotide ratios increase, reach a minimum at an ADR (or DMN)/ribonucleotide ratio in the region of 1/6, and remain constant or increase slightly as the ADR (or DMN)/ribonucleotide ratios approach 2/1. When DMN is combined with poly r(A-U), the ED_{50} of the poly r(A-U) decreases from 6.8 to 0.36 μM (an 18-fold decrease), while the ED_{50} of the DMN decreases from 11.1 to 0.06 μM (a 185-fold decrease). For the ADR/poly r(A-U) combination, the ED_{50} of the poly r(A-U) decreases 7-fold, while the ED_{50} of the ADR decreases 54-fold.

The fractional inhibitory concentration index (Sühnel 1990) is employed to evaluate synergism, additivism, or antagonism. An FIC < 1.0 indicates that the combination is synergistic, while an FIC > 1.0 indicates that the combination is antagonistic. An FIC $\simeq 1.0$ indicates that the combination is additive (or indifferent). In our previous work (Jamison et al. 1988, 1989, 1990a-g), only dye/poly r(A-U) combinations that exhibit at least an eightfold enhancement in antiviral activity are considered active. To remain consistent with this criterion, the dye and the poly r(A-U) are considered to exhibit synergy only if the dye/poly r(A-U) combination exhibits an FIC ≤ 0.2. The minor groove intercalating dyes EB, PI, DMN, and ADR exhibit a significant synergism in combination with poly r(A-U) at the concentrations employed in this study, with FICs of 0.065, 0.045, 0.095, and 0.166, respectively.

Three major/minor groove intercalating dyes are tested in this study: 9-aminoacridine (9-AA), N^2,N^6-dimethyl-9-hydroxy-ellipticine (DMHE), and N^2-methyl-9-hydroxy-ellipticine (NMHE). For all three dyes, the ratios ED_{50} decrease as the dye/ribonucleotide ratios increase, reach a minimum at a dye/ribonucleotide ratio in the region of 1/4, and then increase as the dye/ribonucleotide ratios approach 2/1. When 50 μM NMHE is combined with 200 μM poly r(A-U), the ED_{50} of the poly r(A-U) decreases from 6.8 to 0.28 μM (a 25-fold decrease), while the ED_{50} of the NMHE decreases from 17.6 to 0.07 μM (a 251-fold decrease). For the DMHE/poly r(A-U)combination, the ED_{50} of the poly r(A-U) decreases 21-fold, while the ED_{50} of the DMHE decreases 274-fold. In the case of the 9-AA/poly r(A-U) combination, the ED_{50} of the poly r(A-U) decreases 3-fold and the ED_{50} of the 9-AA decreases 22-fold. The minor/major groove intercalating dyes NMHE and DMHE a significant synergism in combination with poly r(A-U) at the concentrations employed in this study, with FIC of 0.045 and 0.051, while the FIC value of the 9-AA/poly r(A-U) combination is less significant (0.399). These similarities of results for NMHE and DMHE suggest that the enhancement phenomenon is not affected by the methylation of the indolic nitrogen of the NMHE (Fig. 1). The diminished antiviral activity of 9-AA may be due to the absence of a sizable substituent at the 9-position of the acridine. All the minor groove intercalating agents tested in our previous work (Jamison et al. 1988, 1989, 1990a-g) possess a sizable substituent group attached to the planar, heterocyclic ring systems, except 9-AA (Fig. 1). These minor groove intercalators all induce a large degree of enhancement of the antiviral activity of poly r(A-U), except 9-AA. Therefore, while a planar, heterocyclic ring system may be sufficient to enhance the antiviral activity of the poly r(A-U), the presence, placement and the nature of the substituents in the side chains of these ring systems may be crucial in modulating the degree of enhancement observed.

Two major groove intercalating dyes are tested: acridine orange (AO) and proflavine (PRO). Neither the AO/poly r(A-U) combination nor the PRO/poly r(A-U) combination displays potentiated antiviral activity. Therefore, there is no apparent synergism between the poly(A-U) and the major groove intercalating dyes at the concentrations employed in this study. These results are consistent with a model for an A-form helix in which the major groove lies on the inner surface of the cylinder and is not accessible for intercalation.

The previous results demonstrate that the Human Foreskin Fibroblast – Vesicular Stomatitis Virus (HSF-VSV) assay can be used to screen putative intercalating substances for their ability to potentiate the antiviral activity of polyribonucleotides by testing only the maximum binding ratio predicted by the neighbor exclusion model. Since the ED_{50} for the 1/4 and the 1/6 dye/nucleotide ratios in the ADR and DMN studies are identical within the error of the technique, it was decided to standardize the screening protocol to a dye/nucleotide ratio of 1/4. When nine additional dyes (3′-methoxy-A-9-aminoacridine, N-phenyl-9-aminoacridine, amsacrine, o-amsacrine, chloroquine diphosphate (CHL), methylene blue, quinacrine, quinine hydrochloride (QUI), and toluidine blue O, whose mode of intercalation is not known, were combined with poly r(A-U) at a dye/nucleotide ratio of 1/4, the ED_{50} of CHL and QUI decreases 65- and 32-fold, respectively. CHL and QUI exhibit a significant synergism in combination with poly r(A-U) at the concentrations employed in this study, with FICs of 0.133 and 0.149.

5 Antiviral Activity of Anthraquinones

Subsequently, the HSF – VSV bioassay has been employed to evaluate over 70 chemical substances for their ability to potentiate the antiviral activity of poly r(A-U). The majority of active substances fall into two categories: anthraquinones and xanthenes. Ten anthraquinones [alizarin, alizarin blue black B, alizarin complexone, alizarin sulfonate, ametantrone (HAQ), carminic acid (CAR), cibacron blue 3GA, emodin, mitoxantrone (DHAQ), and rhein] have been evaluated for their antiviral activity (Jamison et al. 1990c). Five of the ten anthraquinones (alizarin, ametantrone, carminic acid, emodin, and mitoxantrone) are known to form complexes with nucleic acids (Swanbeck 1966; Zunino et al. 1972; Zee-Cheng and Cheng 1978; Lown et al. 1979; Arcamone 1981; Kapuscinski et al. 1981). When poly r(A-U) is combined individually with these ten anthraquinones, the ED_{50} of DHAQ and HAQ decreased 251- and 125-fold, respectively. CAR potentiated the antiviral activity of poly r(A-U) 12-fold, but the decrease in the ED_{50} of CAR could not be quantified because CAR alone is not active. FIC values indicate that both DHAQ (0.126) and HAQ (0.126) interact synergistically with poly r(A-U).

The five active anthraquinones (ADR, CAR, DHAQ, DMN, and HAQ) share several common physical, chemical, and structural parameters. First, all of the active anthraquinones possess planar-conjugated ring systems with molecular surface areas of 40 – 45 $Å^2$ (Pullman 1986). Second, all of the active compounds possess a moderate to high aqueous solubility (Traganos 1986). Third, they possess a pKa values greater than 6 and usually less than 8.5 and thus are neutral or slightly cationic at pH 7 (Plumbridge et al. 1980; Traganos 1986). Fourth, they all have either sizable 2-(hydroxyethyl)aminoethylamino substituents or bulky sugar or amino sugar moieties (Zee-Cheng and Cheng 1978). Fifth, all of the active anthraquinones, except carminic acid, possess amine groups (Traganos 1986). Sixth, all the active anthraquinones possess hydroxyl groups or amino groups in positions 1 and 4 or possess a 5,8-dihydroxyanthraquinone nucleus (Zee-Cheng and Cheng 1978; Plumbridge et al. 1980). These parameters encourage the formation and stabilization of complexes (especially intercalative complexes) of these anthraquinones with the double-stranded nucleic acids (Zee-Cheng and Cheng 1978; Traganos 1986).

6 Antiviral Activity of Xanthenes

Ten xanthenes [eosin B, fluoresceinamine isomers I and II, pyronin B and Y, rhodamine B (RB), rhodamine 6G (R6G), rhodamine 123 (R123), sulforhodamine B (SB), and sulforhodamine 101] (Fig. 2) have been evaluated for their antiviral activity (Jamison et al. 1990e). Four of the ten xanthenes (pyronin B and Y, rhodamine B, and rhodamine R6G) are known to form complexes with nucleic acids (Dutt 1973, 1980; Müller and Crothers 1975; Darzynkiewicz et al. 1986; Kapuscinski and Darzynkiewicz 1987). When poly r(A-U) is combined individually with these ten xanthenes, the ED_{50} of R123 decreased 350-fold. The decrease in the ED_{50} of RB, R6G, and SB could not be quantified because none of these dyes alone are active. The FIC value for R123 (0.097) indicates that R123 interacts synergistically with poly r(A-U).

While all of the xanthene dyes possess tricyclic, conjugated xanthylium chromophores or have the xanthylium chromophore embedded in their ring systems, only four of the xanthenes (rhodamine B, rhodamine 6G, rhodamine 123, and sulforhodamine B) significantly enhance the antiviral activity of poly

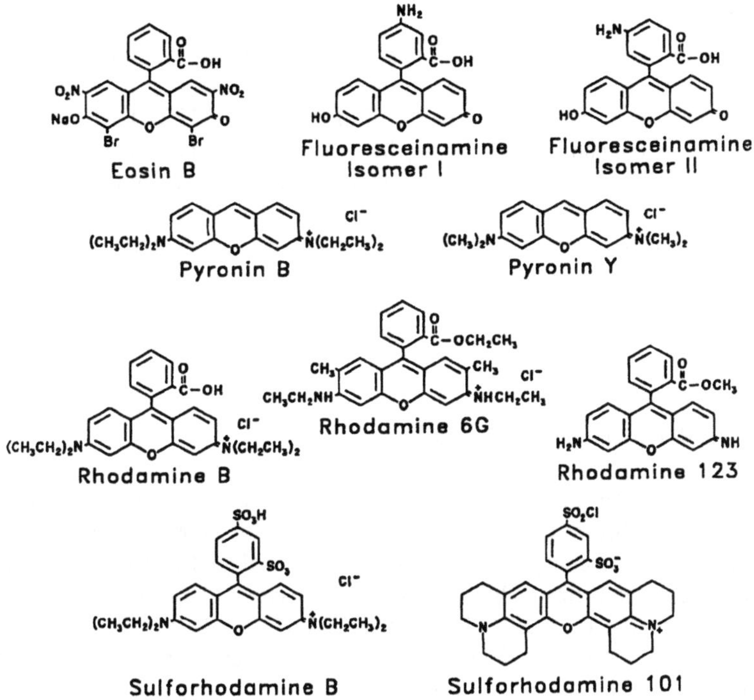

Fig. 2. Xanthenes

r(A-U). The antiviral activities of the four active xanthene/poly r(A-U) combinations are approximately twice those of the pyronin/poly r(A-U) combinations. The major structural difference between the pyronin dyes which are inactive and the rhodamine dyes which are active is that the rhodamines (sulforhodamines) possess a 9-substituted phenyl group that contains an ortho carboxyl substituent. This suggests that the 9-substituted phenyl group may be responsible for the additional enhancement observed for the rhodamines and sulforhodamines. Comparison of the antiviral activities of the four rhodamine (or sulforhodamine)/poly r(A-U) combinations illustrates that their antiviral activities are essentially equivalent. Thus, the enhancement phenomenon appears to be relatively insensitive to methyl or ethyl substitutions on the carboxyl group, to the nature and position of small substituents on the phenyl ring, and to ethyl substitutions on the amine or imine groups.

7 Toxicity of Dye/RNA Combinations

Initial assessments of cytotoxicity are made by microscope examination of HSF cell morphology and by trypan blue exclusion. Alteration of normal cell morphology, such as cell rounding or detachment of HSF cells from the microtiter plate, is considered as exhibiting cytotoxicity. Employing these criteria, cytotoxity is observed for R6G at concentrations ≥ 6.25 μM; DHAQ or NMHE alone at concentrations ≥ 25 μM; ADR, DMHE, DMN, EB, R123, or SB alone at concentrations ≥ 50 μM; HAQ alone at concentrations ≥ 20 μM; and for ADR (or DHAQ, DMN, HAQ)/poly (A-U) combinations with D/P ratios ≥ 1/2 and ADR (or DHAQ, DMN, HAQ) concentrations ≥ 50 μM. No cytotoxicity is observed when cell monolayers are treated with poly r(A-U), 9-AA, CAR, CHL, PI, RB, or QUI alone; for any of the 9-AA (or CAR, CHL, PI, QUI, RB, R123, SB)/poly r(A-U) combinations; or for ADR (or DHAQ, DMN, HAQ)/poly r(A-U) combinations with D/P ratios ≤ 1/4 (Jamison et al. 1990g).

8 Dye/RNA Combinations and HIV-1

The results of our preliminary studies conducted in collaboration with J. Oxford and K. Broadhurst of Retroscreen, Ltd. (London, UK) and Dr. C.-c. Tsai of Kent State University demonstrate that several of these dyes and dye/poly r(A-U) combinations exhibit anti-HIV activity at concentrations which are nontoxic to the host cells. The anti-HIV (HIV-1 3B, MOI = 0.02) activity of CAR and CAR/poly r(A-U) combinations in peripheral blood mononuclear cells (PBMCs) has been tested because CAR exhibits a therapeutic index of 8300 in the HSF – VSV assay (Fig. 3). Aliquots of supernatant are removed from each dilution on days 6, 11, and 16 and tested for p24 antigen using the Coulter ELISA assay system. Zidovudine treated PBMCs and sham-treated, HIV-1 infected PBMCs serve as controls. A test agent is considered active if it reduces the supernatant p24 antigen to less than 50% of the positive control at day 11 and is nontoxic at that concentration. Poly r(A-U) alone is not active at any of the concentrations employed in this study. A CAR concentration of 400 μM reduces the level of P24 antigen by 50%. When CAR is combined with poly r(A-U), the CAR concentration required to reduce the level of p24 antigen decreases to 25 μM (a 16-fold decrease). CAR, poly r(A-U), and the CAR/poly r(A-U) combination are nontoxic at concentrations that exhibit antiviral activity. These results indicate that the ability of intercalative dyes to potentiate the antiviral activity of poly r(A-U) is not VSV-specific nor HSF-specific.

9 Interferon Induction and Direct Viral Inactivation of Dye/RNA Combinations

As an initial step in the process of elucidating the mechanism responsible for the potentiated antiviral activity, 200 μM poly r(A-U), each 50 μM dye solution, and the 50 μM dye/200 μM poly r(A-U) combinations have been tested for their ability to induce interferon (Jamison et al. 1990g). Poly (rI)·poly (rC) was employed as the positive control while MEM served as the negative control. All of the active dyes (except the xanthenes), the poly r(A-U), and the dye/poly (A-U) combinations with a dye/ribonucleotide (D/P) ratio of 1/4 induce β-interferon. The results from the interferon neutralization assay [1:1600 dilution of anti-β IFN antibody (G-028-501-568)] demonstrate that, except for CHL/poly r(A-U) and the QUI/poly r(A-U) combinations, the interferon-inducing capability of the dye/poly r(A-U) combinations closely approximates the sum of the interferon-inducing capabilities of the poly r(A-U) and the dyes employed. These results suggest that the majority of the dyes tested potentiate the antiviral activity of the poly r(A-U) without affecting the amount of interferon induced.

The direct viral inactivation of VSV has been determined for 200 μM poly r(A-U), 50 μM dye alone, and the 50 μM dye/200 μM poly r(A-U) combinations as well as for dye, poly r(A-U) and dye/poly r(A-U) combinations at concentrations four times greater than the ED_{50} of the dye/poly r(A-U) combination. The results of these experiments indicate that the dyes, poly r(A-U), and the dye/poly r(A-U) combinations did not inactivate VSV at concentrations near the 50% viral inhibitory dose (Jamison et al. 1990a, b, d,-f).

10 Subcellular Localization of Dyes and Dye/RNA Combinations

The fact that the enhanced antiviral activity of the dye/poly r(A-U) combinations is not due to increased interferon induction, direct viral inactivation, or host cell cytotoxicity suggests that the enhanced antiviral activity of the dye/poly r(A-U) combinations may result from the modulation of one or more cellular processes. Therefore, we have designed experiments to determine the subcellular distribution of the dyes and dye/poly r(A-U) combinations. Identical fields of observation of HSF cells subjected to 3 h exposure of 50 μM dye, 200 μM poly r(A-U), or a 50 μM dye/200 μM poly r(A-U) combination have been studied by phase contrast and UV microscopy. HSF cells are also co-incubated with dye alone, poly r(A-U) alone, and the dye/poly r(A-U) combinations at concentrations four times greater than the ED_{50} of the dye/poly r(A-U) combination in an effort to elucidate the subcellular distribution of the dye and poly r(A-U) at concentrations at which the dye/poly r(A-U) combinations exhibit antiviral activity, but the dye alone and the poly r(A-U) alone do not (Jamison et al. 1990g). For the major groove intercalator (AO), which does not enhance the antiviral activity of poly r(A-U), the dye readily enters the nucleus, but the nuclear accumulation of the dye is not enhanced by poly r(A-U). Minor or major/minor groove intercalating dyes enhance the antiviral activity of poly r(A-U) and are concentrated in the nucleoli and in the chromatin near the periphery of the nuclear membrane. Some of these dyes (9-AA, DMN, EB) readily enter the nucleus and their nuclear accumulations are potentiated by poly r(A-U), while others (DMHE, NMHE) are sequestered in the cytoplasm and only accumulate in the nucleus in the presence of poly r(A-U). Phase contrast and fluorescence photomicrographs of HSF monolayers co-incubated with active anthraquinones and xanthenes, poly r(A-U) alone, or with the dye/poly r(A-U) combinations exhibit two basic patterns: (1) the anthraquinones

Poly r (A-U) versus HIV-1 in PBMC's

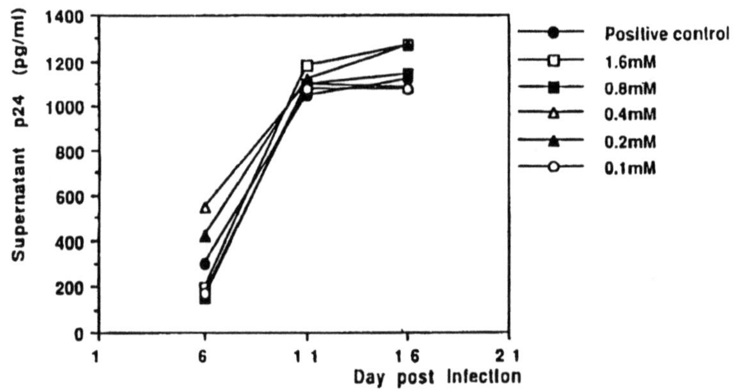

Carminic acid versus HIV-1 in PBMC's

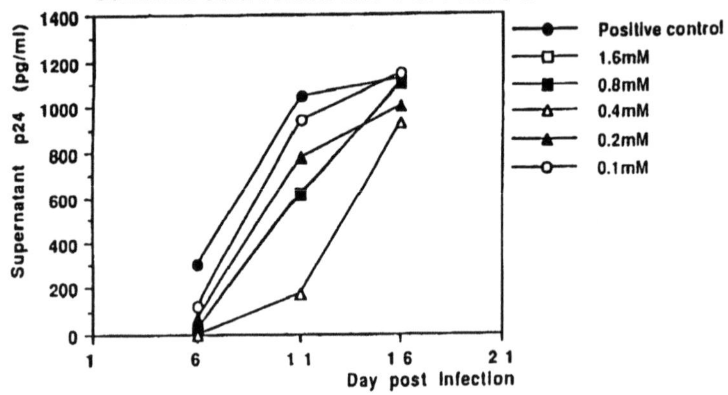

Carminic acid/Poly r (A-U) Combination versus HIV-1 in PBMC's

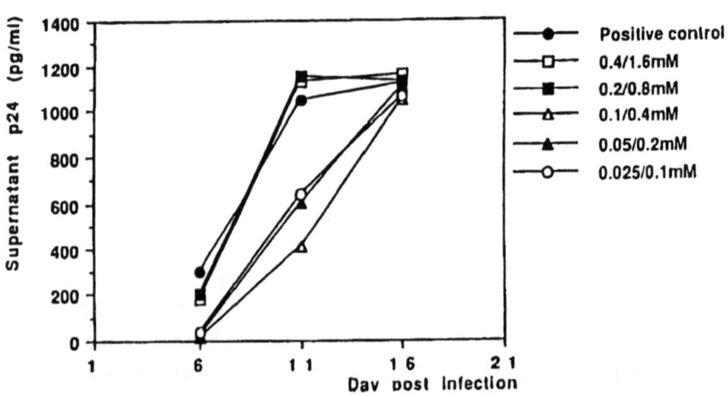

enter the cell as well as the nucleus and the accumulation of the anthraquinones in the nucleus is greatly potentiated by the poly r(A-U); (2) the xanthenes enter the cell and are sequestered in cytoplasmic organelles (mitochondria) and only the poly r(A-U) and minute quantities of the xanthenes accumulate in the nucleus when the dye is combined with poly r(A-U). These results suggest that the potentiated antiviral activity of the intercalative dye/poly r(A-U) combinations is due to the altered kinetics of uptake and subcellular distribution and the subsequent modulation of one or more cellular processes (Jamison et al. 1991).

11 Dye-Induced Condensation of RNA

Transmission electron microscopy has been employed to determine if dye-induced changes in the topology of the poly r(A-U) may have fostered the enhanced uptake of the polymer. Either 200 μM poly r(A-U) or a 50 μM EB/200 μM poly r(A-U) combination is spotted on carbon-coated formvar grids and then visualized by electron microscopy using negative staining techniques. Poly r(A-U) alone exhibits an elongated conformation that possesses a number of hairpin loops as well as single- and double-stranded domains (Fig. 4). The double-stranded domains are found predominantly at the base of 30-nm hairpin loops. Micrographs of the EB/poly r(A-U) combination illustrate the presence of toroidal (doughnut-shaped) structures with inside diameters ranging from 35 to 75 nm and outside diameters ranging from 80 to 225 nm. These toroids are due to the EB-induced condensation of the poly r(A-U) (Fig. 4B; Jamison et al. 1992, 1993).

Fig. 3. Anti-HIV-1 activity of carminic acid and carminic acid/poly r(A-U) combinations. Test solutions were created by adding 1.5×10^6 peripheral blood mononuclear cells (PBMCs) to 300 μm of 1.6 mM poly r(A-U), 1.6 mM carminic acid, or a carminic acid/poly r(A-U) (0.4/1.6 mM) combination. These test solutions were then diluted in 5 twofold dilutions with PBS containing PBMCs to create 5 different concentrations. After a 3-h incubation at 37°C, 750 μl RPMI 1640 (supplemented with L-glutamine, Hepes, and 10% FCS) was added to each dilution. Following a 17-h incubation at 37°C, HIV-1 (3B, MOI = 0.02) was added to each dilution. After an additional 5-h incubation at 37°C, cell-free virus was removed and the PBMCs were resuspended in 2.5 ml of drug-free RPMI 1640 and incubated at 37°C for 16 days. At days 6, 11, and 16, 200 μl of supernatant was removed from each dilution and tested for p24 antigen using the Coulter ELISA assay system. Zidovudine-treated PBMCs and sham-treated, HIV-1 infected PBMCs served as controls. A test agent was considered active if it reduced the supernatant p24 antigen to less than 50% of the positive control at day 11 and was nontoxic at that concentration. Poly r(A-U) alone was not active at any of the concentrations employed in this study. A carminic acid concentration of 0.4 mM reduced the level of p24 antigen by 50%. When carminic acid was combined with poly r(A-U), the carminic acid concentration required to reduce the level of p24 antigen decreased to 0.025 mM (a 16-fold decrease). Toxicity was evaluated on days 1, 2, and 5 using trypan blue exclusion. Carminic acid, poly r(A-U), and the carminic acid/poly r(A-U) combination were nontoxic at concentrations exhibiting antiviral activity (data not shown)

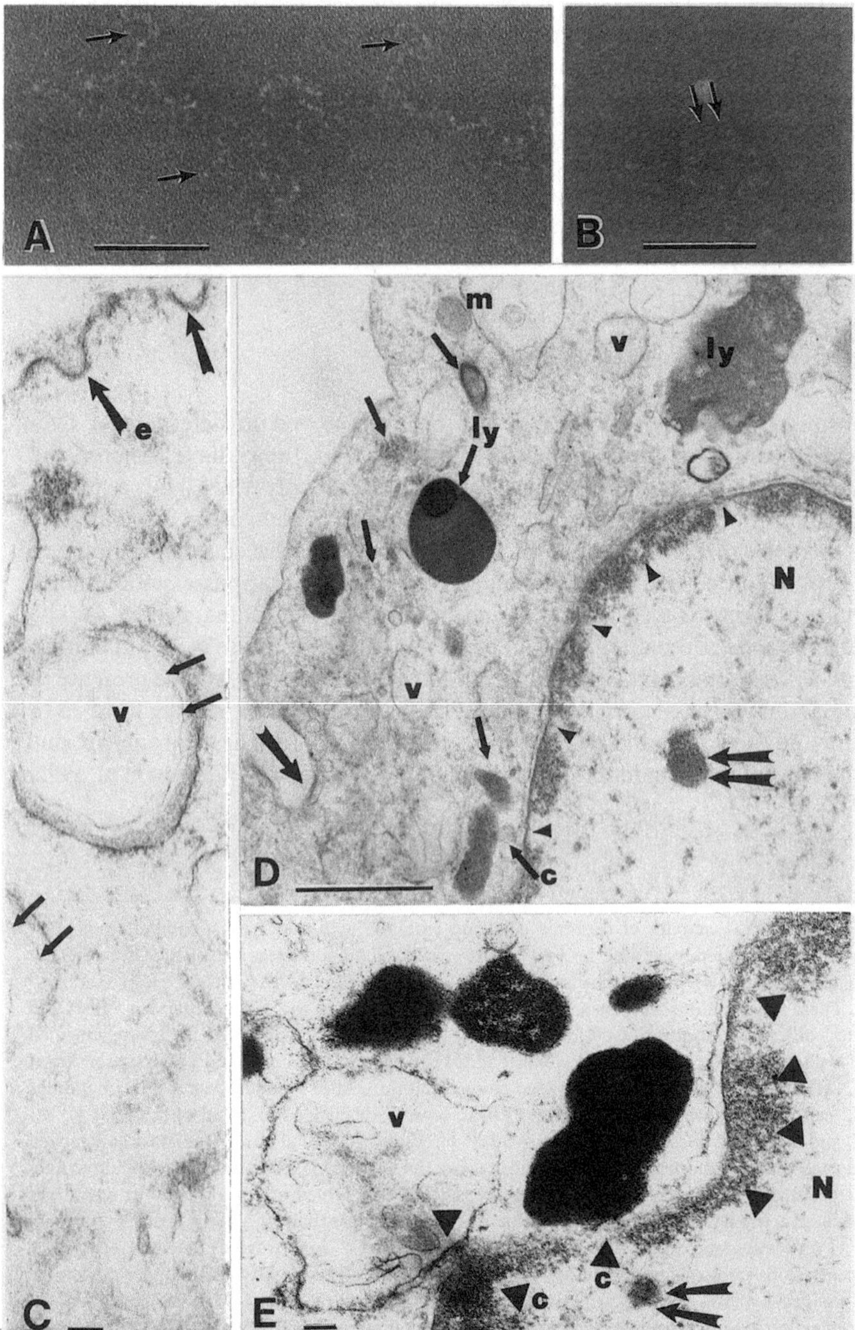
A
B
C
D
E
e
v
m
ly
N
c

Birnstiel and his associates (Wagner et al. 1991a, b; Cotten et al. 1992) have incorporated the ability of polycations to condense DNA into a strategy which exploits receptor-mediated endocytosis to deliver genes into eukaryotic cells. One of their approaches employs transferrin-polylysine conjugates which serve to condense the DNA into compact, electroneutral, ligand-coated toroids with diameters of 80–100 nm. These ligand-coated DNA particles bind to transferrin receptors on the surface of target cells and are endocytosed. Endosomal exit is promoted by agents, such as chloroquine or quaternary amines, that either induce endosomal or lysosomal swelling or by the endosomolytic activity of replication-defective adenovirus. Once the DNA has escaped from the acidic compartment, it finds its way into the nucleus. In our study (Jamison et al. 1993), EB has been shown to condense poly r(A-U) into toroids with dimensions that could be accommodated by coated endocytotic pits (Fig. 4C-D). Taken together these results suggest that the enhanced antiviral activity and altered subcellular distribution of EB/poly r(A-U) combinations may be related to the formation of toroids which enter and escape from the endosomal (Fig. 4C) and/or lysosomal (Fig. 4D-E) compartments and then enter the nucleus via a process similar to that described by Birnstiel and coworkers. Putative delivery events are suggested by the contact of rimmed vacuoles and lysosomes with the nuclear envelope (Fig. 4E).

Fig. 4. Montage of electron micrographs depicting poly r(A-U), EB/poly r(A-U), and EB/poly r(A-U)-treated HSF cells. **A**, **B**, negatively stained 200 µM poly r(A-U) (in **A**) and a 50 µM EB/200 µM poly r(A-U) combinations (in **B**) were prepared in 0.9% NaCl, pH 7.4 and mounted on carbon-coated formvar grids. Poly r(A-U) exhibits an elongated conformation consisting of single- and double-stranded domains. Double-stranded domains are detected at the base of 30-nm-diameter hairpin loops (*single arrows*). In **B**, the EB/poly r(A-U) combination shows a condensed topology in the shape of toroids. The example depicted here has a 100-nm outside diameter and a 40-nm inside diameter. **C–E** Transmission electron micrographs of HSF cells treated with a 50 µM EB/200 µM poly r(A-U) combination for 3 h at 37 °C and 5% CO_2 and subsequently fixed in cacodylate-buffered glutaraldehyde and embedded in epoxy. The HSF cells were then cut in 500-nm-thick sections and either contrasted with uranyl acetate (**D**) or left untreated (**C**, **E**). Untreated cells can be examined because EB is electron-dense. **C** *Thick arrows* depict endocytotic pits (*e*) near 0.5 to 1 µM vacuoles (*v*). These vacuoles are rimmed by the EB/poly r(A-U) and contain membranous material denoted by *small arrows*. **D** The *large arrow* denotes endocytosis, while the *small arrows* depict several uptake sites for the EB/poly r(A-U) including an endocytotic vesicle, rimmed vacuoles (*v*), a mitochondrion (*m*), and heterolysosomes (*ly*). Electron-dense material appears to be located adjacent to the nuclear envelope where numerous contacts (*c*) can be found, especially near the nuclear pores (*arrow heads*). The heterochromatin is often disrupted and constitutes a perinuclear band, while small, condensed, chromatic bodies (*double arrows*) appear in the nucleus (*N*). The enlarged nucleolus (not shown) exhibits exaggerated electron density. **E** A rimmed vacuole (**v**) is in contact with electron-dense lysosomes which are adjacent to or in contact with the nuclear envelope. The *arrow heads* denote the nuclear pores. *Double arrows* in the nucleus (*N*) depict the putative delivery of an electron-dense material. The electron-dense material appears to be encircled by a delicate meshwork. Scales in **A**, **B**, **C** and **E** are 100 nm and 1 µM in **D**

12 Biological Consequences of Dye/RNA Combinations

The antiviral effects of the dye/poly r(A-U) combinations are the composite of the interferon-, dye-, and dsRNA-induced effects. Two subcellular compartments appear to be affected: the mitochondria and the nucleus/nucleolus. Lewis (1992) has shown that interferon-α or-β treatment downregulates the expression of many of the 13 genes encoded by mammalian mitochondria, including the ATP synthetase gene and genes which encode essential members of the electron transport chain. IFN treatment also reduces the rate of synthesis of certain mitochondrial polypeptides. The mechanism(s) responsible for these effects are not fully understood and may be due to a number of effects including: interferon-mediated interference with RNA splicing (Shan et al. 1990), the inability to maintain normal levels of ATP, and interferon-mediated changes in the stability of mitochondrial rRNA and mRNA (Lewis 1992). In contrast, two mitochondrial mRNAs and both the mitochondrially encoded ribosomal RNAs are not affected by IFN treatment. Some of the cationic xanthene dyes employed in our work (Jamison et al. 1990e), like rhodamine 123, inhibit the import and processing of the precursors of mitochondrial matrix proteins (Kolarov and Nelson 1984); mitochondrial protein synthesis (Abou-Khalil et al. 1986); and several mitochondrial energy-transducing activities (Emaus et al. 1986; Ronot et al. 1986; Wieker et al. 1987). Other intercalating dyes (actinomycin D, PRO, and EB) inhibit the synthesis and processing of ribosomal RNA in human mitochondria. The processing step which entails removing the tRNA-Phe sequence from the 5′ end of the ribosomal RNA precursor is uniquely sensitive to PRO and EB (Gaines and Attardi 1984). These observations suggest that cationic dyes may potentiate the antiviral activity of interferon by inhibiting mitochondrial functions not affected by interferon. Because most mitochondrial proteins are encoded by nuclear genes, synthesized on cytoplasmic ribosomes, and imported into the mitochondria, mitochondria are also susceptible to the nuclear effects of IFN, the dyes, and the dsRNA (Schatz 1987; Lewis 1992).

Double-stranded RNAs are known to modulate many nuclear and cytoplasmic processes. For example, they are known to induce multiple species of interferons (Nosik et al. 1988); to possess intrinsic antiviral activity (Wathelet et al. 1988); to co-ordinately induce and modulate the activity of at least 23 genes (Lammers et al. 1988); and to act as systemic immunomodulators (Wong and Goeddel 1986). Montefiori and coworkers (1989) reported the importance of the pleiotropic effects of dsRNA when they demonstrated that ampligen potentiated the anti-HIV activities of AZT, foscarnet, and other antiviral agents as well as all three species recombinant interferon by a mechanism other than interferon induction. In addition, dsRNA has been shown to selectively inhibit the growth of tumors in immunocompetent as well as in immunocompromised hosts by inhibiting tumor transfer RNA and ribosomal RNA methylases and RNA polymerases; by inhibiting ribosome production; by disaggregating polysomes; and by inhibiting DNA, RNA, and protein synthesis of the tumor cells without killing the host cells (Snyder et al. 1971; Liau et al. 1973, 1975). Tumor-specific processes are more tightly coupled to cellular processes than are most virus-specific processes. The fact that dsRNA can discriminate between tumor-specific and host cell-specific processes suggests that dsRNA can also discriminate between virus-specific and host cell-specific processes.

Intercalation of cationic dyes into nucleic acids has been implicated in a wide variety of pharmacological effects including: inhibition of DNA transcription and RNA processing, inhibition of the cell cycle, inhibition of thymidine incorporation into DNA and uridine incorporation into RNA, formation of DNA-protein cross-links as well as the formation of protein-associated and non-protein-associated DNA breaks, the inhibition of helicases, and condensation of nucleic acids in solution and in situ (Kapuscinski and Darzynkiewicz 1986; Traganos 1986; Bachur et al. 1992). Intercalating dyes (AO, PRO , EB, and DMN) have also been shown to decrease the rate of degradation of heterologous nuclear RNA (Brinker et al. 1973); to inhibit both RNA splicing and transcription termination in isolated nucleoli; to induce transcription termination in vivo under conditions where the growth rate is only slightly reduced (Nielsen et al. 1984); and to inhibit the self-catalyzed cyclization of the intervening sequence of pre-rRNA by intercalating into functionally important secondary or tertiary structures of the intervening sequence (Tanner and Cech 1985). EB has been shown to intercalate into RNA hairpins similar to those involved in HIV-1 mRNA splicing. The intercalation produces a concerted conformational change in the entire helix backbone as well as in the hairpin itself (White and Draper 1987).

13 Summary

The results of our previous studies (Jamison et al. 1988, 1989, 1990 a, b, c, d, e) have shown that the ability of intercalative dyes to modulate the antiviral activity of poly r(A-U) is related to the groove through which the dyes intercalate into the poly r(A-U). When poly r(A-U) is combined with the minor groove intercalating dyes or the minor/major groove intercalating dyes, optimum enhancement of antiviral activity is observed at the dye/ribonucleotide ratio predicted by the neighbor exclusion model (usually 1/4 or 1/6). No enhancement is observed when poly r(A-U) is combined with major groove intercalating dyes. When poly r(A-U) is combined with additional intercalative dyes to produce a dye/ribonucleotide ratio of 1/4 and a ribonucleotide concentration of 200 μM, the antiviral activity of poly r(A-U) is enhanced 8- to 20-fold, while 50% effective doses of the poly r(A-U) and the dyes decreases 18- to 347-fold. Interferon neutralization assays demonstrate that the interferon-inducing capability of the dye/poly r(A-U) combinations approximates the sum of the interferon-inducing capabilities of the poly r(A-U) and the dyes employed and suggests that the dyes potentiate the antiviral activity of poly r(A-U) without affecting the amount of interferon induced. Direct viral inactivation studies demonstrate that the dyes, poly r(A-U), and the dye/poly r(A-U) combinations do not inactivate VSV at concentrations near the 50% viral inhibitory dose. Assessment of cytotoxicity by microscope examination of HSF cell morphology and trypan blue exclusion indicates that the dye/poly r(A-U) combinations exhibit antiviral activity at concentrations well below those that induce cyto-toxicity. Several of the dyes and the dye/poly r(A-U) combinations exhibit anti-HIV-1 activity, suggesting that the enhancement phenomenon is not virus-specific nor host cell-specific. The enhancement phenomenon is sensitive to the base sequence of the polynucleotide with dye/poly r(A-U) and dye/poly r(G-C) combinations displaying enhanced antiviral activity, while dye/poly (rI)·poly (rC) and dye/poly d(A-T) combinations do not. These results suggest that while intercalation of the dye and interferon induction are necessary for enhanced antiviral activity, neither intercalation nor interferon induction alone is sufficient to potentiate the antiviral activity of polyribonucleotides.

Transmission electron micrographs illustrate that one of the most active dyes, EB, induces the condensation of poly r(A-U) into toroids which may be endocytosed into and subsequently escape from the acidic compartments (endosome, lysosome) of the cell. After escaping from the acidic compartment, the dyes and dye/poly r(A-U) combinations become localized in either the mitochondria, nucleus, or both. Since the dye and the poly r(A-U) form reversible complexes, the potentiated antiviral activity may be a composite of the effects of the dye alone, the poly r(A-U) alone, and the dye/poly r(A-U) combination as well as the interferon-mediated effects. Each of these factors [interferon, dye, and poly r(A-U)] induce a set of pharmacological effects which perturb nuclear (nucleolar) and mitochondrial structure as well as modulate gene activity. RNA synthesis, processing, and transport appear to be susceptible to these dye-, dsRNA-, and dye/dsRNA-induced effects. Messenger RNA and rRNA production are affected at the transcriptional and posttranscriptional levels as well as at the level of translation. The reversibility and toxicity of these effects are dose- and time-dependent with a short exposure time (2–3 h) favoring reversibility and inhibition of viral processes and longer times of exposure 24–30 h) resulting in irreversible host cell toxicity.

Host cells exhibit a range of conditions under which they can survive and replicate their niche. Likewise, the virus exhibits a niche of its own which is a subset of the host cell niche (Fig. 5). IFN/dsRNA activates "blocks" of host cell genes whose products reversibly perturb host cell structure and function so that the host cell niche temporarily becomes partially or completely exclusive of the viral niche. However, some viruses have developed mechanisms to thwart this process. Numerous studies demonstrate that short-term administration of dyes and/or dye/RNA combinations reversibly perturb host cell structure and function. Our results demonstrate that these perturbations result in enhanced antiviral activity at concentrations which are not toxic to the host cells. We believe that administration of dye/RNA combinations modulates additional host processes, returns the competitive advantage to the host cell, and induces the exclusion of the viral niche. Ideally, naturally occurring intercalative dyes may exhibit antiviral activity themselves or may interact synergistically with

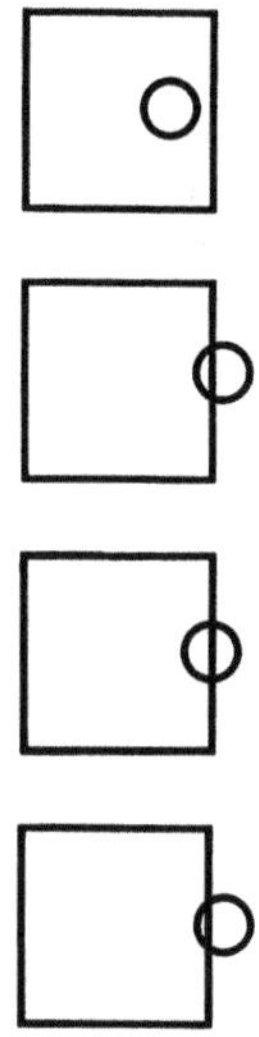

Fig. 5. Virus (○)–host cell (□) interactions

dsRNA by forming reversible, intercalative complexes which act as "binary weapons" whose virus to host cytotoxicity differences create a "therapeutic window" that is more amenable to prevention and/or treatment of viral infection than either constituent alone. Since poly r(A-U) alone and some of the intercalative dyes alone are relatively labile in living systems, they will be rapidly degraded and thus inactivated before they irreversibly affect host cell processes.

Acknowledgements. This work was supported in part by grants from the Akron City Hospital Foundation and the Office of Geriatric Medicine and Gerontology, Northeastern Ohio Universities College of Medicine. We would like to thank them for their support. We would also like to thank Keith Krabill, Katrina Broadhurst, John Oxford, Chun-che Tsai, and all the other graduate students, medical students, and technicians who participated in the collection of data.

References

Abou-Khalil WH, Arimura GK, Yunis AA, Abou-Khalil S (1986) Inhibition by rhodamine 123 of protein synthesis in mitochondria of normal and cancer tissues. Biochem Biophys Res Commun 137:759–765

Arcamone F (1981) Doxorubicin anticancer antibiotics. Academic Press, New York

Bachur NR, Yu F, Johnson R, Hickey R, Wu Y, Malkas L (1992) Helicase inhibition by anthracycline anticancer agents. Mol Pharmacol 41:993–998

Becker Y (ed) (1984) Antiviral drugs and interferon: the molecular basis of their activity. Nijhoff, Boston

Bennett DD (1985) Drugs that fight cancer...naturally. Sci News 128:58–60

Berman HM, Young PR (1981) The interaction of intercalating drugs with nucleic acids. Annu Rev Biophys Bioeng 10:87–114

Blough HA, Ray EK (1980) Glucose analogues in chemotherapy of herpesvirus infections. Pharmacol Ther 10:669–681

Bobst AM, Torrence PF, Kouidou S, Witkop B (1976) Dependence of interferon induction on nucleic acid conformation. Proc Natl Acad Sci USA 73:3788–3792

Brinker JM, Madore HP, Bello LJ (1973) Stabilization of heterogeneous nuclear RNA by intercalating drugs. Biochem Biophys Res Commun 52:928–934

Carter WA, O'Malley JA, Beeson M, Cunningham P, Kelvin A, Vere-Hodge A, Alderfer JL, Ts'o POP (1976) An integrated and comparative study of the antiviral effects and other biological properties of the polyinosinic · polycytidylic acid duplex and its mismatched analogues. III. Chronic effects and immunological features. Mol Pharmacol 12:440–453

Content J, Cogniaux-LeClercq J (1968) Comparison of the in vitro action of ethidium chloride on animal viruses with that of other photodyes. J Gen Virol 3:63–75

Cotten M, Wagner E, Zatloukal K, Phillips S, Curiel DT, Birnstiel ML (1992) High-efficiency receptor-mediated delivery of small and large (48 kilobase) gene constructs using the endosome-disruption activity of defective or chemically inactivated adenovirus particles. Proc Natl Acad Sci USA 89:6094–6098

Crothers DM (1968) Calculation of binding isotherms for heterogeneous polymers. Biopolymers 6:575–584

Darzynkiewicz Z, Kapuscinski J, Carter SP, Schmid FA, Melamed MR (1986) Cytostatic and cytotoxic properties of pyronin Y: relation to mitochondrial localization of the dye and its interaction with RNA. Cancer Res 46:5760–5766

De Clercq E, Torrence PF (1977) Comparative study of various double-stranded RNAs as inducers of human interferon. J Gen Virol 37:619–623

De Clercq E, Wells R, Grant R, Merigan T (1971) Thermal activation of the antiviral activity of synthetic double-stranded polyribonucleotides. J Mol Biol 56:83–100

Dutt MK (1973) In situ localization of DNA with basic dyes of the xanthene group. Curr Sci 42:354–356

Dutt MK (1980) Absorption spectra of nuclei stained with two different basic dyes after selective extraction of RNA. Indian J Exp Biol 18:57–58

Eckstein F, Gish G (1989) Phosphorothioates in molecular biology. Trends Biochem Sci 14:97–100

Elion GB (1984) Acyclovir. In: Becker Y (ed) Antiviral drugs and interferon: the molecular basis of their activity. Nijhoff, Boston, pp 71–88

Emaus RK, Grunwald R, Lemasters JJ (1986) Rhodamine 123 as a probe of transmembrane potential in isolated rat-liver mitochondria: spectral and metabolic properties. Biochim Biophys Acta 850:436–448

Field HJ, Goldthorpe SE (1989) Antiviral drug resistance. Trends Pharmacol Sci 10:333–337

Gaines G, Attardi G (1984) Intercalating drugs and low temperatures inhibit the processing of ribosomal RNA in isolated human mitochondria. J Mol Biol 172:451–466

Grigorian SS, Ershov FI, Poverennyi AM, Podgorodnichenko VK, Popov GA (1987) Interferon-inducing activity of liposome-incorporated double-stranded polynucleotides and the means for its enhancement. Vopr Virusol 32:352–357

Helène C, Thuong NT (1988) Oligo-[α]-deoxyribonucleotides covalently linked to intercalating agents. A new family of sequence-specific nucleic acid reagents. In: Eckstein F, Lilley DMJ (eds) Nucleic acids and molecular biology 2. Springer Berlin Heidelberg New York, pp 105–123

Iliescu R, Repanovici R, Mutiu A, Sahnazarov N, Danielescu G, Popa LM, Cajal N (1983) Investigations of the effect of cellular and viral nucleic acids on certain virus infections. Note 2. Effect of nucleic acids on virus multiplication in cell cultures. Rev Roum Med Virol 34:191–196

Jamison JM, Flowers DG, Jamison E, Kitareewan S, Krabill K, Rosenthal KS, Tsai C-c (1988) Enhancement of the antiviral and interferon-inducing activities of poly r(A-U) by carminic acid. Life Sci 42:1477–1483

Jamison JM, Flowers DG, Kitareewan S, Krabill K, Tsai C-c (1989) Potentiation of the antiviral activity of poly r(A-U) by riboflavin, FAD and FMN. Cell Biol Int Rep 13:215–222

Jamison JM, Krabill K, Flowers DG, Tsai C-c (1990a) Enhancement of the antiviral activity of poly r(A-U) by ametantrone and mitoxantrone. Life Sci 46:653–661

Jamison JM, Bonilla PJ, Tsai C-c (1990b) Modulation of the antiviral activity of poly (A-U) by ethidium bromide and propidium iodide. Antiviral Chem Chemother 1:53–60

Jamison JM, Krabill K, Flowers DG, Tsai C-c (1990c) Polyribonucleotide–anthraquinone interactions: in vitro antiviral activity studies. Cell Biol Int Rep 14:219–228

Jamison JM, Krabill K, Flowers DG, Tsai C-c (1990d) In vitro antiviral activity of poly r(A-U) and ellipticines. Biochimie 72:235–243

Jamison JM, Krabill K, Hatwalkar A, Jamison E, Tsai C-c (1990e) Potentiation of the antiviral of poly r(A-U) by xanthene dyes. Cell Biol Int Rep 14:1075–1084

Jamison JM, Bonilla PJ, Tsai C-c (1990f) Enhancement of the antiviral activity of poly r(A-U) by adriamycin and daunomycin. Antiviral Chem Chemother 1:285–292

Jamison JM, Krabill K, Allen KA, Stuart SH, Tsai C-c (1990g) RNA-intercalation agent interactions: in vitro antiviral activity studies. Antiviral Chem Chemother 1:333–347

Jamison JM, Krabill K, Gilloteaux J, Flowers DG, Tsai C-c (1991) Subcellular localization and antiviral activity of cationic dye/poly r(A-U) combinations. Antiviral Res 15: 15

Jamison JM, Krabill K, Gilloteaux J, Tsai C-c, Summers JL (1992) Antiviral and antitumor activity of RNA-dye combinations. J Interferon Res 12:S152

Jamison JM, Gilloteaux J, Summers JL (1993) Effect of ethidium bromide on the topology, antiviral activity and subcellular distribution of poly r(A-U). Cell Biol Int (in press)

Kapuscinski J, Darzynkiewicz Z (1984) Condensation of nucleic acids by intercalating aromatic cations. Proc Natl Acad Sci USA 81:7368–7372

Kapuscinski J, Darzynkiewicz Z (1986) Relationship between the pharmacological activity of antitumor drugs Ametantrone and mitoxantrone (Novatrone) and their ability to condense nucleic acids. Proc Natl Acad Sci USA 83:6302–6306

Kapuscinski J, Darzynkiewicz Z (1987) Interactions of pyronin Y (G) with nucleic acids. Cytometry 8:129–137

Kapuscinski J, Darzynkiewicz Z, Traganos F, Melamed MR (1981) Interactions of new antitumor agent, 1,4-dihydroxy-5,8-bis[[2-[(2-hydroxyethylamino]-ethyl]amino]-9,10-anthracenedione, with nucleic acids. Biochem Pharmacol 30:231–240

Kolarov J, Nelson BD (1984) Import and processing of cytochrome b-c_1 complex subunits in isolated hepatoma ascites cells: inhibition by rhodamine 6G. Eur J Biochem 144:387–392

Lammers R, Gross G, Mayr U, Collins J (1988) Alternative mechanisms for gene activation induced by poly (rI)·poly (rC) and Newcastle disease virus. Eur J Biochem 178:93–99

Lewis JA (1992) Effects of interferons on mitochondria: regulation of mitochondrial gene expression. In: Baron S, Coppenhaver DH, Dianzani F, Fleischmann WR Jr, Klimpel GR, Niesel DW, Stanton GJ, Tyring SK (eds) Interferon: principles and medical applications. The University of Texas Medical Branch at Galveston, Department of Microbiology, Galveston, pp 251–260

Liau MC, Hunt JB, Smith DW, Hurlbert RB (1973) Inhibition of transfer and ribosomal RNA methylases by polyinosinate. Cancer Res 33:323–331

Liau MC, Smith DW, Hurlbert RB (1975) Preferential inhibition by homopolyribonucleotides of the methylation of ribosomal acid and disruption of the production of ribosomes in a rat tumor. Cancer Res 35:2340–2349

Lodemann E, Diederich J, Sattler V, Wacker A (1973) Induction of interferon in L cells by poly I poly C in the presence of cationic compounds. Arch Ges Virusforsch 40:87–92

Lown JW, Chen H-h, Sim S-k, Plambeck JA (1979) Reactions of the antitumor agent carminic acid and derivatives with DNA. Bioorg Chem 8:17–24

Montefiori DC, Sobol RW Jr, Li SW, Reichenbach NL, Suhadolnik RJ, Charubala R, Pfleiderer W, Modliszewski A, Robinson WE Jr, Mitchell WM (1989) Phosphorothioate and cordycepin analogues of 2', 5'-oligoadenylate: inhibition of human immunodeficiency virus type 1 reverse transcriptase and infection in vitro. Proc Natl Acad Sci USA 86:7191–7194

Müller W, Crothers DM (1975) Interactions with nucleic acids. 1. The influence of heteroatoms and polarizability on the base specificity of intercalating ligands. Eur J Biochem 54:267–277

Nelson JW, Tinoco I Jr (1984) Intercalation of ethidium ion into DNA and RNA oligonucleotides. Biopolymers 23:213–233

Nielsen OF, Carin M, Westergaard O (1984) Studies on transcription termination and splicing of the rRNA precursor in vivo in the presence of proflavine. Nucleic Acids Res 12:873–886
Nosik NN, Bopegamage SA, Saiitkulov AM, Ershov FI (1988) The mosaic nature of the interferons induced by polyribonucleotides. Vopr Virusol 33:760–763
Plumbridge TW, Knight V, Patel KL, Brown JR (1980) Mechanism of DNA-binding of some aminoakylamino-derivatives of anthraquinone and naphthacenequinone. J Pharm Pharmacol 32:78–80
Pullman B (1986) A few considerations on quinones as antitumor agents. In: Lowdin P-O, Sabin JR, Zerner MC (eds) International Journal of Quantum Chemistry, Quantum Biology Symposium 13. Wiley, New York, pp 95–105
Ronot X, Benel L, Adolphe M, Mounolou JC (1986) Mitochondrial analysis in living cells: the use of rhodamine 123 and flow cytometry. Biol Cell 57:1–8
Saenger W (ed) (1984) Principles of nucleic acid structure. Springer, Berlin Heidelberg New York
Santoro MG, Amici C, Elia G, Benedetto A, Garaci E (1989) Inhibition of virus protein glycosylation as the mechanism of the antiviral action of prostaglandin A in Sendai virus-infected cells. J Gen Virol 70:789–800
Schatz G (1987) Signals guiding proteins to their correct locations in mitochondria. Eur J Biochem 165:1–6
Sehgal PB, Tamm I, Vilcek J (1975) Enhancement of human interferon production by neutral red and chloroquine: analysis of inhibition of protein degradation and macromolecular synthesis. J Exp Med 142:1283–1300
Sen GC, Ransohoff RM (1993) Interferon-induced antiviral actions and their regulation. In: Maramorosch K, Murphy FA, Shatkin JA (eds) Advances in virus research, vol 42. Academic Press, San Diego, pp 57–102
Shan B, Vazquez E, Lewis JA (1990) Interferon selectively inhibits the expression of mitochondrial genes: a novel pathway for interferon-mediated responses. EMBO J 9:4307–4314
Snyder AL, Kahn HE, Kohn KW (1971) Inhibition of the processing of ribosomal precursor RNA by intercalating agents. J Mol Biol 58:555–565
Sühnel J (1990) Evaluation of synergism or antagonism for the combined action of antiviral agents. Antiviral Res 13:23–40
Swanbeck G (1966) Interaction between deoxyribonucleic acid and some anthracene and anthraquinone derivatives. Biochim Biophys Acta 123:630–633
Tanner NK, Cech TR (1985) Self-catalyzed cyclization of the intervening sequence RNA of Tetrahymena: inhibition by methidiumpropyl EDTA and localization of the major dye binding sites. Nucleic Acids Res 13:7759–7779
Traganos F (1986) Dihydroxyanthraquinone and related *bis* (substituted) aminoanthraquinones: a novel class of antitumor agents. In: Dethlefsen LA (ed) International encyclopedia of pharmacology and therapeutics, Section 121. Cell cycle effects of drugs Pergamon Press, Elmsford, NY, pp 269–286
Tsai C-c (1978) Stereochemistry of drug-nucleic acid interactions and its biological implications. In: Clarke FH (ed) Annual reports in medical chemistry. Academic Press, New York, pp 316–326
Wagner E, Cotten M, Foisner R, Birnstiel ML (1991a) Transferrin-polycation-DNA complexes: the effect of polycations on the structure of the complex and DNA delivery to cells. Proc Natl Acad Sci USA 88:4255–4259
Wagner E, Cotten M, Mechtler K, Kirlappos H, Birnstiel ML (1991b) DNA-binding transferrin conjugates as functional gene-delivery agents: synthesis by linkage of polylysine or ethidium homodimer to the transferrin carbohydrate moiety. Bioconjugate Chem 2:226–231
Warlo DG, Rosenthal KS (1988) Inhibition of phosphorylation and viral protein kinase activity by D609 (trycyclodecan-9-yl-xanthogenate). Antiviral Res 9:95
Wang AH-J, Ughetto G, Quigley GJ, Rich A (1987) Interactions between an anthracycline antibiotic and DNA: molecular structure of daunomycin complexed to d(CpGpTpApCpGp) at 1.2-Å resolution. Biochemistry 26:1152–1163
Waring MJ (1981) DNA modification and cancer. Annu Rev Biochem 50:159–192

Wathelet MG, Clauss IM, Content J, Huez GA (1988) Regulation of two interferon-inducible human genes by interferon, poly (rI)·poly (rC) and viruses. Eur J Biochem 174:323–329

White SA, Draper DE (1987) Single base bulges in small RNA hairpins enhance ethidium binding and promote an allosteric transition. Nucleic Acids Res 15:4049–4064

Wieker H-J, Kuschmitz D, Hess B (1987) Inhibition of yeast mitochondrial F_1-ATPase, F_0F_1-ATPase and submitochondrial particles by rhodamines and ethidium bromide. Biochem Biophys Acta 892:108–117

Wong GW, Goeddel DV (1986) Tumour necrosis factors α and β inhibit virus replication and synergize with interferons. Nature (Lond) 323:819–821

Zee-Cheng RKY, Cheng CC (1978) Antineoplastic agents. Structure-activity relationship study of *bis* (substituted aminoethylamino) anthraquinones. J Med Chem 21:291–294

Zunino F, Gambetta R, di Marco A, Zaccara A (1972) Interaction of daunomycin and its derivatives with DNA. Biochim Biophys Acta 277:489–498

Chemical Synthesis of 2′,5′-Oligoadenylate Analogues

R. Charubala and W. Pfleiderer[1]

1 Introduction

The interferons (IFN) are a family of proteins secreted by mammalian cells in response to various inducers (Baglioni 1979; Revel 1979; Stewart 1979; Friedman 1981; Pestka 1981; Torrence 1982; Sehgal et al. 1983). The interferons are divided into three different types α, β, γ, whereby each class can be summarized from different species for which a separate IFN gene is coded. The interferons are pleotropic modifiers of cell functions as seen from their antiviral properties (Sehgal et al. 1983), the inhibition of cell growth, the inhibition of tumor cell proliferation, their augmentation of natural killer (NK) activity, the modulation of immune response, the interaction of nuclear formation and the delay of cells entering into S-phase of the cell cycle (Suhadolnik et al. 1983).

2 Biochemical Mechanism of Interferon Activity

Different activities of interferon have been summarized in a general scheme (Torrence 1982; Sen 1984), indicating that dsRNAs are important mediators of various phenomena in cell-free extracts of interferon-treated cells (Fig. 1). The central molecule, dsRNA, should be seen as a source of primary biochemical effects inducing two different reaction cascades, both of them finally leading to the inhibition of protein synthesis. The upper part of the circle is regarded as the *2-5-OligoA cascade*, whereas the lower half represents the *protein kinase pathway*.

In a nutshell interferon induces, on the one hand, the enzyme 2-5-OligoA synthetase, which is activated by dsRNA to synthesize from ATP a series of most interesting 2′,5′-connected oligoadenylate-5′-triphosphates (Kerr and Brown 1978). The trimer possesses the best features to activate the latent RNase L which degrades finally viral and cellular mRNA thereby inhibiting protein synthesis. These potentially important antimetabolites are, however, quickly deactivated particularly in cell-free systems by enzymatic degradation by phosphodiesterases. On the other hand, interferon increases the amount of

[1]Fakultät für Chemie, Universität Konstanz, Postfach 5560, D-78434 Konstanz, Germany

Progress in Molecular and Subcellular Biology, Vol. 14
W.E.G. Müller/H.C. Schröder (Eds.)
© Springer-Verlag Berlin Heidelberg 1994

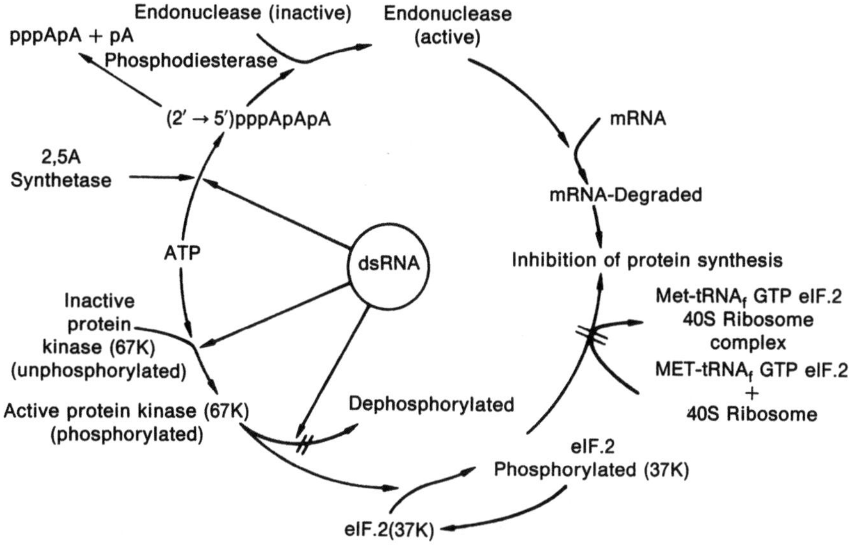

Fig. 1. dsRNA-mediated phenomena of cell-free extracts of interferon-treated cells

active protein kinase (67 K) by inducing phosphorylation in the presence of dsRNA followed by phosphoryl transfer to the elF-2 factor which interferes with the MET-tRNA$_f$-GTP-40S ribosome complex to stop protein synthesis.

3 The 2-5-OligoA System

The discovery of the 2-5-oligoA system can be traced back to Friedman et al. (1972) and also includes various other laboratories (Kerr et al. 1974; Roberts et al. 1976; Hovanessian et al. 1977), climaxing finally in the structure elucidation of the low molecular weight inhibitor (LMWI) by Kerr and Brown (1978). The proposed structure of this interesting molecule (Fig. 2) was quite different from the natural 3′,5′-connected oligonucleotides known thus far. A new type of 2′,5′-internucleotidic linkage was the most striking structural feature of chemical interest. The major component of most preparations of the LMWI is the trimer, but other oligonucleotides of the general formula (2′,5′)ppp$(A_p)_nA$, where n = 1–15, may be produced.

These findings simplified the complex activity spectrum of interferons to a large extent and encouraged synthetic chemists to prepare the LMWI trimer, pppA2′p5′A2′p5′A (Den Hartog et al. 1979, 1981; Ikehara et al. 1979; Jones and Reese 1979; Martin et al. 1979; Sawai et al. 1979; Imai and Torrence 1981; Ohtsuka et al. 1982), its core (Engels and Krahmer 1979; Markham et al. 1979; Ogilvie and Theriault 1979; Charubala and Pfleiderer 1980a, 1981; Chattopadhyaya 1980; Gioeli et al. 1981; Karpeisky et al. 1982; Takaku and Ueda 1983;

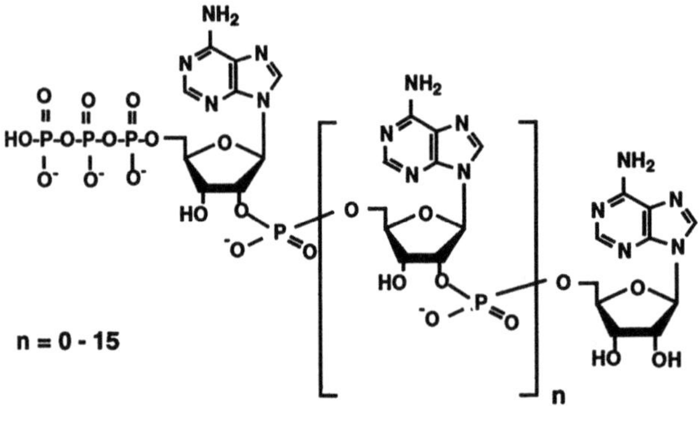

Fig. 2. 5′-O-triphosphoryl-oligoadenylyl- (2′,5′)-adenosines

Hayakawa et al. 1985; Kvasyuk et al. 1987; Noyori et al. 1992) and structurally modified analogues by chemical means. The various chemical syntheses were performed either by the phosphotriester method or, later, by the phosphoramidite approach and using, in general, different blocking group combinations for transient protection. The most common protecting groups for the base moieties are the benzoyl and isobutyryl group, for the 2′-OH group of the sugar moiety the tert-butyldimethylsilyl, the 4-methoxytetra-hydropyranyl and the o-nitro-benzyl group, whereas phosphate protection is achieved by the o-chlorophenyl, 2.5-dichlorophenyl, tribromoethyl, β-cyanoethyl and the 2-(4-nitrophenyl)ethyl(npe) group. 2.4.6-Triiso-propyl-benzenesulfonyl-, mesitylsulfonyl- and quinoline-sulfonyl-nitrotriazolide have been used as condensing reagents. Furthermore, polymerization of N^6-benzoyl-3′-O-(o-nitrobenzyl)-adenosine-5′-monophosphate leads also to the formation of 2′,5′-oligoadenylates (Sawai et al. 1979).

An early synthesis of the 2-5A trimer core at a preparative scale starting from adenosine (**1**) is illustrated by the complex multistep approach (Charubala et al. 1981; Fig. 3).

Adenosine (**1**) was first converted into N^6-benzoyl-5′-O-monomethoxytrityl-adenosine (**2**) which gave upon silylation with tert-butyldimethylsilyl chloride the three silyl derivatives **3-5** separable by chromatography. Compound **4** was then converted into the phosphotriester **6** which can be regarded as a universal monomeric building block for oligonucleotide syntheses. It can be converted into the components **8** and **9** either by oximate cleavage (Reese and Zard 1981), forming a phosphodiester or by detritylation to liberate a free 5′-OH group. Condensation of **8** and **9** produced **10** which can be selectively deprotected to the 2′-terminal phosphodiester **11** due to the improved properties of the npe blocking group. The fully protected trimer **12** resulted from the condensation of **11** + **7** at 70% yield. The cleavage of the various protective groups was then done successively, first, by using DBU to eliminate the npe groups, ammonia to cleave the benzoyl groups, fluoride ion to remove the silyl groups and finally

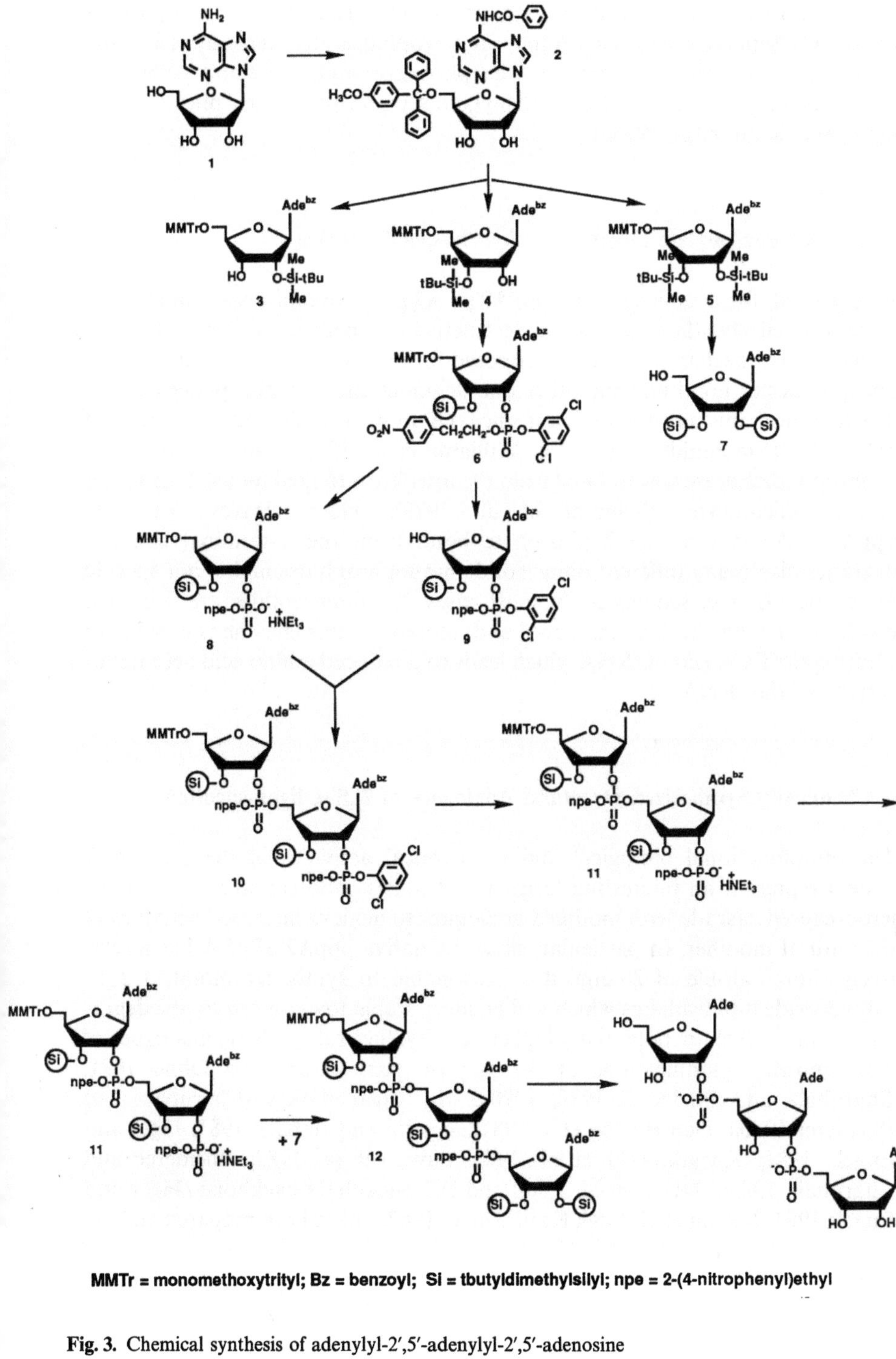

MMTr = monomethoxytrityl; Bz = benzoyl; Si = tbutyldimethylsilyl; npe = 2-(4-nitrophenyl)ethyl

Fig. 3. Chemical synthesis of adenylyl-2′,5′-adenylyl-2′,5′-adenosine

acid to remove the monomethoxytrityl group. The crude product was purified by DEAE Sephadex chromatography and lyophilization. Recently, the same trimer has also been prepared by the phosphoramidite method using bis-(N,N-diisopropylamino)-2-(4-nitrophenyl)ethoxy-phosphane as phosphitylating agent (Sobol et al. 1993a).

4 2′,5′-Oligoadenylate Degradation by Phosphodiesterase

As expected, for a molecule like pppA2′p5′A2′p5′A revealing such interesting biological effects, there is also a destructive mechanism to neutralize this activity. Two mechanisms can be proposed for its inactivation: one cleaves the phosphodiester bond and the other one degrades the 5′-triphosphate residue. The first mechanism is predominant since the products of degradation are ATP and 5′-AMP (Schmidt et al. 1978; Williams et al. 1978; Minks et al. 1979). A phosphodiesterase was isolated from the interferon-treated mouse L-cells and mouse reticulocytes (Schmidt et al. 1979), which cleaves not only pppA2′p5′A2′p5′A at the 2′-phosphate bond from the 2′-terminal end but degrades also many different dinucleotide monophosphates and is not specific for particular base sequences. In most cases 2′,5′-dinucleotides are degraded much faster than the 3′,5′-analogues and, moreover, this enzyme is capable of splitting the CCA end of tRNA which leads to a reduced amino acid acceptance activity of the tRNA.

5 Chemically Synthesized Structural Analogues of 2′,5′-Oligonucleotide

The multifunctional biological and biochemical activities of the 2-5-oligoA system represent an interesting target for chemists, who can attack the interferon-caused cascade with modified analogues to achieve increased activities of the natural modifier. In particular, since the native pppA2′p5′p5′A has a relatively short half-life of 20 min, it is reasonable to synthesize modified 2′,5′-trinucleoside diphosphates which will be more stable towards enzymatic degradation and will show potentially higher activity. Several synthetic analogues of 2-5A showing modifications in the aglycon part (Sawai and Ohno 1981; Charubala and Pfleiderer 1982, 1990a), the sugar moiety (Charubala and Pfleiderer 1980a; Den Hartog et al 1981;Gosselin and Imbach 1981; Jäger and Engels 1981; Kwiatkowski et al. 1982; Sawai et al. 1983; Pfleiderer and Charubala 1985a; Nyilas et al. 1986) and the phosphate backbone (Jäger and Engels 1981; Nelson et al. 1984; Kariko et al. 1987a) have been prepared and are summarized partly as **13–19** in Fig. 4. Since the 2-5A trimer core reveals in most in vitro tests the same biological activity as its 5′-triphosphate, the chemical syntheses are therefore commonly directed towards the core molecules which are to some extent more easily prepared.

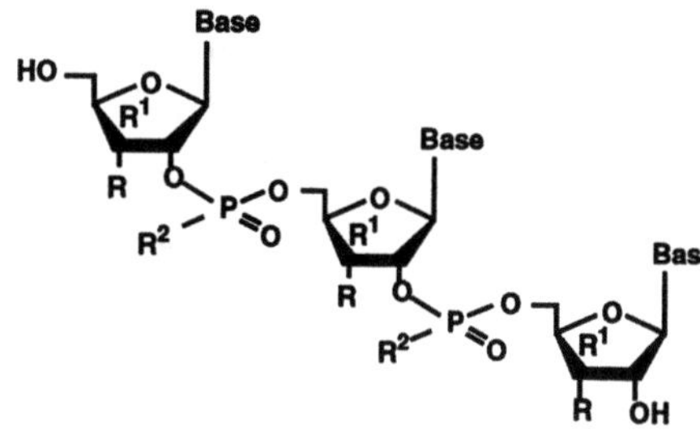

	Base	R	R^1	R^2
13	Ade	H	H	O^-
14	Ade	H	OH	O^-
15	Ade	OCH_3	H	O^-
16	Hyp	OH	H	O^-
17	Ade	OH	H	S^-
18	Ade	OH	H	CH_3

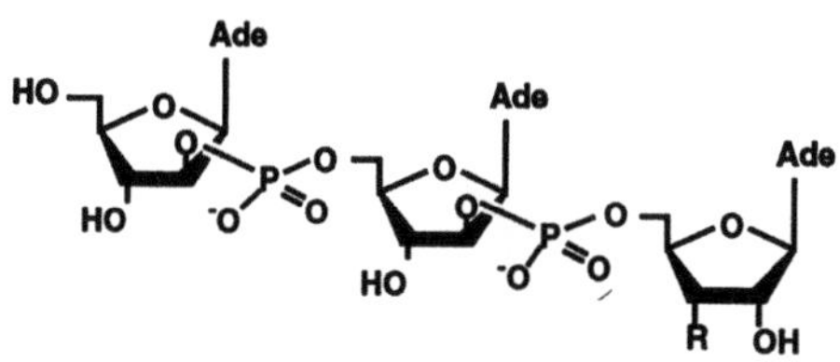

19

Fig. 4. Structural analogues of the 2-5A trimer core

5.1 Modification at the Sugar Moiety

The synthesis of the 2′,5′-cordycepin trimer core (**13**) was one of the first aims in this series due to the closely related structure of the natural core inhibitor (Fig. 5). The synthetic approach to **13** makes use of a more simplified blocking group strategy (Himmelsbach et al. 1984; Pfleiderer et al. 1986) using, in principle, with the exception of the 5′-OH group, one type of protection for the amino-, hydroxy-and phosphate groups in applying the 2-(4-nitrophenyl)ethyl (npe) and 2-(4-nitrophenyl)- ethoxycarbonyl (npeoc) group to block these functionalities.

3′-Deoxyadenosine (cordycepin) (**20**) was transformed in a one-pot reaction via trimethylsilylation and acylation with 1-methyl-3-[2-(4-nitrophenyl) ethoxycarbonyl)]-imidazolium chloride into N^6-2(4-nitrophenyl)ethoxycarbonyl-cordycepin (**21**). Monomethoxytritylation gave **22** which was again acylated to **23** and finally deblocked by acid to the monomeric building block **24**. Compound **22** was then also phosphorylated to give the mixed phosphotriester **25** which was the starting material for the other two building blocks **26** and **27** derived from oximate and acid treatment, respectively.

Condensation of **26** and **27** in the presence of 2.4.6-triisopropylbenzenesulfonyl chloride and N-methyl-imidazole led at 88% yield to the dimer **28**. Subsequent oximate treatment to **29** and another condensation with **24** afforded the fully protected trimer **30** at 90% yield. The final deblocking now needs only subsequent treatment with DBU and acetic acid to produce the required trimer **13** which was purified by DEAE-Sephadex chromatography.

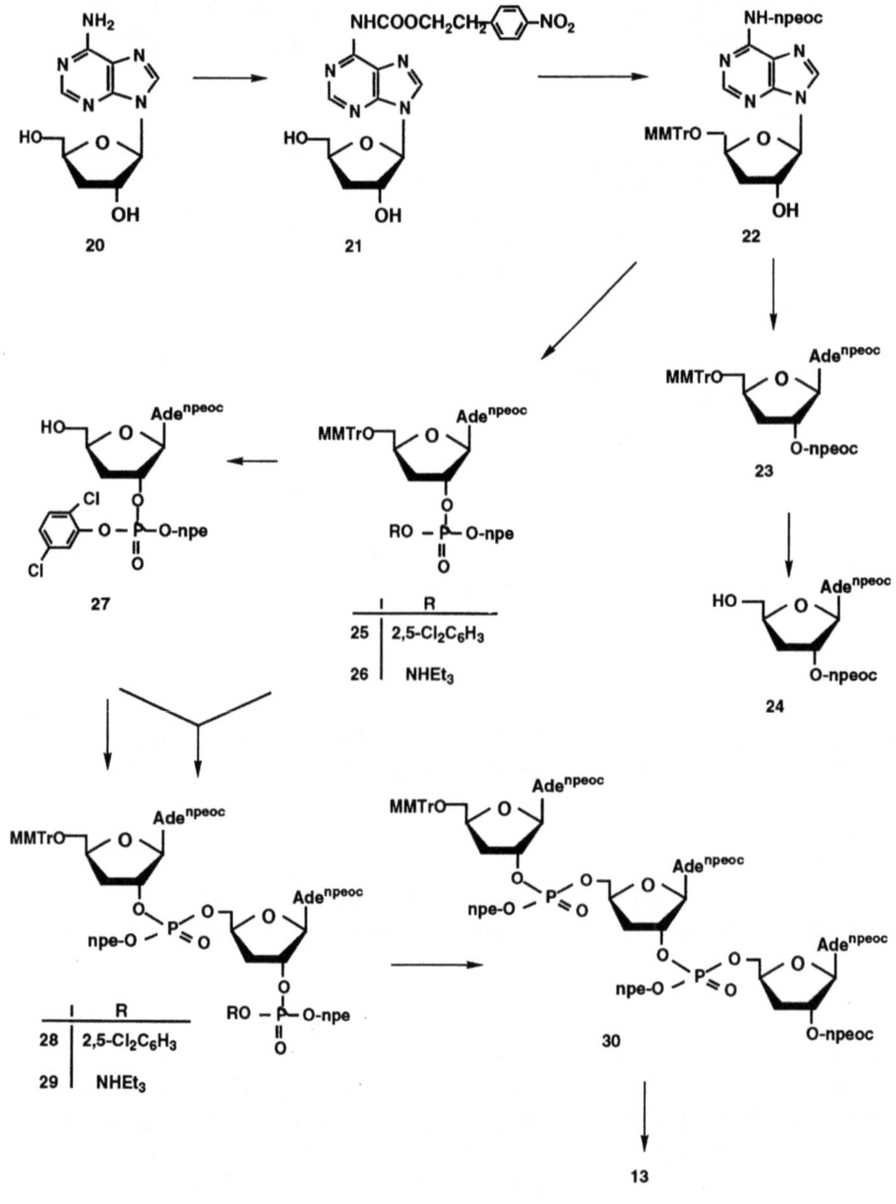

Fig. 5. Chemical synthesis of the cordycepin trimer core (**13**)

Screening of the cordycepin trimer indicated extended metabolic stability, without toxicity to cells, and a broad spectrum of biological activities characterized by a more potent inhibition of protein synthesis in lysed rabbit reticulocytes (Doetsch et al. 1981a), the prevention of the transformation of Epstein-Barr virus-infected lymphocytes (Doetsch et al. 1981b), the synthesis of EBV-induced nuclear antigen (Henderson et al. 1982), tobacco mosaic virus replication in tobacco plants (Devash et al. 1984) and chondrosarcoma growth in animals (Willis et al. 1983). Other sugar-modified analogues prepared were p5′(3′dA)2′p5′A2′p5′A(pCordy-A-A) (**31**), p5′(3′dA)2′p5′A2′p5′(3′dA) (pCordy-A-Cordy) (**32**), p5′A2′p5′A2′p5′(3′dA) (p5′A-A-Cordy) (**33**) (Torrence et al. 1988) which are degraded by the 2′-phosphodiesterase at a rate comparable to p5′A2′p5′A2′p5′A itself (Fig. 6). On the other hand, the trimer p5′A2′p5′(3′dA)2′p5′A (pA-Cordy-A) (**34**), like the cordycepin trimer 5′-monophosphate (pCordy-Cordy-Cordy), was completely resistant to degradation. These data imply that sensitivity to the 2′,5′-phosphodiesterase activity of mouse L-cells requires the presence of the 3′-OH group in the penultimate nucleoside (Alster et al. 1986). The same structural dependence is observed with the corresponding 5′-triphosphates (Torrence et al. 1988) regarding their binding to and activation of the 2-5A-dependent endonuclease (RNase L). Lack of the penultimate 3′-OH group causes here an eight-fold decrease in binding and an even more dramatic 500- to 1000- fold drop in activating the enzyme. This led to the conclusion that only the 3′-hydroxy group of the second nucleotide residue, counted from the 2′-terminal end, is needed for effective activation of RNase L.

Furthermore, the syntheses of more severely modified 2-5A and cordycepin trimers carrying an acyclonucleoside (**35a–d** and **36a–d**) at the 2′-terminal end or consisting even of fully acyclic analogues (**37, 38**) (Mikhailov and Pfleiderer 1985; Mikhailov et al. 1991) have been performed according to established procedures (Fig. 7). Also, a new type of trimeric oligonucleotide derived from the 2′,5′-adenylate dimer and bearing an antitumor or antiviral agent, such as 5-[(E)-2-bromovinyl)]-2′-deoxyuridine (bvdU) (**39**), 9-[(2-hydroxyethoxy)methyl]-guanine (acyclovir) (**41**), 9-β-D-arabinofuranosyl-adenine (araA) (**42**) and 2′-deoxy-5-fluoro-uridine (fdU) (**40**), at the 2′-end (Fig. 7), has been synthesized by the phosphotriester approach (Herdewijn et al. 1989a) to achieve a potential prodrug form. The antiviral screening of **39–41**

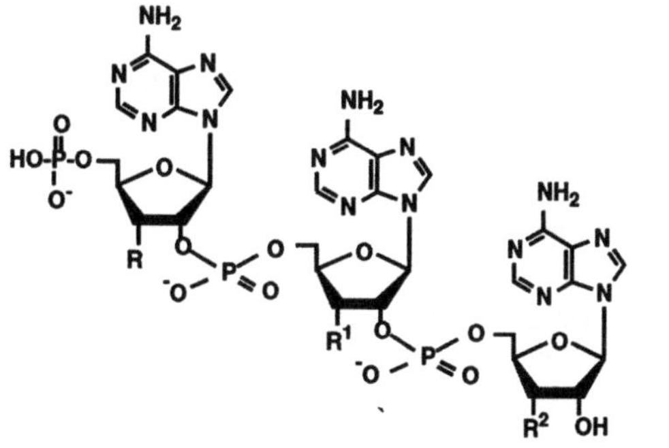

	R	R1	R2
31	H	OH	OH
32	H	OH	H
33	OH	OH	H
34	OH	H	OH

Fig. 6. Mixed 2′,5′-adenosine-cordycepin trimer 5′-monophosphates

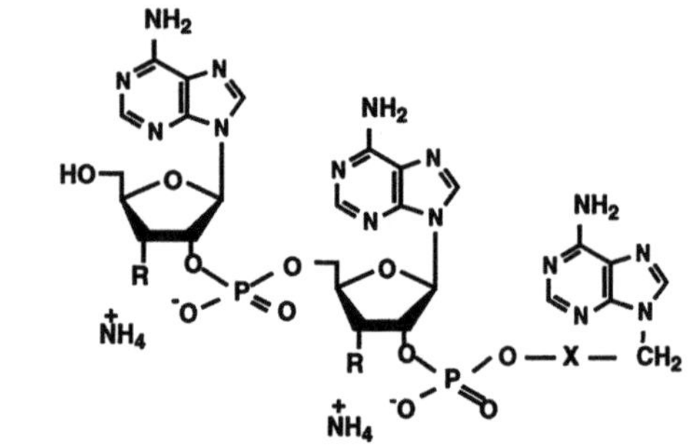

35	R	X	R	36
a	OH	CH_2	H	a
b	OH	CH_2CH_2	H	b
c	OH	$CH_2CH_2CH_2$	H	c
d	OH	CH_2CH_2O	H	d

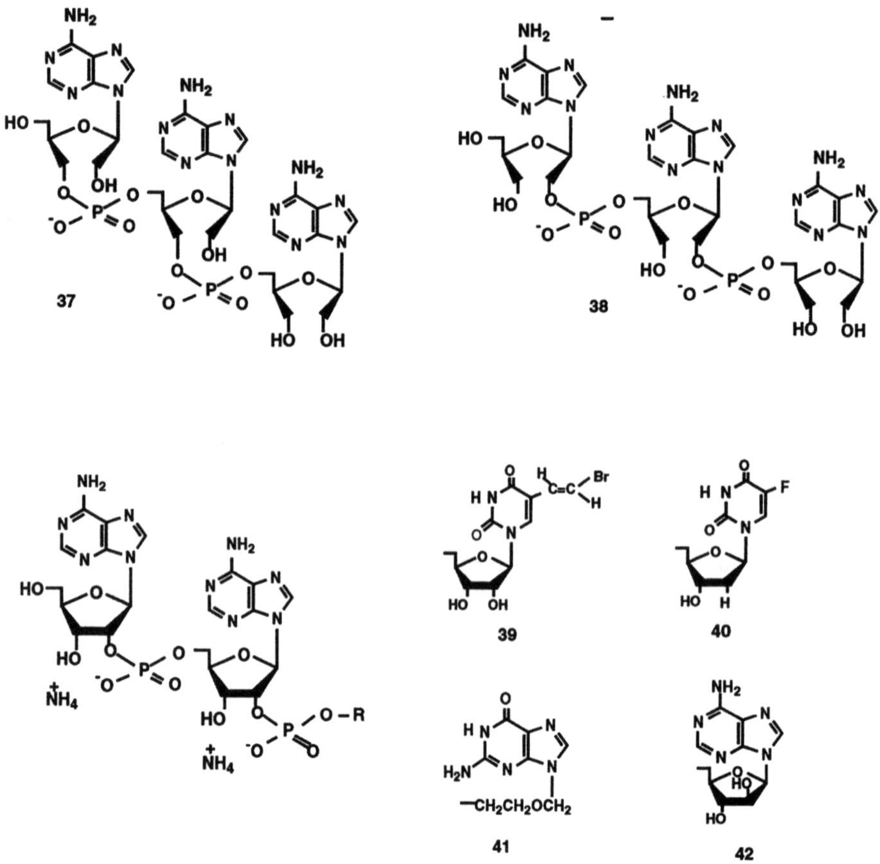

Fig. 7. New types of modified 2′,5′-oligonucleotide trimers

(Herdewijn et al. 1989a) showed biological activities closely related to the parent nucleosides and acyclonucleosides, indicating their potential release by enzymatic cleavage from the oligomers (Fig. 7).

Another chemical synthesis leading to the 3′-azido- (**43**) and 3′-amino-3′-deoxy-β-D-xylofuranosyl-adenine trimer (**44**) in a stepwise approach from the appropriately protected monomeric building blocks via the phosphotriester method demonstrates again the versatility and advantages of the npe- and npeoc-protective groups over other blocking group combinations (Herdewijn et al. 1989b). Studies on the chemical and enzymatic stability of the free oligonucleotides revealed complete resistance towards 0.33 N NaOH for 24 h at 37 °C due to the absence of any 3′-neighbouring *cis*-located function capable of catalyzing the internucleotide hydrolysis, whereas snake-venom phosphodiesterase achieves complete digestion from the 2′-end.

An extended synthetic program also dealt with the preparation of the corresponding 3′-fluoro-, 3′-chloro- and 3′-bromo-3′-deoxy-β-D-xylofuranosyladenine trimers (**45–47**) (Herdewijn et al. 1989a, 1991; Ruf et al. 1987). In this case, selection of a suitable protecting group for the various functionalities was very crucial and depended mainly upon the stability of the modified xyloadenosine moiety in the final deblocking steps. The npe and npeoc group could be applied in the case of 3′-azido- and 3′-fluoro-series, respectively, since all common deblocking steps using DBU in pyridine, aqueous ammonia, 1 M tetrabutylammonium fluoride in THF and acid did not harm the xylonucleoside derivatives. However, the 3′-chloro- (**46**) and 3′-bromo-3-deoxyfuranosyl adenines (**47**) required a different blocking group strategy due to epoxide formation during removal of the 2′-O-tert-butyldimethylsilyl group with fluoride ion and the introduction of a 3′,4′-double bond on DBU treatment. The use of the 2,5-dichloro-phenyl group for phosphate protection and the monomethoxytrityl group for the 2′-OH function turned out to be suitable combinations leading to the free trimers without further intramolecular interconversions.

There are also several other modified 2′,5′-oligonucleotide trimers and tetramers described in the literature such as the A2′p5′A2′p5′-(3′-amino-3′-deoxyA) (Pfleiderer et al. 1985b; Visser et al. 1986), the A2′p5′A′p′A2′p5′-L-riboA (Visser et al. 1986) and the araA2′p5′araA2′p5′araA (**19**) (Kwiatkowski et al. 1982). Furthermore, there are the 2-5A trimer core analogues known to bear at the 2′-terminal end a 2′deoxy-, 2′,3′dideoxy- (Engels 1980) and a 3′-O-methyladenosine unit as well as the corresponding trimer built up from 3′-O-methyl-3′deoxyadenosine (Jäger and Engels 1981; Sharma et al. 1983). Finally, the *cis*-diol function in the 2-5-oligoA trimer was converted into the (2-carboxyethyl)-ethylidene group to prepare an appropriate affinity sorbent for the isolation of endonuclease L (Kvasyuk et al. 1984).

5.2 Modification of the Aglycon in 2-5A Analogues

A variety of base substituted analogues of 2-5-oligoA have been proven valuable in determining the structural and conformational factors that govern the interaction of the 2-5A-dependent endonuclease RNAase L with its activator (Luxembourg 1988). For instance, 2-5A oligonucleotides in which all adenosines were replaced by either uridine, cytidine or inosine showed the vital role of one or more adenosine components (Torrence et al. 1984) to restore biological activity. Sequence specific 2-5A analogues in which each adenosine was systematically replaced by inosine provided evidence that the N^6-amino/N^1 functionality of the adenosine at the 5′-terminus was crucial for endonuclease binding, while the middle part played a minor role in binding and activation; the N^6-amino/N^1 grouping of the 2′-terminal adenosine was not critical for effective binding but was a requirement for activation of RNase L (Imai et al. 1985).

The synthesis of the 2′,5′-inosinate trimer (**16**) was done in a straigthforward manner following either the phosphotriester approach (Charubala and Pfleiderer 1982, 1990a) or by polymerization of inosine 5′-phosphorimidazolide in the presence of Pb^{2+} ions (Sawai and Ohno 1981). The 2′,5′-tubercidin trimer in which the N-7 in the purine nucleus is replaced by CH has also been reported (Seela et al. 1984) as well as the 5.6-dichloro-1-β-D-ribo-furanosyl-benzimidazole (DBR) analogue (Ruf and Pfleiderer 1991). Since the nucleoside DBR inhibits partially the synthesis of hnRNA and to a larger extent the production of RNA and triggers also the production of interferon in human fibroblasts indirectly stimulating antiviral activity, the combination of some of the structural features of bioactive molecules was realized in the 2′,5′-linked trimer of DBR to form a new type of oligomer (Fig. 8).

The successful enzymatic synthesis of the 2′,5′-linked (^{32}P)ppp5′-(8-azido-A) trimer from (^{32}P)-8-azido-ATP (Suhadolnik et al. 1988) allowed covalent

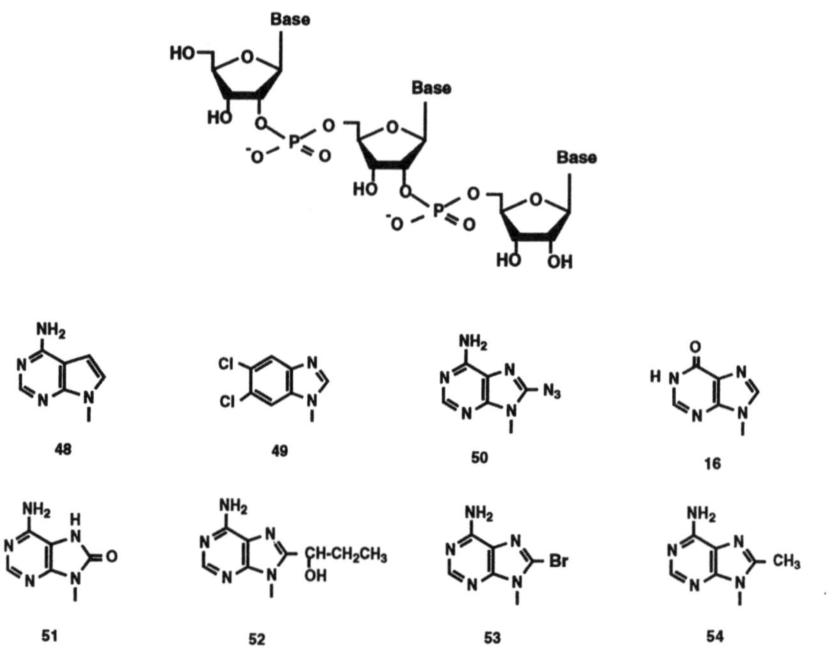

Fig. 8. Base-modified 2′,5′-oligonucleotides

photolabelling of specific proteins of unfractionated extracts of interferon-treated L 929 cells. Since six proteins have been labelled, it was also of interest to elucidate which of the 8-azido groups is actually interacting with the 2′,5′-binding domain. One of the three possible trimers, A2′p5′A2′p5′-(8-azido-A) (**50**), was prepared by the phosphotriester method; its subsequent phosphorylation at the 5′-end by (^{32}P)-phosphate transfer from ATP led to a photoprobe which links covalently to a protein of M_r 80 K known to be RNase L (Charubala et al. 1989a). Various other sequence-specific oligonucleotides containing different C-8 substituents like hydroxy (**51**), hydroxypropyl (**52**) (Kanaou et al.1990), bromo (**53**) (Lesiak and Torrence 1986) and methyl (**54**) (Kitade et al. 1991) have been synthesized by polycondensation reactions of the corresponding 5′-monophosphates in a uranyl- or lead ion-catalyzed ligation process. The 5′-monophosphates of **51** and **52** as well as p5′A2′p5′A2′p5′-(8-hydoxy-A) showed some resistance to snake venom phosphodiesterase cleavage, while **53** and its tetramer possess about 5% of the protein synthesis inhibitory potency of 2-5A which has to be seen under the aspect of a 0.2% translational inhibitory activity of 8-bromoadenosine 5′-monophosphate (Yoshida and Takaku 1986). Oligomers carrying the 8-methyladenosine moiety at different positions showed that the modification at the 2'-terminal end causes increased stability towards SV phosphodiesterase digestion, effects a stronger binding affinity and can be regarded as a more effective inhibitor of translation (Kitade et al. 1991).

5.3 Modification of the Internucleotidic Bonds in 2′,5′-Oligoadenylates

Backbone modification of 2′,5′-oligonucleotides at the site of the internucleotidic bond present another possibility of influencing biological activities by structural alterations. The primary source of conformational flexibility in the 2-5A molecules is determined primarily by a backbone similar to the normal 3′,5′-linked DNA and RNA oligonucleotides (Srinivasan and Olson 1986). However, theoretical and experimental analyses have revealed that the conformations of 2′,5′-linked di- and polynucleotide chains are significantly different from those of 3′,5′-connected analogues.

Our interest in a structural variation of the internucleotidic linkage was derived from the fact that we are dealing here with a crucial determinant in the interaction between a 2-5A molecule and RNase L. The few reports on these lines found in the literature and were focused mainly on the 2′,5′-methylphosphonate and methylphosphotriester analogues (Jäger and Engels 1981; Eppstein et al 1982), which turned out to be completely inactive in these “uncharged” forms.

Trimeric adenylates with (3′,5′)(3′,5′) and mixed (2′,5′)(3′,5′) and (3′,5′)(2′,5′) linkages were synthesized via the phosphotriester approach (Charubala and Pfleiderer 1980a; Charubala et al. 1981), leading to the conclusion that the replacement of only one 2′,5′-linkage resulted in at least one order of magnitude less activity, while alteration of all 2′,5′ to 3′,5′-bonds, the biological activity vanished completely (Lesiak et al. 1983). In addition to their wide use in the stereochemical analysis of enzyme reaction mechanisms (Eckstein 1985), phosphorothioate analogues of nucleotides and polynucleotides have been proven to be extremely valuable probes in studies of DNA conformations, DNA/protein interactions and antiviral activities.

Stereospecific syntheses of phosphorothioate analogues of 2-5A were performed from (SP)ATPαS with 2-5A synthetase from L 929 cell extracts (Lee and Suhadolnik 1985). On the basis of enzymatic degradations and the known preference of SVPD for the Rp configuration (Burgers and Eckstein 1978), it was suggested that the 2-5A synthetase proceeds by inversion of the configuration of (SP)ATPαS to yield the Rp configuration at the 2′,5′-internucleotidic linkages. Absolute proof of configuration must await chemical synthesis of the pure diastereoisomers, because only one of the four possible stereoisomers of the 2′,5′-adenylate phosphorothioate trimer can be prepared enzymatically. A first attempt to synthesize this new type of 2′,5′-oligoadenylate led to the isolation of the two diastereomeric dimers, but the trimer core analogues could not be separated into the pure components (Nelson et al. 1984). However, some time later we succeeded in the synthesis, separation, purification and characterization of the two diastereoisomeric Rp and Sp dimers, the four RpRp, SpRp, RpSp and SpSp trimers (Charubala and Pfleiderer 1988, 1992) as well as the eight RpRpRp, SpRpRp, RpSpRp, SpSpRp, RpRpSp, RpSpSp, SpRpSp and SpSpSp tetramers Charubala and Pfleiderer unpubl. results 1993). The success of these complicated studies was dependent on a special blocking group strategy in applying the N^6-benzoyl-instead of the N^6-2-(4-nitro-phenyl)-ethoxycarbonyl derivatives in the phosphoramidite approach to achieve effective separations of the diastereoisomer pairs. The long sequence of reactions was performed with two starting components -N^6-benzoyl-3′-O-tert-butyldimethylsilyl-5′-O-monomethoxytrityl-adenosine-2′-[2-(4-nitrophenyl)ethyl, N,N-diisopropyl-phosphoramidite] (**55**) (Sobol et al. 1993a) and N^6-benzoyl-2′,3′-di-tert-butyl-dimethylsilyl-adenosine (**7**). Condensation of **7** and **55** under tetrazole catalysis and subsequent oxidation by sulfur led to the diastereoisomeric phosphorothioate-triester mixture **56** + **57** which was separated by silica-gel chromatography into the pure components in 42 and 48% yield, respectively. After detritylation to **58** and **59** each dimer block could be coupled again with the phosphoramidite **55** forming after sulfur treatment two new pairs with the RpRp (**60**) + SpRp (**61**) and RpSp (**62**) + SpSp (**63**) configuration (Fig. 9). This time separation was achieved on silica-gel plates which allow better performance than commonly used columns. Finally, the eight tetramers resulted from subsequent analogous reaction – detritylation, condensation, sulfur oxidation and separation – (Fig. 9). Furthermore, detritylation of **60–63** followed by phosphorylation produced the corresponding 5′-phosphotriesters. The final deprotection of the various blocking groups from the dimers, trimers and tetramers proceeded subsequently by acid, DBU, ammonia and fluoride ions to form the free oligonucleotides **64–75** which were purified by DEAE-Sephadex chromatography.

The configurational assignment of the new chiral centers on phosphorus was accomplished by HPLC, charge separation, ^{31}P-NMR spectra and enzymatic hydrolyses (Kariko et al. 1987a; Suhadolnik et al. 1987; Charubala and Pfleiderer 1992). It was also demonstrated that the configuration of the internucleotidic phosphorothioate linkages does not affect the binding to RNase L, but markedly controls the activation process (Kariko et al. 1987a). Activation decreases in the order RpRp > SpRp > RpSp, whereas the SpSp 2-5A trimer core and its monophosphate (**67**) can be considered as effective inhibitors which bind strongly to RNase L but are unable to activate the enzyme up to concentrations as high as 10^{-3} and 10^{-5} M, respectively. It should also be mentioned that a recent report describes an effective and stereoselective synthesis of the Sp dimeric and the SpSp trimeric 2′,5′-adenylates via the hydrogen phosphonate method (Battistini et al. 1992).

Furthermore, the four diastereoisomeric phosphorothioate trimers RpRp, RpSp, SpRp and SpSp of the cordycepin series have been made available by a similar strategy which again applies the phosphoramidite approach (Charubala and Pfleiderer 1990b; Sobol et al. 1993c). It has been established that the RpRp and SpRp trimer cores are able to bind to and activate RNase L, but the RpSp and SpSp isomers do not induce activity in the enzyme.

The most exciting recent findings with the 2′,5′-oligoadenylate phosphorothioate 5′-mono- and triphosphates, however, deal with the inhibitory effect on HIV-1 reverse transcriptase and HIV-1 replication mediated at least in part by interaction with the template/primer binding complex in a non-competitive mechanism (Montefiori et al. 1989; Sobol et al. 1990, 1993b).

We have also shown that the cordycepin trimer 5′-monophosphate is also an effective inhibitor of HIV-1 replication when encapsulated in liposomes targeted by antibodies for the T-cell receptor molecule CD3 (Müller et al. 1991; Schröder et al. 1992). It was also noted that the 2′,5′adenylate phosphorothioate tetramer 5′-monophosphates are much more effective inhibitors of HIV-1 RT activity than the cordycepin analogues. The broad spectrum activity of these types of compounds is further seen from the activation of RNase L from HeLa cells when the three diastereoisomeric 2′,5′-adenylate phosphorothioate 5′-monophosphates with an RpRpRp, SpRpRp and RpSpSp configuration are microinjected into the cytoplasm of HeLa cells. The SpSpSp 5′-monophosphate, however, binds again to Rnase L but does not activate the enzyme to cleave rRNA (Suhadolnik et al. 1989; Charachon et al. 1990).

The specificity of RNase L for the 2′,5′-internucleotidic linkage implicates first a phosphodiester moiety and, secondly, an Rp configuration in a chiral phosphorothioate function for activation. These facts initiated the synthesis of another series of 2′,5′-oligoadenylate analogues designated as phosphodiester/phosphorothioate hybrids (Vroom et al. 1987; Sobol et al. 1993a) by a combined phosphotriester/phosphoramidite approach, leading to the following compositions: 2′-5′-$pA_{Rp}ApA$, $pA_{Sp}ApA$, $pApA_{Rp}A$, $pApA_{Sp}A$, $pA_{Rp}ApApA$, $pA_{Sp}ApApA$, $pApA_{Rp}ApA$, $pApA_{Sp}ApA$, $pApApA_{Rp}A$ and $pApApA_{Sp}A$. Biological studies established that the internucleotidic linkages of the 2-5A trimer are

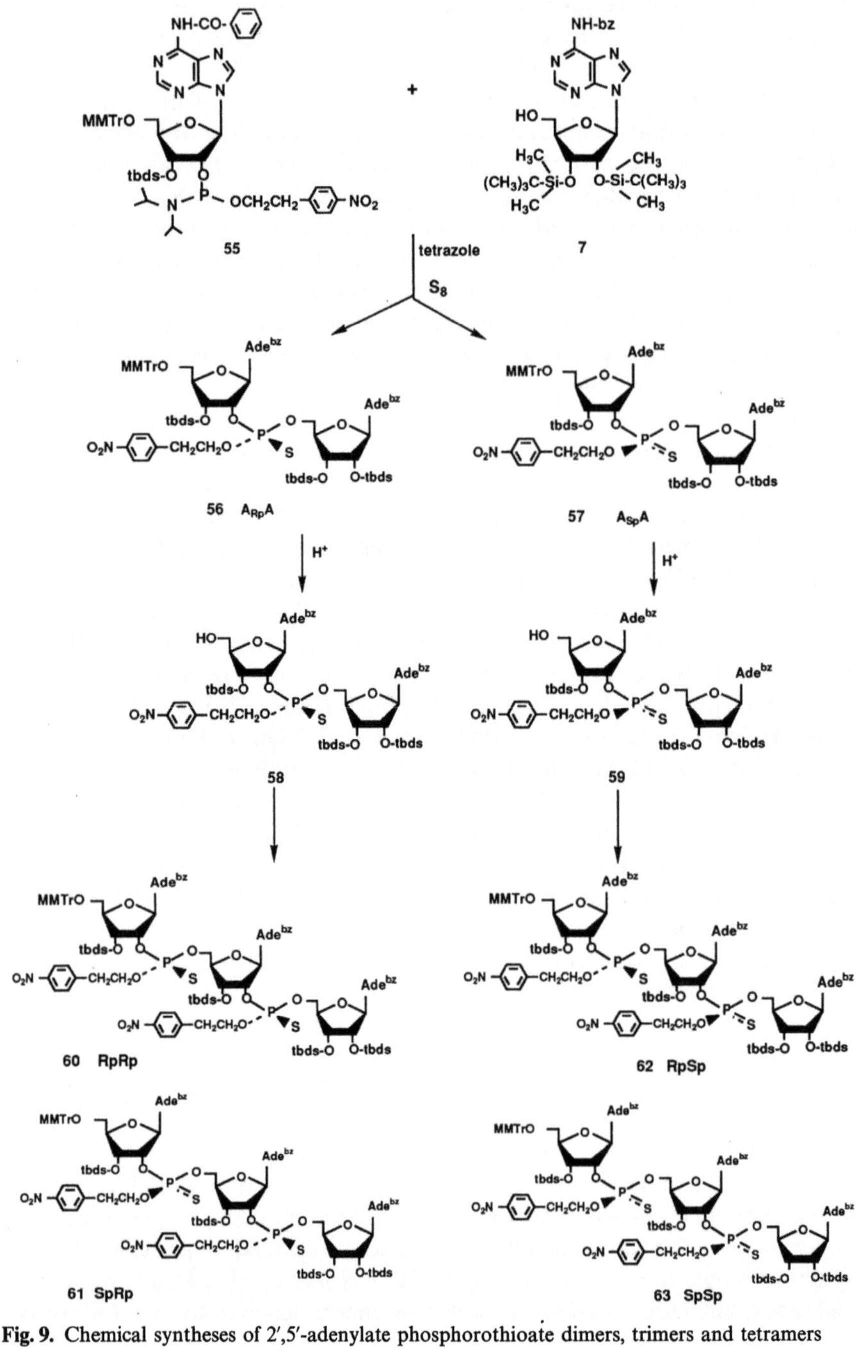

Fig. 9. Chemical syntheses of 2′,5′-adenylate phosphorothioate dimers, trimers and tetramers

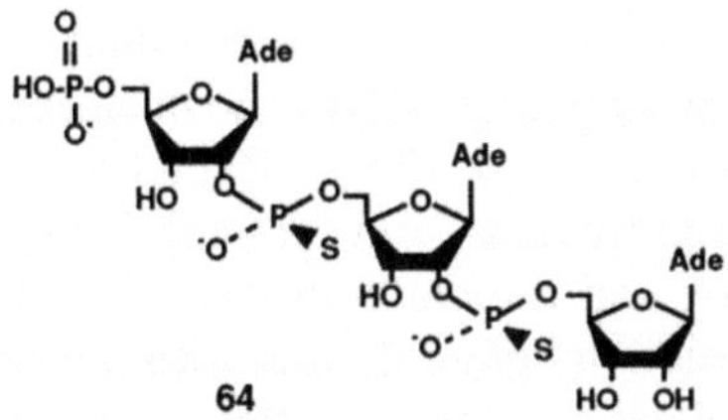

64

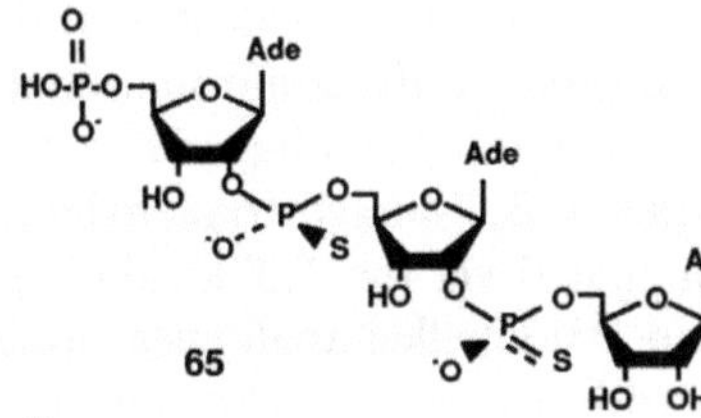

65

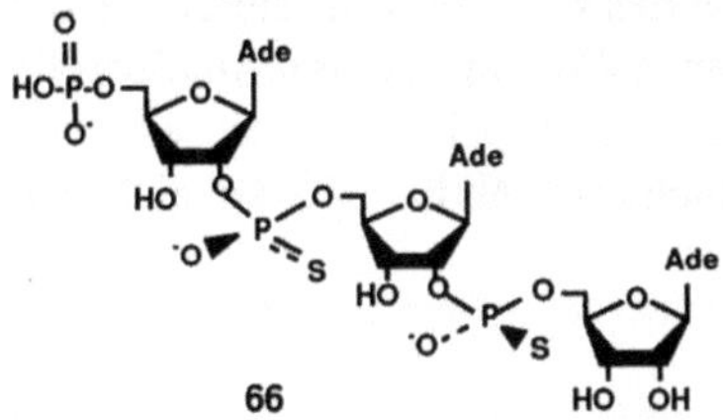

66

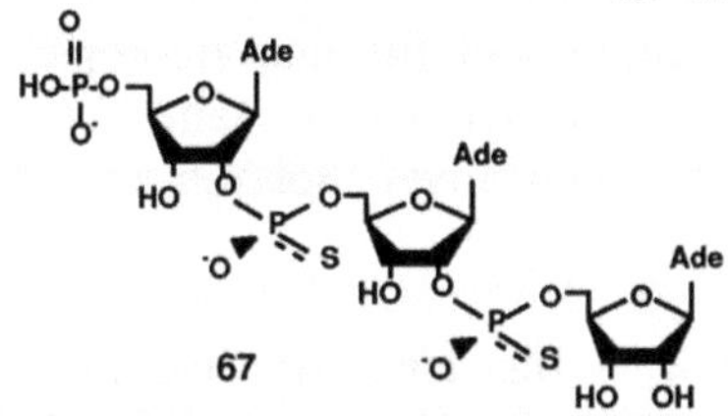

67

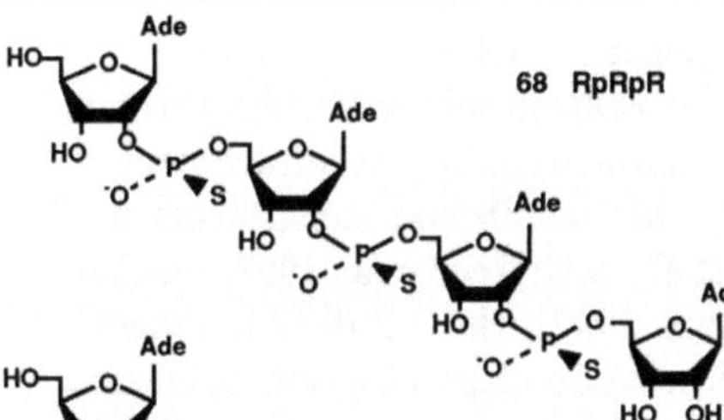

68 RpRpR

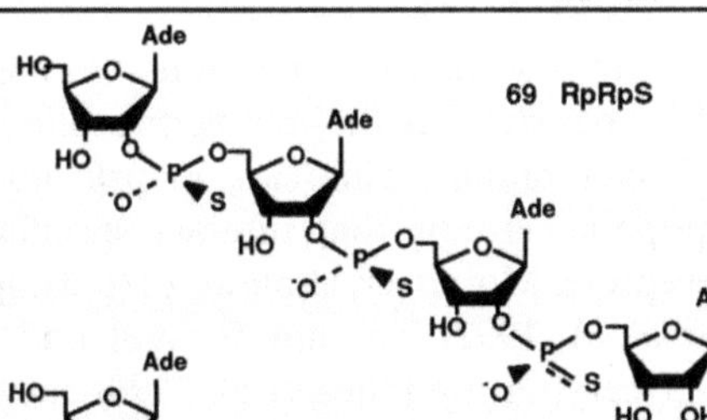

69 RpRpS

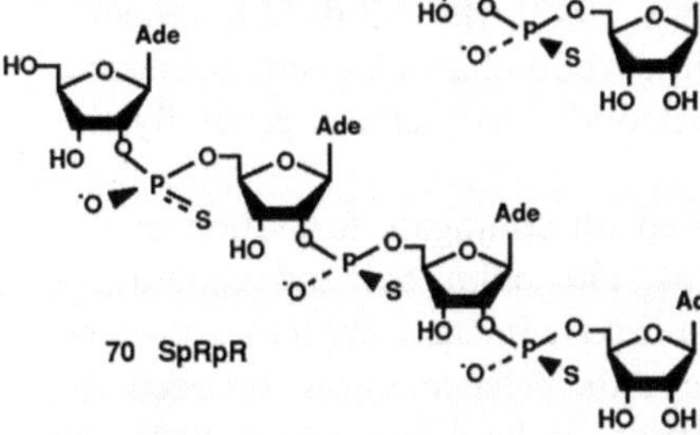

70 SpRpR

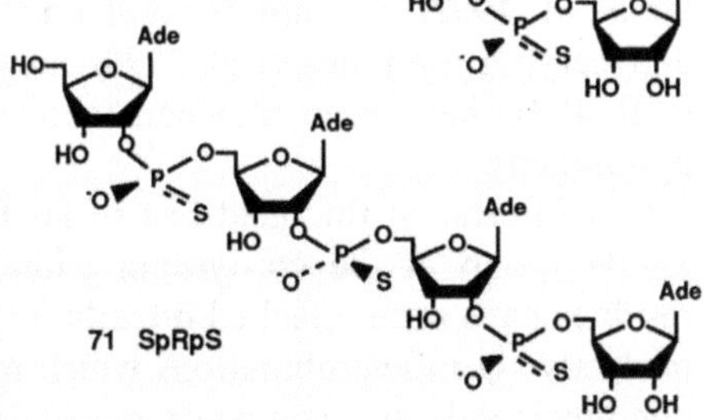

71 SpRpS

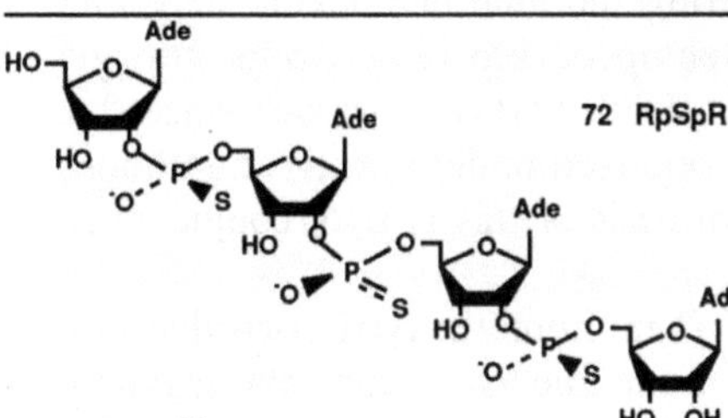

72 RpSpR

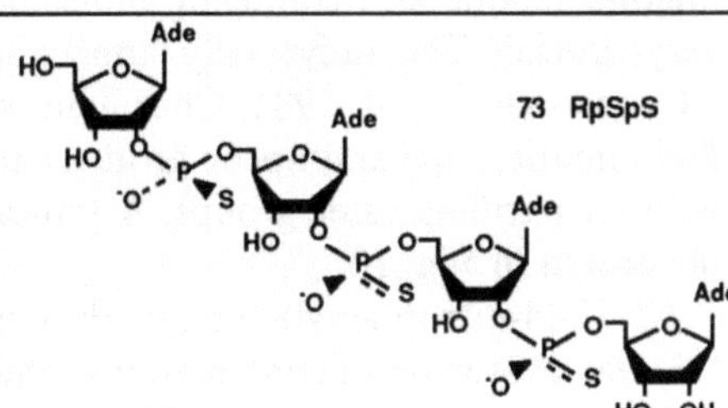

73 RpSpS

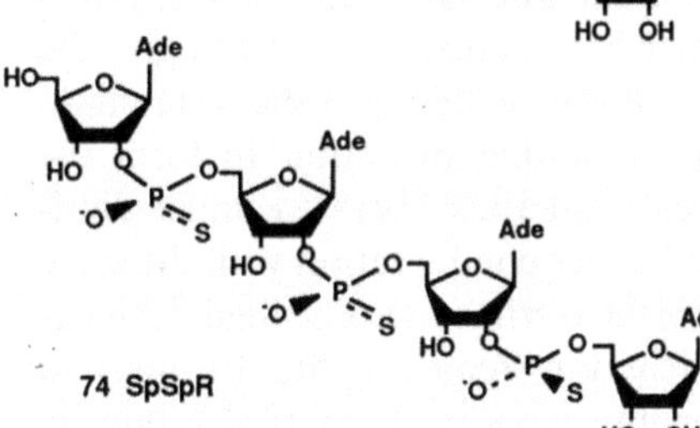

74 SpSpR

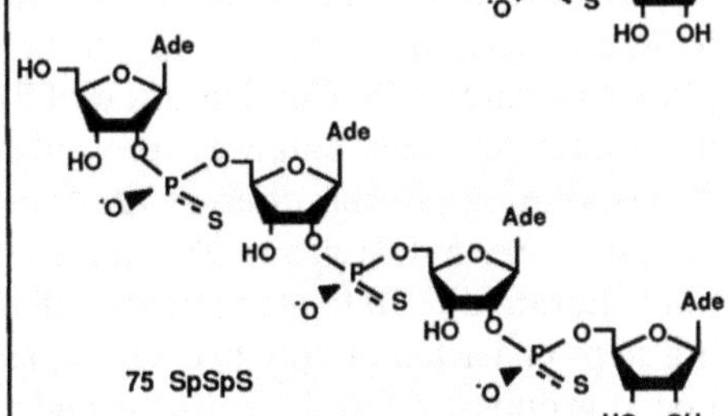

75 SpSpS

Fig. 9. (*Cont.*)

not functionally equivalent since the second internucleotidic linkage is primarily responsible for the conformational and allosteric changes in RNase L which are induced upon 2′,5′-oligoadenylate activation (Sobol et al. 1993a). Rabbit antibodies to 2′,5′-linked triadenylate were prepared by immunization with 2-5A_3 conjugated via the 2′,3′-levulinic group to BSA and radioimmunoassay based upon ^{125}I-labelled analogues. Studies on the reactivity of antibodies with phosphorothioate and seco analogues of oligoadenylates show that the stereospecific substitution of the diastereotopic oxygens by sulfur in the internucleotidic phosphodiester linkage change the immunoreactivity of such analogues, whereby the seco derivatives display in some cases high reactivity (Mikhailopulo et al. 1991).

5.4 2′,5′-Oligoadenylate Conjugates

One of the major limitations in the application of most of these 2′,5′-oligoadenylates is the low permeability of the polyanionic molecules through cell membranes. This can, in principle, be improved by the attachment of lipophilic groups that interact specifically with membrane substituents and receptors. Moreover, cholesterol (Letsinger et al. 1989; Reed et al. 1991; Mackellar et al. 1992), vitamin E (Will and Brown 1992), the 1.2-di-O-hexadecyl-3-glyceryl group (Shea et al. 1990) and hydrocarbon chains (Saison-Behmoras et al. 1991) have been attached to oligonucleotides to increase generally the lipophilicity.

Our interest in this field has so far focused on conjugate formation at the 2′-OH group of the cordycepin trimer (**13**). The palmitoyl and cholesteryl residues have been attached directly or via a linker. In these syntheses suitable protective group combinations which preserve the ester functions between the oligonucleotide and conjugate molecule during the final deblocking steps are very crucial. The universally applicable npe/npeoc blocking group strategy (Himmelsbach et al. 1984; Charubala et al. 1987) worked very well since the β-eliminating npe and npeoc groups can be removed under aprotic conditions, without harming ester groups. A typical synthesis of this type of conjugate is illustrated in Fig. 10.

N^6-[2-(4-nitrophenyl)ethoxycarbonyl]-5′-O-monomethoxytrityl-cordycepin (**22**) was chosen as starting material which, on the one hand, was palmitoylated to **76** and detritylated to **77** and, on the other hand, phosphitylated to give the phosphoramidite **78**. Condensation of **77** + **78** was achieved in the usual manner under tetrazole catalysis and subsequent iodine oxidation to form the corresponding phosphotriester **79**. Acid treatment then liberated under detritylation a free 5′-OH group (**80**) suitable for further condensation with the same phosphoramidite **78** to generate the fully protected trimer **81**. The final deblocking steps consisted of only two subsequent reactions removing first the npe and npeoc groups by DBU treatment and then the monomethoxytrityl group by acid treatment. The free cordycepin conjugate (**82**) was finally isolated in a pure state without ion exchange chromatography just by washing the solid with ether

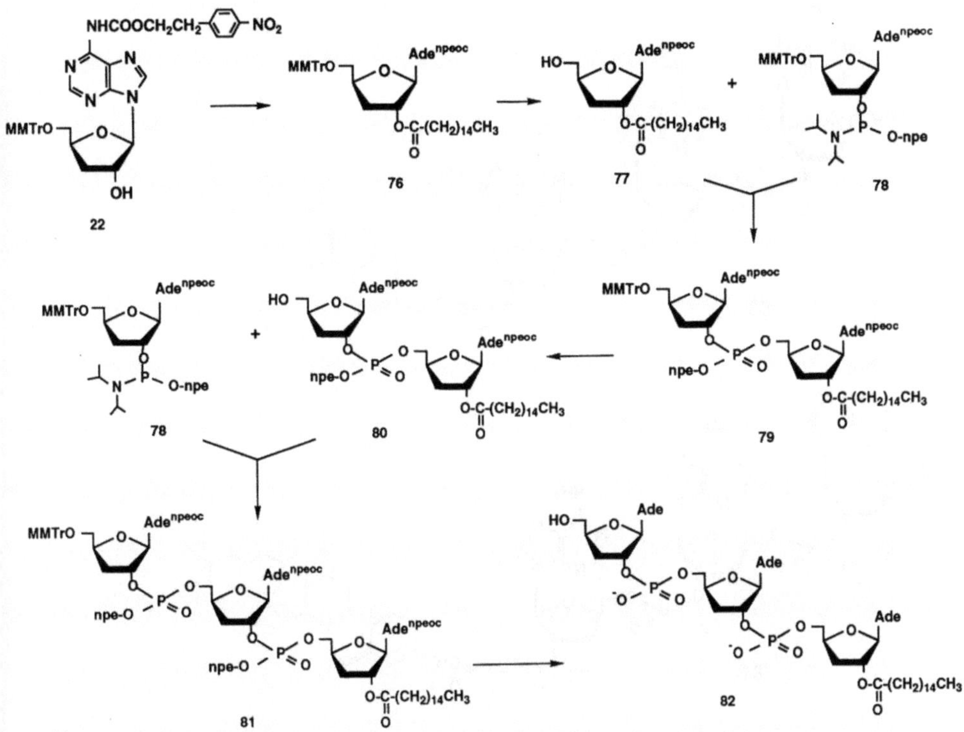

Fig. 10. Chemical synthesis of cordycepinyl-cordycepinyl-2′-O-palmitoyl-cordycepin

and water (Fig. 10; Charubala and Pfleiderer, unpubl. results 1993). In a similar manner cholesterol was attached via a carbonate function (**83**) and a succinate linker (**84**) to the 2′-terminal end (Fig. 11; Wasner and Pfleiderer unpubl. results 1993).

The presence of a lipophilic group at the terminal 2′-OH group can compensate to some extent the polar character of the oligonucleotide, thereby enhancing the penetration through eukaryotic cell membranes as observed in recent screening experiments.

2-5A analogues in which the ribose moiety of the 2′-terminal oligoribonucleotide was transformed into N-substituted morpholino residues proved to be resistant to 2′-PDE degradation, still functioning as potent activators of RNase L (Imai et al. 1982). There is, furthermore, a report on the synthesis of a 2′,5′-adenylate trimer carrying a tetradecanoyl group at the 3′-OH group of the 2′-terminal ribose moiety (Visser et al. 1986).

New oligonucleotide conjugates of various types have a great potential from a structural point of view and will reveal, undoubtedly, new biological effects if close collaboration between synthetic chemists and biochemists, physiologists and biologists is guaranteed.

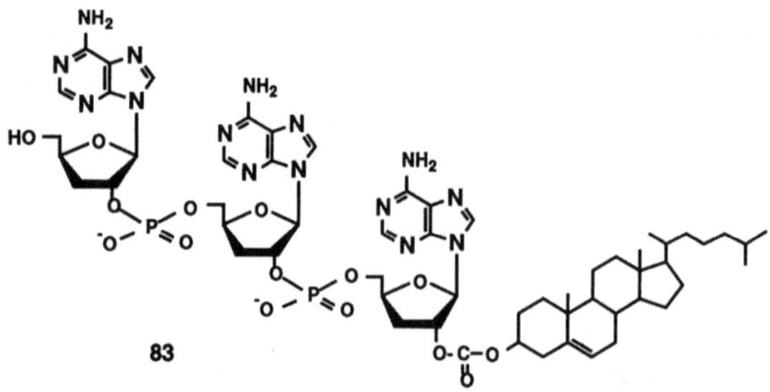

83

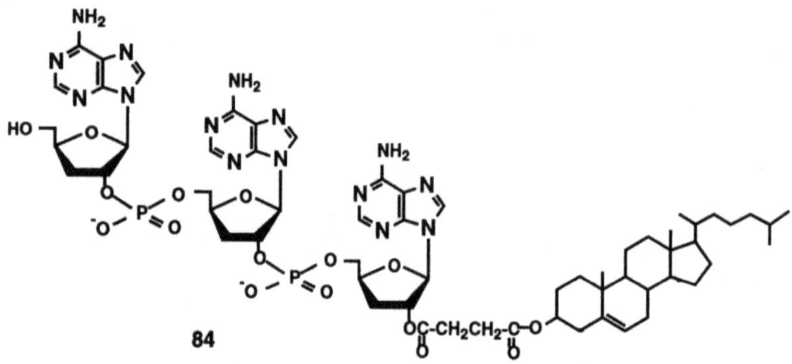

84

Fig. 11. Cholesterol-oligonucleotide conjugates

References

Alster D, Brozda D, Kitade Y, Wong A, Charubala R, Pfleiderer W, Torrence PF (1986) 2′,5′-Phosphodiesterase activity depends upon the presence of a 3′-hydroxyl moiety in the penultimate position of the oligonucleotide substrate. Biochem Biophys Res Commun 141:551–561

Baglioni C (1979) Interferone-induced enzymatic activities and their role in the antiviral state. Cell 17:255–264

Battistini C, Brasca MG, Fustinoni S, Lazzari E (1992) An efficient and stereoselective synthesis of 2′,5′-oligo-(Sp)-thioadenylates. Tetrahedron 48:3209–3226

Burgers PMJ, Eckstein F (1978) Absolute configuration of the diastereomers of adenosine 5′-O-(1-thiophosphate): consequences for the stereochemistry of polymerization by DNA-dependent RNA polymerase from *Escherichia coli.* Proc Natl Acad Sci USA 75:4798–4800

Charachon G, Sobol RW, Bisbal C, Salchzada T, Silhol M, Charubala R, Pfleiderer W, Lebleu B, Suhadolnik RJ (1990) Phosphorothioate analogues of (2′-5′) (4): agonist and antagonist activities in intact cells. Biochemistry 29:2550–2556

Charubala R, Pfleiderer W (1980a) Synthesis and properties of adenylate trimers A2′p5′A2′p5′A, A2′p5′A3′p5′A, and A3′p5′A2′p5′A. Tetrahedron Lett 21:1933–1936

Charubala R, Pfleiderer W (1980b) Synthesis and properties of 3′-deoxyadenylate trimer dA2′p5′A2′p5′A. Tetrahedron Lett 21:4077–4080

Charubala R, Pfleiderer W (1982) Synthesis of inosinate trimer I2′p5′I2′p5′I and tetramer I2′p5′I2′p5′I2′p5′I. Tetrahedron Lett 23:4789–4792

Charubala R, Pfleiderer W (1988) The chemical synthesis of (2′-5′)-P-thioadenylate dimers, trimers and tetramers. Nucleosides Nucleotides 7:703–706

Charubala R, Pfleiderer W (1990a) Synthesis and properties of inosinate trimer I2′p5′I2′p5′I and inosinate tetramer I2′p5′I2′p5′I2′p5′I. Heterocycles 30:1141–1153

Charubala R, Pfleiderer W (1990b) Chemical synthesis, characterisation and biological properties of four possible trimeric 3′-deoxyadenosine (cordycepin) phosphorothioates and the 3′-5′-adenylate analogues. Collect Czech Chem Commun Spec Issue 55:181–184

Charubala R, Pfleiderer W (1992) Syntheses and characterisation of phosphorothioate analogues of (2′-5′) adenylate dimer, trimer and their monophosphates. Helv Chim Acta 75:471–479

Charubala R, Uhlmann E, Pfleiderer W (1981) Synthese und Eigenschaften vom Adenylyl-adenylyl-adenosinen. Liebigs Ann Chem:2392–2406

Charubala R, Uhlmann E, Himmelsbach F, Pfleiderer W (1987) Chemical synthesis of the 2′-5′-cordycepin trimer core. Helv Chim Acta 70:2028–2038

Charubala R, Pfleiderer W, Sobol RW, Li SW, Suhadolnik RJ (1989a) Chemical synthesis of adenylyl-(2′-5′)-adenylyl-(2′-5′)-8-azidoadenosine, and activation and photoaffinity labelling of RNase L by (^{32}P) p5′A2′p5′A2′p5′$N_3{}^8$A. Helv Chim Acta 72:1354–1361

Charubala R, Pfleiderer W, Alster D, Brozda D, Torrence PF (1989b) Synthesis and biological activity of a bis-substituted 3′-deoxyadenosine analog of 2-5 A. Nucleosides Nucleotides 8:273–284

Charubala R, Pfleiderer W, Suhadolnik RJ, Sobol RW (1991) Chemical synthesis and biological activity of 2′-5′-phosphorothioate tetramer cores. Nucleosides Nucleotides 10 (1–3):383–388

Chattopadhyaya JB (1980) Synthesis of adenylyl(2′-5′)adenylyl-(2′-5′)adenosine (2-5 A core). Tetrahedron Lett 21:4113–4116

Creasey AA, Bartholowmew JC, Merigen TC (1980) The importance of G0 in the site of action of interferons in the cell cycle. Exp Cell Res 134:155–160

Den Hartog JAJ, Doornbos J, Crea R, Van Boom JH (1979) Synthesis of the 5′-triphosphate of a trimer containing 2′-5′internucleotide linked riboadenosines. Recueil 98:469–470

Den Hartog JAJ, Wijnands RA, Van Boom JH, Crea R (1981) Chemical synthesis of pppA2′p5′A2′p5′A, an interferon induced inhibitor of protein synthesis, and some functional analogues. J Org Chem 46:2242–2251

Devash Y, Gera A, Willis DH, Reichmann M, Pfleiderer W, Charubala R, Sela I, Suhadolnik RJ (1984) 5′-Dephosphorylated 2′,5′-adenylate trimer and its analogs. J Biol Chem 259:3482–3486

Doetsch PW, Wu JM, Sawada Y, Suhadolnik RJ (1981a) Synthesis and characterisation of (2′-5′)ppp 3′dA(p3′dA)$_n$, an analogue of (2′-5′)ppA(pA)$_n$. Nature (Lond) 291:355–358

Doetsch PW, Suhadolnik RJ, Sawada Y, Mosca JD, Flick MB, Reichenbach NL, Dang AQ, Wu JD, Charubala R, Pfleiderer W, Henderson EE (1981b) Core (2′-5′) oligoadenylate and the cordycepin analog: inhibitors of Epstein-Barr virus induced transformation of human lymphocytes in the absence of interferon. Proc Natl Acad Sci USA 78:6699–6703

Drocourt JL, Dieffenbach CW, Ts'o PO, Justesen J, Thang MN (1982) Structural requirements of (2′-5′)-oligoadenylate for protein synthesis inhibition in human fibroblasts. Nucleic Acids Res 10:2163–2174

Eckstein F (1985) Nucleoside phosphorothioates. Annu Rev Biochem 54:367–402

Engels J (1980) Synthesis of 2′-end modified 2′,5′-adenylate trimers. Tetrahedron Lett 21:4339–4342

Engels J, Krahmer U (1979) Gezielte Synthese des trimeren Isoadenylats A2′p5′A2′p5′A. Angew Chem 91:1007

Eppstein DA, Marsh VY, Schryver BB, Larsen MA, Barnett JW, Verheyden JPH, Prisbe EJ (1982) Analogs of (A2′p)$_n$A correlation of structure of analogs of ppp(A2′p)$_2$A and (A2′p)$_2$A with stability and biological activity. J Biol Chem 257:13390–13397

Friedman RM (1981) A primer. Academic Press, New York

Friedman RM, Metz DH, Esteban RM, Tovell DR, Ball LA, Kerr IM (1972) Mechanism of interferon action: inhibition of viral messenger ribonucleic acid translation in L cell extracts. J Virol 10:1184–1198

Gioeli C, Kwiatskowski M, Öberg B, Chattopadhyaya JB (1981) The tetraisopropyldisiloxane-1,3-diyl: a versatile protecting group for the synthesis of adenylyl(2′-5′)adenylyl(2′-5′)adenosine (2-5 A core). Tetrahedron Lett 22:1741–1744

Gosselin G, Imbach JL (1981) Synthese du Trimere de la β-D-Xylofuranosyl-9-adenine a Liaisons Internucleotidiques 2′-5′. Tetrahedron Lett 22:4699–4702

Gresser I (1977) Commentary: on the varied biological effects of interferon. Cell Immunol 34:406–415

Hayakawa Y, Nobori T, Noyori R (1985) Synthesis of 2′-end- modified 2′,5′-oligoadenylates. Nucleic Acids Res Symp Ser 16:129–132

Henderson EE, Doetsch PW, Charubala R, Pfleiderer W, Suhadolnik RJ (1982) Inhibition of Epstein-Barr virus-associated nuclear antigen (EBNA) induction by (2′-5′) oligoadenylate and the cordycepin analog: mechanism of action for inhibition of EBV-induced transformation. Virology 122:198–201

Herberman RB, Ortaldo JR, Mantovani A, Hobbs DS, King H, Pestka S (1982) Effect of human recombinant interferons on cytotoxic activity of natural killer cells and monocytes. Cell Immunol 67:160–167

Herdewijn P, Charubala R, De Clercq E, Pfleiderer W (1989a) Synthesis of 2′-5′ connected oligonucleotides. Prodrugs for antiviral and antitumoral nucleosides. Helv Chim Acta 72:1739–1748

Herdewijn P, Charubala R, Pfleiderer W (1989b) Modified oligomeric 2′-5′A analogues: synthesis of 2′-5′ oligonucleotides with 9-(3′-azido-3′-deoxy-β-D-xylofuranosyl) adenine and 9-(3′-amino-3′-deoxy-β-D-xylofuranosyl)adenine as modified nucleosides. Helv Chim Acta 72:1729–1738

Herdewijn P, Ruf K, Pfleiderer W (1991) Synthesis of modified oligomeric 2′-5′ A analogues: potential antiviral agents. Helv Chim Acta 74:7–23

Himmelsbach F, Schulz BS, Trichtinger T, Charubala R, Pfleiderer W (1984) The p-nitrophenylethyl (NPE) group. Tetrahedron 59:59–72

Hovanessian AG, Brown RE, Kerr IM (1977) Synthesis of low molecular weight inhibitor of protein synthesis with enzyme from interferon treated cells. Nature 268:537–539

Ikehara M, Oshie K, Ohtsuka E (1979) Synthesis of a protein biosynthesis inhibitor, 5′-triphosphoryladenylyl (2′-5′)adenylyl-(2′-5′) adenosine. Tetrahedron Lett 20:3677–3680

Imai J, Torrence PF (1981) Bis (2,2,2-trichloroethyl) phosphorodichloridite as a reagent for the phsophorylation of oligonucleotides: preparation of 5′-phosphorylated 2′-5′-oligoadenylates. J Org Chem 46:4015–4021

Imai J, Johnston I, Torrence PF (1982) Chemical modification potentiates the biological activities of 2-5 A and congeners. J Biol Chem 257:12739–12741

Imai J, Lesiak K, Torrence PF (1985) Respective role of each of the purine N^6-amino groups of 5′-O-triphosphoryladenylyl (2′-5′)adenylyl(2′-5′)adenosine in binding to and activation of RNase L. J Biol Chem 260:1390–1393

Itkes AV, Karpeisky M Ya, Kartasheva ON, Mikhailov SN, Mosieyev GP, Pfleiderer W, Charubala R, Yakaovlev GI (1988) A route to 2′,5′-oligoadenylates with increased stability towards phosphodiesterases. FEBS Lett 236:325–328

Jäger A, Engels J (1981) Synthesis of methylphosphonate and methylphosphotriester analogues of 2′,5′-adenylate trimers. Nucleic Acids Res Symp Ser 9:149–152

Jones SS, Reese CB (1979) Chemical synthesis of 5′-O-triphosphoryladenylyl(2′-5′)adenylyl(2′-5′)adenosine (2-5 A). J. Am Chem Soc 101:7399–7401

Kanaou M, Ohmori H, Takaku T, Yokoyama S, Kawai G, Suhadolnik RJ, Sobol RW Jr (1990) Chemical synthesis and biological activities of analogues of 2′,5′-oligoadenylates containing 8-substituted adenosine derivatives. Nucleic Acids Res 18:4439–4446

Kariko K, Li SW, Sobol RW Jr, Suhadolnik RJ, Charubala R, Pfleiderer W (1987a) Phosphorothioate analogues of 2′,5′-oligoadenylate activation of 2′,5′-oligoadenylate dependent endoribonuclease by 2′,5′-phosphorothioate cores and 5′-monophosphates. Biochemistry 26:7136–7142

Kariko K, Sobol RW, Suhadolnik L, Li SW, Reichenbach NL, Suhadolnik RJ, Charubala R, Pfleiderer W (1987b) Phosphorothioate analogues of 2′-5′-oligoadenylate. Enzymatically synthesized 2′-5′-phosphorothioate dimer and trimer: unequivocal structural assignment and activation of 2′-5′-oligoadenylate-dependent endoribonuclease. Biochemistry 26:7127–7135

Karpeisky MYu, Beigelman LN, Mikhailov SN, Padyukova NSh, Smrt J (1982) Synthesis of adenylyl-(2′-5′)adenylyl-(2′-5′)adenosine. Collect Czech Chem Commun 47:156–166

Kerr IM, Brown RE (1978) pppA2′p5′A2′p5′A: an inhibitor of protein synthesis synthesised with an enzyme from interferon-treated cells. Proc Natl Acad Sci USA 75:256–260

Kerr IM, Brown RE, Ball LA (1974) Increased sensitivity of cell-free protein synthesis to double-stranded RNA after interferon treatment. Nature (Lond) 250:57–59

Kitade Y, Nakata Y, Hirota K, Maki Y, Pabuccuglu A, Torrence PF (1991) 8-Methyladenosine substituted analogues of 2-5 A: synthesis and their biological activities. Nucleic Acids Res 19:4103–4108

Kvasyuk EI, Kulak TI, Zaitseva GT, Mikhailopulo IA, Charubala R, Pfleiderer W (1984) Synthesis of a modified 2′,5′-adenylate trimer with a 2′,3′-di-O-(2-carboxyethyl)-ethylidene terminal group. Tetrahedron Lett 25:3683–3686

Kvasyuk EI, Kulak TI, Khirpach NB, Mikhailopulo IA, Uhlmann E, Charubala R, Pfleiderer W (1987) Preparative synthesis of trimeric (2-5) oligoadenylic acid. Synthesis 4:535–541

Kwiatkowski M, Gioelli C, Chattopadhyaya JB, Öberg B, Drake AF (1982) Chemical synthesis and conformation of "arabino analogues" of (2′-5′)-isooligoadenylates and their application as probes to determine the structural requirements of cellular exonucleases. Chem Scr 19:49–56

Lee C, Suhadolnik RJ (1985) 2′,5′-Oligoadenylates chiral at phosphorus: enzymatic synthesis, properties and biological activities of 2′,5′-phosphorothioate trimer and tetramer analogues synthesised from (Sp)-ATPaS. Biochemistry 24:551–555

Lesiak K, Torrence PF (1986) Synthesis and biological activities of oligo (8-bromoadenylates) as analogues of 5′-O-triphosphoadenylyl-(2′-5′) adenylyl (2′-5′) adenosine. J Med Chem 29: 1015–1022

Lesiak K, Imai J, Floyd-Smith G, Torrence PF (1983) Biological activities of phosphodiester linkage isomers of 2-5 A. J Biol Chem 258:13082–13088

Letsinger RL, Zhang G, Sun DK, Ikeuchi T, Sarin PS (1989) Cholesteryl-conjugated oligonucleotides: synthesis, properties and activity as inhibitors of replication of human immunodeficiency virus in cell culture. Proc Natl Acad Sci USA 86:6553–6556

Luxembourg A (1988) 2-5 A mediator in search of a function. Bull Inst Pasteur 86:373–417

Mackellar C, Graham D, Will DW, Burgess S, Brown T (1992) Synthesis and physical properties of anti HIV antisense oligonucleotides bearing terminal lipophilic groups. Nucleic Acids Res 20:3411–3417

Markham AK, Porter RA, Gait MJ, Sheppard RC, Kerr IM (1979) Rapid chemical synthesis and circular dichroism properties of some 2′-5′-linked oligoadenylates. Nucleic Acids Res 6:2569–2582

Martin EM, Bridsall NJM, Brown RE, Kerr IM (1979) Enzyme synthesis, characterisation, and NMR spectra of pppA2′p5′a2′p5′A and related oligonucleotides: comparison with chemically synthesized material. Eur J Biochem 95:245–257

Mikhailopulo IA, Kvasyuk EI, Kulak TI, Shulyakovskaya SM, Makarenko MV, Mikhailov SN, Charubala R, Pfleiderer W (1991) Specificities of rabbit anti-(2′-5′)oligoadenylate antibodies towards phosphorothioate and seco analogs of oligoadenylate. Nucleic Acids Res Symp Ser 24:67–70

Mikhailov SN, Pfleiderer W (1985) Synthesis of a new class of acyclic 2′,5′ and 3′,5′ oligonucleotide analogs based on 9 (1,5-dihydroxy-4 (S)-hydroxymethyl-3-oxapent-2 (R)-yl)-adenine. Tetrahedron Lett 26:2059-2062

Mikhailov SN, Charubala R, Pfleiderer W (1991) 3′-Deoxyadenylyl-(2′-5′)-3′-deoxyadenylyl-(2′w)-9-(w-hydroxyalkyl) adenines. Helv Chim Acta 74:887–891

Minks MA, West DK, Benvin S, Baglioni C (1979) Metabolic stability of 2′-5′-oligo (A) and activity of 2′-5′-oligo (A) dependent endonuclease in extracts of interferon-treated and central HeLa cells. J Biol Chem 254:10180–10183

Montefiori DM, Sobol RW Jr, Li SW, Reichenbach NL, Suhadolnik RJ, Charubala R, Pfleiderer W, Modliszewski A, Robinson WE Jr, Mitchell WM (1989) Phosphorothiate and cordycepin analogues of 2′-5′-oligoadenylate: inhibition of human immunodeficiency virus type 1 reverse transcriptase and infection in vitro. Proc Natl Acad Sci USA 86:7191–7194

Müller WEG, Weiler BE, Charubala R, Pfleiderer W, Leserman L, Sobol R W, Suhadolnik RJ, Schröder HC (1991) Cordycepin analogues of 2'.5'-oligoadenylate inhibit human immunodeficiency virus infection via inhibition of reverse transcriptase. Biochemistry 30:2027–2033

Nelson PS, Bach CT, Verheyden JPH (1984) Synthesis of P-thioadenylyl-(2'-5') adenosine and P-thioadenylyl-(2'-5')-P-thioadenylyl-(2'-5') adenosine. J Org Chem 49:2314-2317

Noyori R, Uchiyama M, Nobori T, Hirose M, Hayakawa Y (1992) Practical synthesis of 2'-5'-linked oligoadenylates (2-5A oligomers). Aust J Chem 45:205–225

Nyilas A, Vrang C, Drake A, Öberg B, Chattopadhyaya JB (1986) The cordycepin analogue of 2,5A and its threo isomer. Chemical synthesis, conformation and biological activity. Acta Chem Scand B 40:687–688

Ogilvie KK, Theriault NY (1979) The synthesis of 2'-5'-linked oligoribonucleotides. Tetrahedron Lett 20:2111–2114

Ohtsuka E, Yamane A, Ikehara M (1982) A new synthetic method for 5'-triphosphoryl-adenylyl (2'-5')-adenylyl (2'-5') adenosine and 2-5 A core using 3'-O-tetrahydropyranyl-adenosine derivatives. Chem Pharm Bull 30:376–378

Pestka S (ed) (1981) The interferons. Methods in enzymology, vols 78 and 79. Academic Press, New York

Pfleiderer W, Charubala R (1985a) in: Böger P (ed) Wirkstoffe in Zellgeschehen: 2'-5'-verknüpfte Oligonucleotide eine neue Gruppe antiviral-antitumor aktiver Wirkstoffe. Univ Verlag, Konstanz, pp 9–31

Pfleiderer W, Himmelsbach F, Charubala R, Schirmeister H, Beiter AH, Schulz BS, Trichtinger T (1985b) The p-nitrophenylethyl group – a universal blocking group in nucleoside and nucleotide chemistry. Nucleosides Nucleotides 4:81–94

Pfleiderer W, Schwarz M, Schirmeister H (1986) New developments in nucleotide chemistry. Chem Scr 26:147–154

Pfleiderer W, Schirmeister H, Reiner T, Pfister M, Charubala R (1987) New protecting groups in nucleoside and nucleotide chemistry. In: Bruzik KS, Stec WJ (eds) Biophosphates and their analogues – synthesis, structure metabolism and activity. Elsevier, Amsterdam, pp 133–142

Reed MW, Adams AD, Nelson JS, Meyer RB Jr (1991) Acridine and cholesterol derivatised solid supports for improved synthesis of 3'-modified oligonucleotides. Bioconjugate Chem 2:217-225

Reese CB, Zard L (1981) Some observations relating to the oximate ion promoted unblocking oligonucleotide aryl esters. Nucleic Acids Res 9:4611–4626

Revel M (1979) Molecular mechanisms involved in the antiviral effects of interferon. In: Gresser J (ed) Interferon 1. Academic Press, New York, pp 101–163

Roberts WK, Clemens MJ, Kerr IM (1976) Interferon-induced inhibition of protein synthesis in L-cell extracts: an ATP-dependent step in the activation of and inhibition by double stranded RNA. Proc Natl Acad Sci USA 73:3136–3140

Ruf K, Pfleiderer W (1991) Synthesis and properties of 5.6-dichlorobenzimidazole 2'-5'-and 3'-5'-nucleotide dimers and trimers. Carbohydr Res 216:421–439

Ruf K, Herdewijn P, Pfleiderer W (1987) Synthesis of 2'-5'-adenylate trimers containing 3'-modified β-D xylofuranosyl adenine derivatives at the 2'-end. Nucleosides Nucleotides 6:527–528

Saison-Behmoras T, Tocque B, Rey I, Chassignol M, Thuong NT, Helene C (1991) Short modified antisense oligonucleotides directed against Haras point mutation induce selective cleavage of the mRNA and inhibit t 24 cells proliferation. EMBO J 10:1111–1118

Sawai H, Ohno M (1981) Synthesis of oligoinosinates with 2'-5' internucleotide linkage in aqueous solution using Pb^{2+}ion. Bull Chem Soc Jpn 54:2759–2762

Sawai H, Shibata T, Ohno M (1979) Synthesis of oligonucleotide inhibitor of protein synthesis: pppA2'p5'A. Tetrahedron Lett 20:4573–4576

Sawai H, Imai J, Lesiak K, Johnston M I, Torrence P F (1983) Cordycepin analogues of 2-5 A and its derivatives. Chemical synthesis, and biological activity. J Biol Chem 258:1671–1677

Schmidt A, Zilberstein A, Shulman L, Federman P, Berissi H, Revel M (1978) Interferon action: isolation of nuclease F, a translation inhibitor activated by interferon-induced (2′-5′) oligoisoadenylate. FEBS Lett 95:257–264
Schmidt A, Chernajovsky Y, Shulman L, Federman P, Berissi H, Revel M (1979) Interferon-induced phosphodiesterase degrading (2-5) oligoisoadenylate and the CCA terminus of tRNA. Proc Natl Acad Sci USA 76:4788–4792
Schröder HC, Suhadolnik RJ, Pfleiderer W, Charubala R, Müller WEG (1992) Mini review: (2′-5′) oligoadenylate and intracellular immunity against retrovirus infection. Int J Biochem 24:55–63
Seela F, Ott J, Hißmann E (1984) (2′-5′) and (3′-5′)-Tubercidylyl-tubercidine-Synthese über Phosphit-Triester und Untersuchungen zur Sekundärstruktur. Liebigs Ann Chem 208:692–707
Sehgal PB, Pfeffer LM, Tamm I (1983) In : Came PG, Caliguiri LA (eds) Chemotherapy of viral infection. Springer, Berlin Heidelberg New York, pp 305–311
Sen GC (1984) Biochemical pathways in interferon – action. Pharmacol Ther 24:235–257
Sharma OP, Engels J, Jäger A, Crea R, Van Boom JH, Goswami BB (1983) 3′-O-methylated analogs of 2-5 A as inhibitors of virus replication. FEBS Lett 158:298–300
Shea RG, Marsters J, Bischofberger N (1990) Synthesis, hybridization properties and antiviral activity of lipid-oligodeoxynucleotide conjugates. Nucleic Acids Res 18:3777–3783
Sobol RW, Wilson SH, Charubala R, Pfleiderer W, Suhadolnik RJ (1990) 2′-5′-Oligoadenylate mediated inhibition of HIV-1 reverse transcriptase. J Interferon Res 10:66
Sobol RW, Charubala R, Schirmeister H, Kon N, Pfleiderer W, Suhadolnik RJ (1993a) Synthesis and characterisation of phosphorothioate/phosphodiester analogues of 2-5 oligoadenylate: functional characterisation of the individual internucleotide linkage with respect to RNase L activation. J Biol Chem (submitted)
Sobol RW, Fisher WL, Reichenbfich NL, Kumar A, Beard WA, Wilson SH, Charubala R, Pfleiderer W, Suhadolnik RJ (1993b) HIV-1 reverse transcriptase inhibition by 2′-5′-oligoadenylates. Biochemistry (in press)
Sobol RW, Charubala R, Pfleiderer W, Suhadolnik RJ (1993c) Chemical Synthesis and biological characterisation of Phosphorothioate analogs of 2′,5′-3′-deoxyadenylate trimer. Nucleic Acids Res 21:2437–2443
Srinivasan AR, Olson WK (1986) Conformational studies of (2′-5′) polynucleotides: theoretical computation of energy, base morphology, helical structure and duplex formation. Nucleic Acids Res 14:5461–5479
Stewart WE II (1979) The interferon system. Springer, Berlin Heidelberg New York
Suhadolnik RJ, Doetsch PW, Devash Y, Henderson EE, Charubala R, Pfleiderer W (1983) 2′-5′-Adenylate and cordycepin trimer cores: metabolioc stability and evidence for antimitogenesis without 5′-rephosphorylation. Nucleosides Nucleotides 2:351–366
Suhadolnik RJ, Lee C, Kariko K, Li SW (1987) Phosphorothioate analogues of 2′,5′-oligoadenylate. Enzymatic synthesis, properties and biological activities of 2′.5′-phosphorothioates from adenosine 5′-O-(2-thiotriphosphate) and adenosine-5′-O-(3-thiotriphosphate). Biochemistry 26:7143–7149
Suhadolnik RJ, Sobol RW, Li SW, Reichenbach NL, Haley BE (1988) 2- and 8-Azido photoaffinity probes: enzymatic synthesis, characterisation and biological properties of 2- and 8-azido photoprobes of 2-5A and photolabelling of 2-5A binding proteins. Biochemistry 27:8840–8846
Suhadolnik RJ, Lebleu B, Pfleiderer W, Charubala R, Montefiori DC, Mitchell WM, Sobol RW, Li SW, Kariko K, Reichenbach NL (1989) Phosphorothioate analogs of 2′-5′-A: activation/inhibition of RNase L and inhibition of HIV-1 reverse transcriptase. Nucleosides Nucleotides 8:987–990
Takaku H, Ueda S (1983) A convenient method for the synthesis of adenylyl-(2′-5′)-adenylyl-(2′-5′)-adenosine using 3′-O-benzoyladenosine derivatives. Bull Chem Soc Jpn 56:1424–1427
Torrence PF (1982) Molecular foundations of interferon action. Mol Aspects Med 5:129–171
Torrence PF, Imai J, Lesiak K, Jamoulle JC, Sawai H (1984) Oligonucleotide structural parameters that influence binding of 5′-O triphosphoryladenylyl-(2′-5′)-adenylyl-(2′-5′)-adenosine to the

5′-O-triphosphoryladenylyl-(2′-5′)-adenylyl-(2′-5′)-adenosine dependent endoribonuclease: chain elongation, phosphorylation state and heterocyclic base. J Med Chem 27:726–733

Torrence PF, Brozda D, Alster D, Charubala R, Pfleiderer W (1988) Only one 3′-hydroxyl group of ppp5′A2′p5′A2′p5′A2′p5′A (2-5 A) is required for activation of the 2-5 A-dependent endonuclease. J Biol Chem 263:1131–1139

Visser GM, Tromp M, Westrenen JV, Schipperus O, Van Boom JH (1986) Synthesis of some modified 2′-5′-linked oligoriboadenylates of 2-5 A core. Recueil 105:85–91

Vroom ED, Fidder A, Saris CP, Van der Marel G, Van Boom JH (1987) Preparation of individual diastereomers of adenylyl-(2′-5′)-P-thioadenylyl-(2′-5′)-adenosine and their 5′phosphorylated derivatives. Nucleic Acids Res 23:9933–2020

Will DW, Brown T (1992) Attachment of vitamin E derivatives to oligonucleotides during solid phase synthesis. Tetrahedron Lett 33:2729–2732

Williams BRG, Kerr IM, Gilbert CS, White CN, Ball LA (1978) Synthesis and breakdown of pppA2′p5′A2′p5′A and transient inhibition of protein synthesis in extracts from interferon-treated and control cells. Eur J Biochem 96:35–41S

Willis DH, Pfleiderer W, Charubala R, Suhadolnik RJ (1983) The cordycepin analog of 2′-5′-adenylate trimer core: inhibition of swarm chondrosarcoma. Fed Proc 42:443

Yoshida S, Takaku T (1986) Synthesis and properties of 2′.5′-adenylate trimers bearing 2′-terminal 8-bromo-8-hydroxyadenosine. Chem Pharm Bull 34:2456–2461

Homologies Between Different Forms of 2-5A Synthetases

E. TRUVE[1], M. KELVE[1], A. AASPOLLU[1], H.C. SCHRÖDER[2], and W.E.G. MÜLLER[2]

1 Introduction

(2′-5′) Oligoadenylate synthetases (2-5A synthetases; EC 2.7.7.19) are present in mammalian cells and tissues and synthesize from ATP a series of oligomers termed 2-5A [general formula: ppp(A2′p)$_n$A; with $1 \leq n < 18$ and usually $1 \leq n < 6$] (Hovanessian 1991). For full enzymic activity of the 2-5A synthetases, binding of double-stranded RNA is required (Sen 1982). Three principal 2-5A synthetase isoenzymes have been described with M_r's of 40–46, 69, and 100 kDa (Chebath et al. 1987; Hovanessian et al. 1987, 1988). In the following they are classified as 2-5A synthetase I [M_r 40–46 000], II [M_r 69 000] and III [M_r 100 000]. All three isoforms are induced in cells by interferon (Cohen et al. 1988; Rutherford et al. 1988). 2-5A synthetases I and II are present in both the nucleus and the mitochondria as well as in the rough/smooth microsomal fraction, while 2-5A synthetase III is associated with the rough microsomal fraction only (Hovanessian et al. 1987). The enzymic product, 2-5A, functions as an activator of the endoribonuclease L. 2-5A is rapidly degraded either by the relatively unspecific phosphodiesterase (Schmidt et al. 1979; Johnston and Hearl 1987) or the specific 2′,3′-exoribonuclease (Müller et al. 1980; Schröder et al. 1980, 1984).

Until now the 2-5A synthetase I has been cloned from humans (1.6 kb mRNA; accession numbers: X04371 and M25352; Wathelet et al. 1986) and mice (X58077: Rutherford et al. 1991; P11928: Ichii et al. 1986; M63849: Ghosh et al. 1991; M63860: Ghosh et al. 1991). Here, we present the nucleotide sequence of rat 2-5A synthetase I cDNA (Z18877). The human 2-5A synthetase II (M87284) has recently been cloned by Marié and Hovanessian (1992). No species of 2-5A synthetase III has so far been cloned.

2 Primary Structure of the Rat 2-5A Synthetase cDNA

The cDNA insert coding for 2-5A synthetase I was isolated from *Rattus norvegicus* (hippocampus cDNA library) and is 1421 nt long; it is termed

[1] Institute of Chemical Physics and Biophysics, Akadeemia tee 23, EE0026 Tallinn, Estonia
[2] Institut für Physiologische Chemie, Abteilung Angewandte Molekularbiologie, Johannes Gutenberg-Universität, Duesbergweg 6, D-55099 Mainz, Germany

Progress in Molecular and Subcellular Biology, Vol. 14
W.E.G. Müller/H.C. Schröder (Eds.)
© Springer-Verlag Berlin Heidelberg 1994

RN25ASYN; the base composition (in %) was A = 25.6, T = 20.1, G = 26.8 and C = 27.0. The open reading frame with the ATG codon for methionine is 1077 bp long and follows downstream the sequence CCGGAGGTC, a consensus sequence for the translational start (Kozak 1984). An inverted repeat GAAAGCTTTC is present close to the initiation codon (nt −13 to nt −4). The stop codon is TGA (nt +1074). The typical signal polyadenylation site AATAAA (Zarkower et al. 1986) is found at nt + 1333 to 1338 (Fig. 1). One inverted repeat each is present in the 5′- and 3′-untranslated region of RN25ASYN. No relevant homologies were found on the nt level for invertebrates in the EMBL Data Bank (CDEM33IN).

3 Amino Acid Sequence of Rat 2-5A Synthetase

The aa sequence of rat 2-5A synthetase was deduced from the nucleotide sequence of RN25ASYN cDNA (PC/Gene: TRANSL) (Fig. 2). The cDNA encodes a 41 582 dalton primary translation product, indicating that the cloned rat 2-5A synthetase belongs to class I. It consists of 12% aromatic, 35% hydrophobic and 25% charged aa, resulting in an aliphatic index of 84.6. The

```
                                       5'-end      - 69    AGCTCCACC
-   60 GCGGTGGCGCCGCGCGAGACACAGGACCTGCAGGCTGCAGAGGCAAAAGCTCCGGAGGTC
+    1 ATGGAGCAGGAACTCAGGAGCACCCCGTCCTGGAAGCTGGACAAGTTCATAGAGGTTTAC
+   61 CTCCTTCCAAACACCAGCTTCCGTGATGATGTCAAATCAGCTATCAATGTCCTGTGTGAT
+  121 TTCCTGAAGGAGAGATGCTTCCGAGATACTGTCCACCCAGTGAGGGTCTCCAAGGTGGTG
+  181 AAGGGTGGCTCCTCAGGCAAAGGCACCACACTCAAGGGCAAGTCAGACGCTGACCTGGTG
+  241 GTGTTCCTTAACAATTTCACCAGCTTTGAGGATCAGTTAAACAGACGGGGAGAGTTCATC
+  301 AAGGAAATTAAGAAACAGCTGTATGAGGTTCAGCGTGAAAAACATTTTAGAGTGAAGTTT
+  361 GAGGTCCAGAGTTCATGGTGGCCCAACCCCCGGGCTCTGAGCTTCAAGCTGAGTGCACCA
+  421 CACCTCCAACAGGAGGTGGAGTTTGATGTGCTTCCAGCCTATGATGTCCTAGGTCATGTT
+  481 AGCCTCTACAGCAATCCTGATCCCAAGATCTACACCATCCTCATCTCCGAATGTATCTCC
+  541 CTGGGGAAGGATGGCGAGTTCTCTACCTGCTTCACGGAGCTCCAGAGGAACTTCCTGAAG
+  601 CAGCGCCCAACCAAGCTGAAGAGTCTCATCCGCCTGGTCAAGCACTGGTACCAACTGTGT
+  661 AAGGAGAAGCTGGGGAAGCCGCTGCCCCCACAGTACGCCCTGGAGCTGCTCACGGTCTAT
+  721 GCCTGGGAACGTGGAAATGGAATTACTGAGTTCAACACAGCTCAGGGCTTCCGGACAATC
+  781 TTGGAACTGGTCACAAAGTACCAGCAGCTTCGAATCTACTGGACAAAGTATTATGACTTT
+  841 CAACACCCAGATGTCTCCAAATACCTACACAGACAGCTCAGAAAATCCAGGCCTGTGATC
+  901 CTGGACCCTGCTGACCCAACAGGGAACGTGGCTGGTGGGAACCAAGAAGGCTGGCGGCGG
+  961 TTGGCCTCAGAGGCGAAGCTGTGGCTGCAGTACCCATGTTTTATGAACACCGGTGGTTCC
+1021 CCAGTGAGTTCCTGGGAAGTGCCGGTGGATGAGGCCTGGTCATGCATCCTGCTGTGA #
+1078 ACCCAGCAGCACCAGCCCAGGAGGCTCCGGAGTCAGGGGCACGTGCTGCTCTGCTGCAGG
+1138 ACCTTGACACAGTGAGGGAGGCCCCACTCGGGATCACAGTCCATGCTTCTGATGCCCGCC
+1198 CGCCATGGTTGAATACTGTCCAATCACAGACAGCCTTCCTCAACAGATTCAGAAGGGGCG
+1258 GAAAGAACTCAAGCTTGACTTCCATCTGACCGTCCACCTGTTGGGAGGTTCTGTCCAACC
+1318 ATGTCTGTCAACAACAATAAAGTACAGCAGGTGCC(A)n       3'-end
```

Fig. 1. Nucleotide sequence analysis of RN25ASYN, a 1421-nt-long cDNA fragment (accession number: Z18877) which encodes rat 2-5A synthetase I. The putative stop codon is indicated (#). The putative polyadenylation signal is *double underlined*. The inverted repeats in the 5′- and 3′-untranslated region are in *boldface* and *underlined*

```
RN25ASYN      MEQELRSTPSWKLDKFIEVYLLPNTSFRDDVKSAINVLCDFLKERCFRDT   50
MM25ASYN      -------------DKFIEDYLLPDTTFGADVKSAVNVVCDFLKERCFQGA   37
O25S-MOUSE    MEHGLRSIPAWTLDKFIEDYLLPDTTFGADVKSAVNVVCDFLKERCFQGA   50
HS25ASYN      M-MDLRNTPAKSLDKFIEDYLLPDTCFRMQINHAIDIICGFLKERCFRGS   49
MMOLISAA      M-MDLRNTPAKSLDKFIEEYLLPDTCFRMQINHAIDIICGFLKERCFRGS   49
MMOLISAB      M-MDLRNTPAKSLDKFIEDYLLPDTCFRMQINHAIDIICGFLKERCFRGS   49
                           ***** ****.* *  ... *....*.*******...

RN25ASYN      VHPVRVSKVVKGGSSGKGTTLKGKSDADLVVFLNNFTSFEDQLNRRGEFI  100
MM25ASYN      AHPVRVSKVVKGGSSGKGTTLKGKSDADLVVFLNNLTSFEDQLNRRGEFI   87
O25S-MOUSE    AHPVRVSKVVKGGSSGKGTTLKGRSDADLVVFLNNLTSFEDQLNRRGEFI  100
HS25ASYN      SYPVCVSKVVKGGSSGKGTTLRGRSDADLVVFLSPLTTFQDQLNRRGEFI   99
MMOLISAA      SYPVCVSKVVKGGSSGKGTTLRGRSDADLVVFLSPLTTFQDQLNRRGEFI   99
MMOLISAB      SYPVCVSKVVKGGSSGKGTTLRGRSDADLVVFLSPLTTFQDQLNRRGEFI   99
               .** ****************.*.*********. .*.*.**********

RN25ASYN      KEIKKQLYEVQREKHFRVKFEVQ-SSWWPNPRALSFKLSAPHLQQEVEFD  149
MM25ASYN      KEIKKQLYEVQHERRFRVKFEVQ-SSWWPNARSLSFKLSAPHLHQEVEFD  136
O25S-MOUSE    KEIKKQLYEVQHERRFRVKFEVQ-SSWWPNARSLSFKLSAPHLHQEVEFD  149
HS25ASYN      QEIRRQLEACQRERAFSVKFEVQ-APRWGNPRALSFVLSSLQLGEGVEFD  148
MMOLISAA      QEIRRQLEACQRERAFSVKFEVQ-APRWGNPRALSFVLSSLQLGEGVEFD  148
MMOLISAB      QEIRRQLEACQRERAFSVKFEVQEAPRWGNPRALSFVLSSLQLGEGVEFD  149
              .**..** . *.*. *.****** ...* *.*.*** **. .* ..****

RN25ASYN      VLPAYDVLGHVSLYSNPDPKIYTILISECISLGKDGEFSTCFTELQRNFL  199
MM25ASYN      VLPAFDVLGHGSINKKPNPLIYTILIWECTSLGKDGEFSTCFTELQRNFL  186
O25S-MOUSE    VLPAFDVLGHVNTSSKPDPRIYAILIEECTSLGKDGEFSTCFTELQRNFL  199
HS25ASYN      VLPAFDALGQLTGSYKPNPQIYVKLIEECTDLQKEGEFSTCFTELQRDFL  198
MMOLISAA      VLPAFDALDQLTGSYKPNPQIYVKLIEECTDLQKEGEFSTCFTELQRDFL  198
MMOLISAB      VLPAFDALGQLTGSYKPNPQIYVKLIEDCTDLQKEGEFSTCFTELQRDFL  199
              ****.*.*.. .   .*.* **. ** .*..* *.*************.**

RN25ASYN      KQRPTKLKSLIRLVKHWYQLCKEKLGKPLPPQYALELLTVYAWERGNGIT  249
MM25ASYN      KQRPTKLKSLIRLVKHWYQLCKEKLGKPLPPQYALELLTVYAWEQGNGCN  236
O25S-MOUSE    KQRPTKLKSLIRLVKHWYQLCKEKLGKPLPPQYALELLTVFAWEQGNGCY  249
HS25ASYN      KQRPTKLKSLIRLVKHWYQNCKKKLGK-LPPQYALELLTVYAWERGSMKT  247
MMOLISAA      KQRPTKLKSLIRLVKHWYQNCKKKLGK-LPPQYALELLTVYAWERGSMKT  247
MMOLISAB      KQRPTKLKSLIRLVKHWYQNCKKKLGK-LPPQYALELLTVYAWERGSMKT  248
              ******************* **.**** ************.***.*.

RN25ASYN      EFNTAQGFRTILELVTKYQQLRIYWTKYYDFQHPDVSKYLHRQLRKSRPV  299
MM25ASYN      EFNTAQGFRTVLELVINYQHLRIYWTKYYDFQHKEVSKYLHRQLRKARPV  286
O25S-MOUSE    EFNTAQGFRTVLELVINYQHLRIYWTKYYDFQHQEVSKYLHRQLRKARPV  299
HS25ASYN      HFNTAQGFRTVLELVINYQQLCIYWTKYYDFKNPIIEKYLRRQLTKPTPV  297
MMOLISAA      HFNTAQGFRTVLELVINYQQLCIYWTKYYDFKNPIIEKYLRRQLTKPRPV  297
MMOLISAB      HFNTAQGFRTVLELVINYQQLCIYWTKYYDFKNPIIEKYLRRQLTKPRPV  298
              .*********.****..**.* *********..  ..***.*** *. **

RN25ASYN      ILDPADPTGNVAGGNQEGWRRLASEAKLWLQYPCFMNTGGSPVSSWEV--  347
MM25ASYN      ILDPADPTGNVAGGNPEGWRRLAEEADVWLWYPCFMKNDGSRVSSWDV--  334
O25S-MOUSE    ILDPADPTGNVAGGNPEGWRRLAEEADVWLWYPCFIKKDGSRVSSWDV--  347
HS25ASYN      ILDPADPTGNLGGGDPKRWRQLAQEAEAWLNYPCFKNWDGSPVSSWILLV  347
MMOLISAA      ILDPADPTGNLGGGDPKGWRQLAQEAEAWLNYPCFKNWDGSPVSSWILLV  347
MMOLISAB      ILDPADPTGNLGGGDPKGWRQLAQEAEAWLNYPCFKNWDGSPVSSWILLT  348
              **********..**... **.** **. ** **** . .**.**** .

RN25ASYN      ----P---------------------------------------------  348
MM25ASYN      ----PTVVP---------------------------VPFE----------  343
O25S-MOUSE    ----PTVVP---------------------------VPFE----------  356
HS25ASYN      RPPASS------------------------------LPF-----------  356
MMOLISAA      RPPASS------------------------------LPF-----------  356
MMOLISAB      QHTPGSIHPTGRRGLDLHHPLNASASWGKGLQCYLDQFLHFQVGLLIQRR  398

RN25ASYN      -VDEAWSCILL------  358
MM25ASYN      QVEENWTCILL------  354
O25S-MOUSE    QVEENWTCILL------  367
HS25ASYN      ----------IPAPLHEA  364
MMOLISAA      ----------IPAPLHEA  364
MMOLISAB      QSSSVSWCIIQDRTQVS  415
```

Fig. 2. Homologies between the deduced aa sequence of RN25ASYN with the following 2-5A synthetases I: *MM25ASYN* (mouse mRNA for 2-5A synthetase L2; accession number: X58077; Rutherford et al. 1991); *O25S-MOUSE* (mouse 2-5A synthetase 1; P11928; Ichii et al. 1986); *HS25ASYN* (human 1.6-kb mRNA for 2-5A synthetase; X04371, M25352; Wathelet et al. 1986); *MMOLISAA* (mouse 2-5A synthetase gene; M63849; Ghosh et al. 1991); *MMOLISAB* (mouse 2-5A synthetase gene; M63860; Ghosh et al. 1991). *Stars* indicate identical aa; *dots* well-conserved aa; *dashes* arbitrary gaps. The putative ATP/GTP-binding motif in RN25ASYN and MM25ASYN is *double underlined*

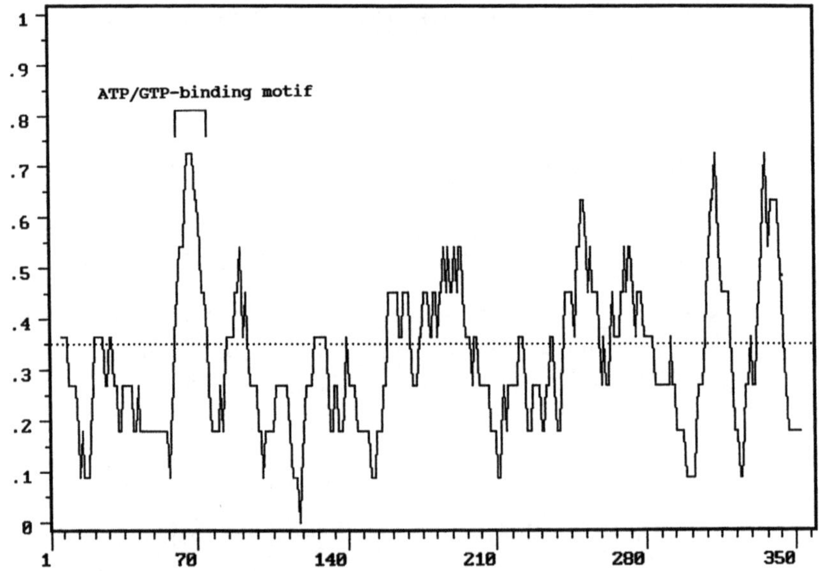

Fig. 3. Plot of C, G, N, Q, S, T, Y (polar/neutral aa) proportion for the aa sequence of RN25ASYN. The ATP/GTP-binding motif is indicated

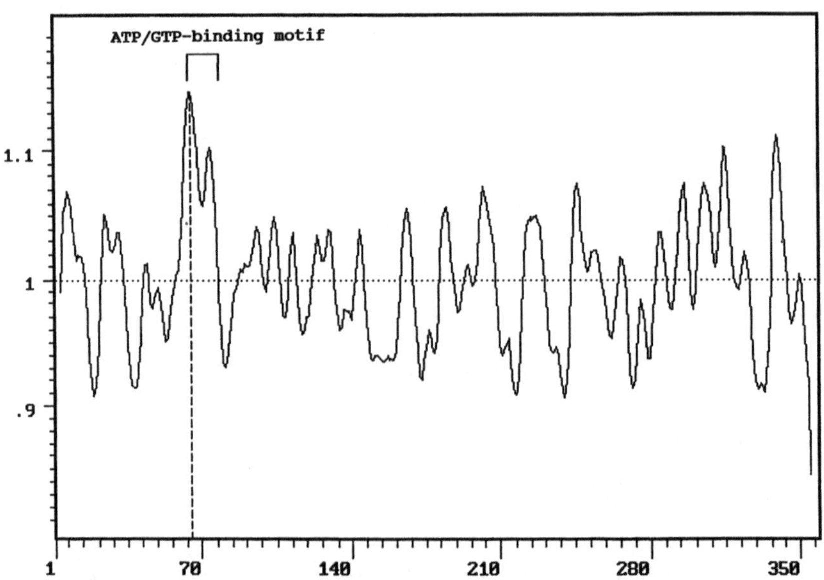

Fig. 4. Plot of the prediction of chain flexibility for the aa sequence of RN25ASYN. The ATP/GTP-binding motif is indicated

```
           60        70        80        90       100
VHPVRVSKVVKGGSSGKGTTLKGKSDADLVVFLNNFTSFEDQLNRRGEFI
------------>**>------>**XX------*****XXXXXXXXXXXX
------------>**>------>**XX------*****XXXXXXXXXXXX
```

Fig. 5. Semigraphic plot of protein secondary structure prediction around the ATP/GTP-binding motif (*double underlined*) of the aa sequence of RN25ASYN. Symbols: ×, helical; –, extended; >, turn; *, coil conformation

isoelectric point (pI) is 9.22. The protein can be considered according to the instability index as a stable protein with a half-life of 36.6 h (PC/Gene: PHYSCHEM); no "good" PEST region (rich in the aa Pro, Glu, Ser and Thr) is present (PC/Gene: PESTFIND). Experimental evidence obtained from Chang and Wu (1991) from human cells indicated that the half-life of the 2-5A synthetase I is >2 h. No strong secretory signal sequence was found. The deduced aa sequence of RN25ASYN displays, like other class I 2-5A synthetases, potential protein kinase C phosphorylation sites, casein kinase II phosphorylation sites and N-myristoylation sites (PC/Gene: PROSITE).

In contrast to human 1.6 kb mRNA for 2-5A synthetase (HS25ASYN; Wathelet et al. 1986) and mouse 2-5A synthetase gene (MMOLISAB; Ghosh et al. 1991), the rat RN25ASYN is provided both with additional N-glycosylation sites and one farnesyl group binding site of the sequence CILL (nt + 355 to + 358). These two types of sites are also present in the mouse 2-5A synthetase L2 mRNA (MM25ASYN; Rutherford et al. 1991) and mouse 2-5A synthetase 1 (O25S-MOUSE; Ichii et al. 1986). The existence of a binding site for a farnesyl group may indicate that rat 2-5A synthetase is membrane-associated (Sinensky and Lutz 1992).

The ATP/GTP-binding motif A (P-loop) (Moller and Amons 1985), GTTLKGKS, is found in RN25ASYN from nt + 68 to + 75 and is present also in the sequence MM25ASYN. Analyses of the aa around the putative ATP/GTP-binding motif displayed the following characteristics. (1) Determination of the clustering of aa groups according to their physico-chemical properties showed that the site of the motif is the most polar neutral part of the total sequence (PC/Gene: PRESIDUE) (Fig. 3). (2) The motif is located in the groove of the protein segment with the highest flexibility and flanked by the segments aa 62 to 68 (GGSSGKG) and aa 71 to 77 (LKGKSDA) (PC/Gene: FLEXPRO) (Fig. 4). (3) In the predicted secondary structure of the rat 2-5A synthetase, the motif which displays extended character is surrounded by turns (Fig. 5) (PC/Gene: GARNIER; Garnier et al. 1978).

Ghosh et al. (1991) have analyzed the region within two mouse 2-5A synthetases I (MMOLISAA-MMOLISAB) to which dsRNA binds. They delimited it to the aa residue 104 to 158, corresponding to the segment aa 105 to 159 in rat 2-5A synthetase RN25ASYN. Analysis of this region by PC/Gene: PRESIDUE revealed a clustering of the aa Arg-Lys, known to be crucial for RNA binding (Fig. 6). The deduced aa sequence of RN25ASYN was further analyzed by the

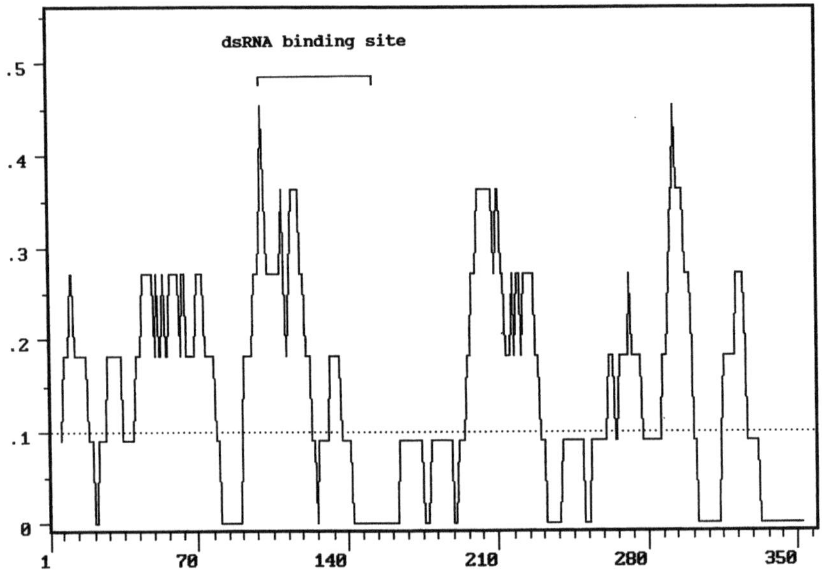

Fig. 6. Putative dsRNA binding site in the Arg-Lys-rich region of the deduced aa sequence of RN25ASYN according to data published by Ghosh et al. (1991)

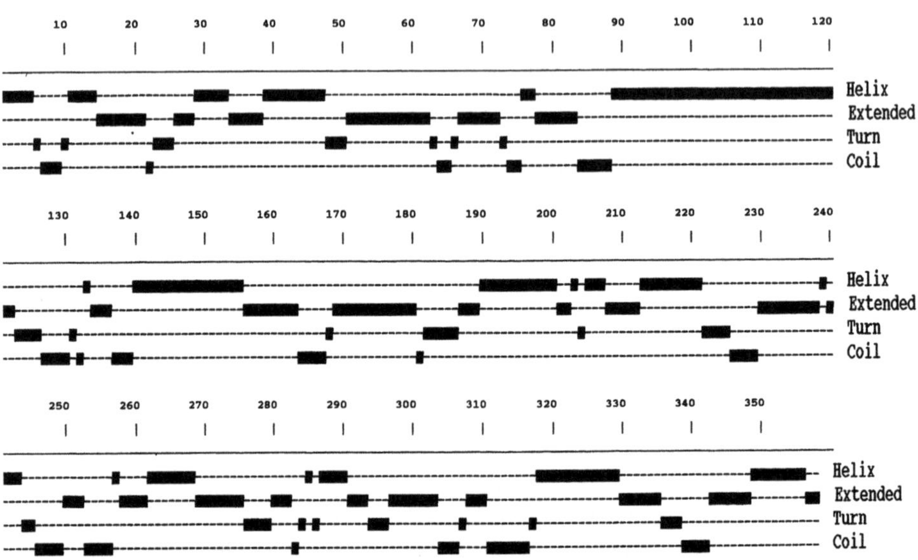

Fig. 7. Secondary structure of rat 2-5A synthetase by Garnier analysis. The helical, extended, turn and coil conformations along the deduced aa sequence are depicted

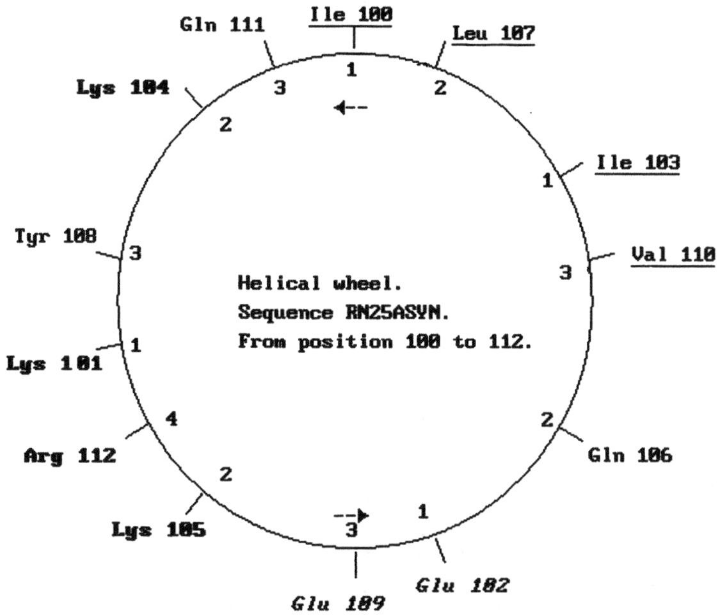

Fig. 8. Helical wheel analysis of the putative dsRNA binding site of the deduced aa sequence of RN25ASYN (aa 100–112). The basic aa Lys and Arg are in *bold face*; the acidic aa Glu in *italics*; the hydrophobic aa are *underlined*; the others are polar aa. The figures within the wheel denote the number of turns; a periodicity of four residues is assumed

procedure of Garnier (PC/Gene: GARNIER). The putative dsRNA binding site was found to be the longest helical stretch, spanning from aa 89 to 120 (Fig. 7). This part of the sequence is interesting because "helical wheel" analysis (PC/Gene: HELWEEL) revealed that the Arg-Lys residues are clustered on one side of the helix, while the hydrophobic aa are located on the opposite side (Fig. 8).

4 Comparison with Other Sequences of 2-5A Synthetases

The rat 2-5A synthetase I, RN25ASYN, was compared with five human and mouse sequences on the deduced aa level (Fig. 2). The comparisons revealed a high homology. The consensus length is 417 aa; among them 220 aa (52.8%) show identity and 75 aa (18%) similarity. If the sequence aa 14 to 347 (using the rat clone RN25ASYN) is selected, the value for identity among the six sequences increases to 66% and similarity to 23%. If the rat RN25ASYN sequence is compared with mouse MM25ASYN, the identity on both the nt and aa level is almost 90%.

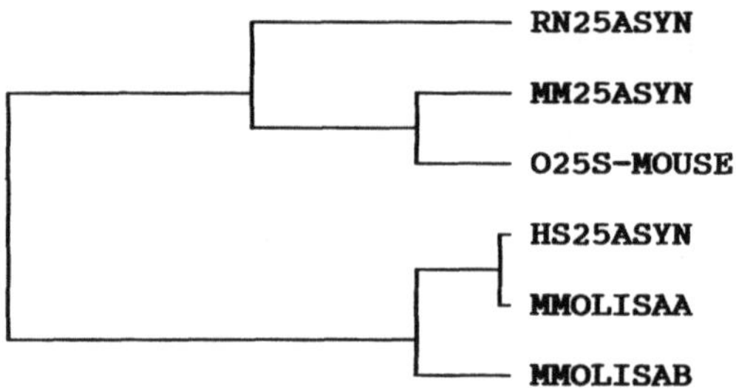

Fig. 9. Dendrogram based on the comparisons of the deduced aa sequences of 2-5A synthetases I shown in Fig. 2

```
RN25ASYN      M---EQELRSTPSWKL----------------------------------   13
HSSYN69KD     MGNGESQLSSVPAQKLGWFIQEYLKPYEECQTLIDEMVNTICDVCRNPEQ   50
              *   *  **  *  **

RN25ASYN      --------------------------------------------------   13
HSSYN69KD     FPLVQGVAIGGSYGRKTVLRGNSDGTLVLFFSDLKQFQDQKRSQRDILDK  100

RN25ASYN      --------------------------------------------------   13
HSSYN69KD     TGDKLKFCLFTKWLKNNFEIQKSLDGSTIQVFTKNQRISFEVLAAFNALS  150

RN25ASYN      --------------------------------------------------   13
HSSYN69KD     LNDNPSPWIYRELKRSLDKTNASPGEFAVCFTELQQKFFDNRPGKLKDLI  200

RN25ASYN      --------------------------------------------------   13
HSSYN69KD     LLIKHWHQQCQKKIKDLPSLSPYALELLTVYAWEQGCRKDNFDIAEGVRT  250

RN25ASYN      --------------------------------------------------   13
HSSYN69KD     VLELIKCQEKLCIYWMVNYNFEDETIRNILLHQLQSARPVILDPVDPTNN  300

RN25ASYN      ------------------------------------------------DK   15
HSSYN69KD     VSGDKICWQWLKKEAQTWLTSPNLDNELPAPSWNVLPAPLFTTPGHLLDK  350
                                                              **

RN25ASYN      FIEVYLLPNTSFRDDVKSAINVLCDFLKERCFRDTVHPVRVSKVVKGGSS   65
HSSYN69KD     FIKEFLQPNKCFLEQIDSAVNIIRTFLKENCFRQSTAKI---QIVRGGST  397
              **   * **  *     ** *    **** ***           * ***

RN25ASYN      GKGTTLKGKSDADLVVFLNNFTSFEDQLNRRGEFIKEIKKQLYEVQREKH  115
HSSYN69KD     AKGTALKTGSDADLVVFHNSLKSYTSQKNERHKIVKEIHEQLKAFWREKE  447
               *** **  ******** *   *   * * *    ***  **    **

RN25ASYN      FRVKFEVQSSWWPNPRALSFKLSAPHLQQEVEFDVLPAYDVLGHVSLYSN  165
HSSYN69KD     EELEVSFEPPKWKAPRVLSFSLKSKVLNESVSFDVLPAFNALGQLSSGST  497
                         *  ** *** *    *   * ******   **  *  *

RN25ASYN      PDPKIYTILISECISLG-KDGEFSTCFTELQRNFLKQRPTKLKSLIRLVK  214
HSSYN69KD     PSPEVYAGLIDLYKSSDLPGGEFSTCFTVLQRNFIRSRPTKLKDLIRLVK  547
              * *  *  **    *     ******** *****   ****** ******

RN25ASYN      HWYQLCKEKLGKP---LPPQYALELLTVYAWERGNGITEFNTAQGFRTIL  261
HSSYN69KD     HWYKECERKL-KPKGSLPPKYALELLTIYAWEQGSGVPDFDTAEGFRTVL  596
              ***  *  ** **   *** ******* **** * *   * ** **** *

RN25ASYN      ELVTKYQQLRIYWTKYYDFQHPDVSKYLHRQLRKSRPVILDPADPTGNVA  311
HSSYN69KD     ELVTQYQQLGIFWKVNYNFEDETVRKFLLSQLQKTRPVILDPGEPTGDVG  646
              **** **** * *   * *    * * *  ** * *******  *** *

RN25ASYN      GGNQEGWRRLASEAKLWLQYPCFMNTGGSPVSSWEVPVDEAWSCILL     358
HSSYN69KD     GGDRWCWHLLDKEAKVRLSSPCFKDGTGNPIPPWKVPVKVI           687
              **    *  *  **   *  ***    * *   * ***
```

Fig. 10. Homologies between the deduced aa sequence of RN25ASYN (2-5A synthetase I) and the aa sequence of the human 69-kDa (2′-5′)oligoadenylate synthetase (p69 2-5A synthetase) (HSSYN69KD; M87284; Marié and Hovanessian 1992)

Multiple sequence alignments were performed to obtain a dendrogram and to estimate the relative relationships between the five 2-5A synthetase clones, applying the CLUSTAL program (Fig. 9). The closest relationship was found between RN25ASYN and mouse MM25ASYN and O25S-MOUSE; common to all three sequences is that they are provided with the N-glycosylation sites and the farnesyl group binding site, none of which is present in any of the other sequences.

The deduced rat 2-5A synthetase aa sequence, RN25ASYN, was compared with the only available sequence of 2-5A synthetase class II, the human 69-kDa 2-5A synthetase (HSSYN69KD; Marié and Hovanessian 1992). The comparison revealed that the first half of the human sequence does not show homology with the rat sequence, while the second half shows a remarkably high homology; on both the nt and aa level the identity was 60 and 51%, respectively (Fig. 10). The high identity of the second half to 2-5A synthetase I has been previously reported leading to the suggestion that 2-5A synthetase II might have two catalytic domains (Marié and Hovanessian 1992).

5 Summary

Sequence analyses of 2-5A synthetases of class I (M_r 40000–46000) revealed high homology among them. The cDNA coding for the M_r 69000 2-5A synthetase of class II displayed in the second half a likewise high homology to the complete sequences of class I enzymes. This high degree of conservation of the 2-5A synthetases supports the assumption that these enzymes play important roles during virus infection (Williams et al. 1979; Coccia et al. 1990) and in the control of growth and differentiation of mammalian cells (Williams and Silverman 1985).

Acknowledgements. We thank Dr. Hakan Persson (Karolinska Institute, Stockholm) for providing us with the rat hippocampus cDNA library and Prof. Mart Saarma (Helsinki University) for thorough and helpful discussions. This work was supported by a grant from the Sandoz Foundation for Gerontological Research (VRF-92-2-3) and the Deutscher Akademischer Austauschdienst.

References

Benech P, Mory Y, Revel M, Chebath J (1985) Structure of two forms of the interferon-induced (2′-5′) oligo A synthetase of human cells based on cDNAs and gene sequences. EMBO J 4:2249–2256

Chang CC, Wu JM (1991) Modulation of antiviral activity of intereferon and 2′,5′-oligoadenylate synthetase gene expression by mild hyperthermia (39.5°C) in cultured human cells. J Biol Chem 266:4605–4612

Chebath J, Benech P, Hovanessian A, Galabru J, Revel M (1987) Four different forms of interferon-induced 2′,5′-oligo(A) synthetase identified by immunoblotting in human cells. J Biol Chem 262:3852–3857

Coccia EM, Romeo G, Nissim A, Marziali G, Albertini R, Affabris E, Battistini A, Fiorucci G, Orsatti R, Rossi GB, Chebath J (1990) A full-length murine 2-5A synthetase cDNA transfected in NIH-3T3 cells impairs EMCV but not VSV replication. Virology 179:228–233

Cohen B, Peretz D, Vaiman D, Benech P, Chebath J (1988) Enhancer-like interferon responsive sequences of the human and murine (2′-5′)oligoadenylate synthetase gene promotors. EMBO J 7:1411–1419

Garnier J, Osguthorpe DJ, Robson B (1978) Analysis of the accuracy and implications of simple methods for predicting the secondary structure of globular proteins. J Mol Biol 120:97–120

Ghosh SK, Kusari J, Bandyopadhyay SK, Samanta H, Kumar R, Sen GC (1991) Cloning, sequencing, and expression of two murine 2′-5′-oligoadenylate synthetases: structure-function relationship. J Biol Chem 266:15293–15299

Hovanessian AG (1991) Interferon-induced and double-stranded RNA-activated enzymes: a specific protein kinase and 2′,5′-oligoadenylate synthetase. J Interferon Res 11:199–205

Hovanessian AG, Laurent AG, Chebath J, Galabru J, Robert N, Svab J (1987) Identification of 69-kd and 100-kd forms of 2-5A-synthetase from interferon-treated human cells by specific monoclonal antibodies. EMBO J 5:1273–1280

Hovanessian AG, Svab J, Marié I, Robert N, Chamaret S, Laurent AG (1988) Characterization of 69- and 100-kDa forms of 2-5A-synthetase from interferon-treated human cells. J Biol Chem 263:4945–4949

Ichii Y, Fukunaga R, Shiojiri S, Sokawa Y (1986) Mouse 2-5A synthetase cDNA: nucleotide sequence and comparison to human 2-5A synthetase. Nucleic Acids Res 14:10117

Johnston MI, Hearl WG (1987) Purification and characterization of a 2′-phosphodiesterase from bovine spleen. J Biol Chem 262:8377–8382

Kozak M (1984) Compilation analysis of sequences upstream from the translational start site in eukaryotic mRNAs. Nucleic Acids Res 12:857–872

Laurent G St, Yoshie O, Floyd-Smith G, Samanta H, Sehgal PB (1983) Interferon action: two (2′-5′)$(A)_n$ synthetases specified by distinct mRNAs in Ehrlich ascites tumor cells treated with interferon. Cell 33:95–102

Marié I, Hovanessian AG (1992) The 69-kDa 2-5A synthetase is composed of two homologous and adjacent functional domains. J Biol Chem 267:9933–9939

Moller W, Amons R (1985) Phosphate-binding sequences in nucleotide-binding proteins. FEBS Lett 186:1–7

Müller WEG, Schröder HC, Zahn RK, Dose K (1980) Degradation of 2′,5′-linked oligoriboadenylates by 3′-exoribonuclease and 5′-nucleotidase. Hoppe-Seyler's Z Physiol Chem 361:469–472

PC/Gene (1991) User and reference manual; release 6.5. IntelliGenetics, Mountain View, CA 2, 1991

Rutherford MN, Hannigan GE, Williams BRG (1988) Interferon-induced binding of nuclear factors to promoter elements of the 2-5A synthetase gene. EMBO J 7:751–759

Rutherford MN, Kumar A, Nissim A, Chebath J, Williams BRG (1991) The murine 2-5A synthetase locus: three distinct transcripts from two linked genes. Nucleic Acids Res 19:1917–1924

Schmidt A, Chernajovsky Y, Shulman L, Federman P, Berissi H, Revel M (1979) An interferon-induced phosphodiesterase degrading (2′-5′)oligoadenylate and the C-C-A terminus of tRNA. Proc Natl Acad Sci USA 76:4788–4792

Schröder HC, Zahn RK, Dose K, Müller WEG (1980) Purification and characterization of a poly (A)-specific exoribonuclease from calf thymus. J Biol Chem 255:4535–4538

Schröder HC, Gosselin G, Imbach J-L, Müller WEG (1984) Influence of the xyloadenosine analogue of 2′,5′-oligoriboadenylate degrading 2′,3′-exoribonuclease and further enzymes involved in poly(A) (+)mRNA metabolism. Mol Biol Rep 10:83–89

Schröder HC, Wenger R, Kuchino Y, Müller WEG (1989) Modulation of nuclear matrix-associated (2′-5′)oligoadenylate metabolism and ribonuclease L activity in H9 cells by human immunodeficiency virus. J Biol Chem 264:5669–5673

Sen GC (1982) Mechanism of interferon action: progress towards its understanding. Prog Nucleic Acid Res Mol Biol 27: 105–156

Sinensky M, Lutz RJ (1992) The prenylation of proteins. Bioessays 14:25–31
Wathelet M, Moutschen S, Cravador A, Dewit L, Defilippi P, Huez G, Content J (1986) Full-length sequence and expression of the 42 kDa 2-5A synthetase induced by human interferon. FEBS Lett 196:113–120
Williams BRG, Silverman RH (eds) (1985) The 2-5A System. Liss, New York
Williams BRG, Golgher RR, Brown RE, Gilbert CS, Kerr IM (1979) Natural occurrence of 2-5A in interferon-treated EMC virus-infected L cells. Nature 282:582–586
Zarkower D, Stephenson P, Sheets M, Wickens M (1986) The AAUAAA sequence is required both for cleavage and for polyadenylation of simian virus 40 pre-mRNA in vitro. Mol Cell Biol 6:2317–2323

2-5A and Virus Infection

N. Fujii[1]

1 Introduction

Many cells die of viral infections in in vitro systems. Even a far larger organism may succumb if its defenses do not function against a virus. Fortunately, cells and living organisms are provided with various defense mechanisms against viral infection. The most important of these is interferon (IFN), which performs a multitude of functions, particularly in the regulation of cell growth, cell differentiation and immune responses (Gastl and Huber 1988).

At the cellular level, the function of IFN appears mainly to be antiviral activity. Consequently, viral infection may be thought of as the endpoint of a long battle between the virus and the antiviral activity of IFN. Viruses, in other words, have attained the ability to replicate despite the hostile activity of IFN. The cells, on the other hand, counteract viral infections through several antiviral systems developed over millions of years of evolution.

IFNs are secreted from virus-infected cells and, in turn, they act on other cells to induce the antiviral state. Indeed, IFNs play a crucial role in the establishment of the antiviral state. They are classified into four major classes, IFN-α, IFN-β, IFN-γ and IFN-ω, based on their biological activity and antigenicity. The location of the IFN genes on the chromosomes and genetic structures (nucleotide sequences) have been elucidated (Capon et al. 1985; Weissmann and Weber 1986; Taniguchi 1988). Studies have been conducted to explore the receptors (Aguet 1990; Uze et al. 1990; Lutfalla et al. 1992) and signal transduction mechanisms (Klein et al. 1990; Reich and Pfeffer 1990; Hannigan and Williams 1991; Pfeffer et al. 1991) of IFN. Molecular biology studies have also continued to explore the mechanism of expression of certain genes by IFN (Reid et al. 1989; Vaiman et al. 1990; Kerr and Stark 1991).

The multiple functions of IFN manifest frequently via several proteins, which are manufactured for the first time or in increased quantity in the cells treated by IFN. Only a few of these proteins have been well characterized (Staeheil 1990). Antiviral activity of IFN is also shown to be due to intracellular induction of 2′,5′-oligoadenylate synthetase (2-5AS), P1 kinase and Mx protein, after IFN

[1] Department of Microbiology, Sapporo Medical College, South 1 West 17, Sapporo 060, Hokkaido, Japan

Progress in Molecular and Subcellular Biology, Vol. 14
W.E.G. Müller/H.C. Schröder (Eds.)
© Springer-Verlag Berlin Heidelberg 1994

binds to certain specific cell surface receptors. Other mechanisms which are thought to be mediated by IFN-induced proteins also appear to operate in the antiviral functions. The multiplication of viruses is often inhibited by the action of some single systems or the cooperative action of several systems. These work to inhibit any of various steps in replication, for example, the uncoating of the viruses, the transcription or translation of virus mRNA or the maturation of the virus particles.

Though most DNA and RNA viruses are sensitive to the antiviral functions of IFN, some are not. It may simply be that the host in one host-viral combination has not yet had the time to evolve a defense against that particular virus, but often, the attacking virus itself has evolved the ability to suppress the 2-5A/RNase L system and the P1 kinase system. Breakdown of the antiviral systems might result in an increase in the sensitivity of host cells to superinfection of other types of exogenous viruses and to activation of the endogenous viruses. It is up to medical science to discover a way of preventing the multiplication of these viruses which have the capacity to resist the antiviral defenses provided by IFN. There are data suggesting that 2′,5′-oligoadenylate (2-5A) oligomers and its analogues introduced into these virus-infected cells restore the antiviral state. We have only begun to explore the mechanisms of suppression of viruses by IFN.

2 Antiviral Action of Interferon

Human antiviral systems induced by each type of IFN consist mainly of the 2-5A/RNase L system, the P1 kinase system and the Mx protein system.

The Mx proteins in humans are MxA (76 kDa) and MxB (73 kDa). Mx1 (72 kDa) and Mx2 (72 kDa) proteins are also found in mice. Furthermore, proteins closely homologous to Mx have been identified in rats, fish and yeasts (Staeheli 1990)

Human MxA protein exhibits antiviral activity against the influenza virus (IFV) and vesicular stomatitis virus (VSV), but not against mengovirus (MGV), encephalomyocarditis virus (EMCV) and semlikiforest virus (SFV) (Pavlovic et al. 1990). Mouse Mx1 protein exerts an almost selective antiviral activity against IFV (Garber et al. 1991). MxA protein is present in the cytoplasm where it can inhibit posttranscriptional steps of virus replication (the transit of mature mRNA from intranuclear sites to the cytoplasm). Mx1 protein exists in the nucleus, where it can prevent the transcription of virus mRNA (Pavlovic et al. 1992).

P1 kinase has a molecular weight of 68 kDa in humans and 65 kDa in mice (Staeheli 1990; Samuel 1991). Although a small amount of P1 kinase is naturally present in cells not treated with IFN, IFN treatment incites copious production of P1 kinase. When this is coexistent with dsRNA (activator RNA) from an invading virus, this enzyme is autophosphorylated, followed by the phosphorylation of the eIF-2 subunit. In other words, both treatment by IFN and viral infection are required for P1 kinase activation in a given cell. The production of activator dsRNA in IFN-treated cells has been proposed in virus-infected cells with either EMCV, IFV, reovirus, vaccinia virus or simian virus 40 (SV 40), because of the observed large increase in the concentration of 2-5A oligomers and the large extent of eIF-2α phosphorylation (Williams et al. 1979a; Nilsen et al. 1982; Hersh et al 1984; Penn and Williams 1984; Samuel 1984; Rice et al. 1985). Phosphorylation of eIF-2 inhibits virus protein synthesis at the initiation step of translation (Pain 1986). This system has been shown to inhibit the multiplication of VSV, picornavirus, reovirus and adenovirus (Whitaker-Dowling and Youngner 1983; Schneider et al. 1985; Imani and Jacobs 1988). Surprisingly, however, substances inhibiting P1 kinase are produced by IFV, poliovirus, human immunodeficiency virus (HIV), adenovirus and vaccinia virus. These substances are proteins which bind to dsRNA, proteins with a high degree of homology with eIF-2 and other substances which degrade P1 kinase or bind to P1 kinase (Paez and Esteban 1984a; Imani and Jacobs 1988; Bischoff and Samuel 1989; Black et al. 1989; Lee et al. 1990; Mellits et al. 1990; Roy et al. 1990, 1991; Davies et al. 1992).

The second of the two IFN-induced, dsRNA-dependent enzymes is 2′,5′-oligoadenylate synthetase (2-5AS). After the activation of 2-5AS by dsRNA, this enzyme catalyzes the synthesis of 2′,5′-oligoadenylate oligomers (2-5A). The latent endoribonuclease RNase L functions to degrade cellular and virus RNA in the presence of 2-5A. The 2-5A/RNase L system, which consists of the two enzymes 2-5AS and RNase L, is thought to participate in the inhibition of multiplication of MGV, EMCV and vaccinia virus (Chebath et al. 1987b; Grun et al. 1987; Rysiecki et al. 1989).

Replication of SV40, hepatitis type-B virus (HBV) and herpes simplex virus (HSV) is inhibited mainly by an unknown system.

It seems likely that these antiviral systems function in concert with each other (Taira et al. 1985; Kumar et al. 1987; Lewis 1988). In fact, the antiviral activity of IFN on EMCV is effectively suppressed by superinfection by vaccinia virus which has the ability to inhibit P1 kinase (Whitaker-Dowling and Youngner 1986; Imani and Jacobs 1988; Davies et al. 1992). Lz cells, in which RNase L activity is limited to 10% of parent L929 cells, show resistance against MGV infection after IFN treatment (Taira et al. 1985). These results indicate that the P1 kinase system participates in IFN-induced antiviral activity against MGV or EMCV infection.

What are the reasons for the assignment of an antiviral function to various systems in the cells? Vaccinia virus may hold an important clue. It produces a protein called K3L, capable of acting as a decoy for P1 kinase and inhibiting the antiviral activity of the P1 kinase system in vaccinia virus-infected cells. Cells, however, are still able to suppress vaccinia virus replication through the 2-5A/RNase L system. It is proposed that cells have developed various antiviral systems for the purpose of stopping the replication of these viruses like vaccinia virus.

3 Antiviral Function of the 2-5A/RNase L System

3.1 Virus Infection and the 2-5A/RNase L System

Our understanding of the 2-5A/RNase L system has considerably widened in recent years, thanks in particular to some successful studies of its mechanisms (Pestka et al. 1987; Staeheli 1990; Samuel 1991).

The functioning of the 2-5A/RNase L system begins only with the induction of enzyme 2-5AS. The levels of 2-5AS actually induced, however, depend on the cell type, for example, it is ca, 10-fold for human KB cells to 100-fold for human FL cells (Fujii et al. 1990b). A 10000-fold increase has been observed in chick embryo cells (Ball 1979).

2-5AS catalyzes the synthesis of the 2′,5′-oligoadenylate [2′,5′-linked oligomers of adenosine; ppp(Ap)nA, n = 2 ~ 18; 2-5A] using ATP as the substrate in the presence of dsRNA, which is supplied in viral infections as described in the previous section. This 2-5A then activates the latent endoribonuclease RNase L present in cells, degrading polysome and virus mRNA, and inhibiting viral protein synthesis. Therefore, this system is thought to work at a localized site of virus multiplication, in which the induced 2-5AS binds to double-stranded replicative forms of virus RNA and synthesizes 2-5A molecules. RNase L is activated by the 2-5A oligomers at the localized site. For the reasons described above, the 2-5A/RNase L system has the tendency to induce antiviral activity with high specificity toward several viruses, such as picornaviruses. The synthesized 2-5A oligomers are relatively unstable, and are degraded into ATP and AMP by the activity of 2′,5′-phosphodiesterase (Schmidt et al. 1979). This enzyme is also thought to block protein synthesis by disrupting the CCA sequence at the 3′-terminus of transfer RNA, in addition to the hydrolysis of 2-5A molecules (Schmidt et al. 1979). The 2-5A/RNase L system is intimately related not only to antiviral activity but also to key physiological functions such as regulation of development, differentiation and proliferation. Detailed descriptions of 2-5AS and RNase L are available in other chapters of this volume.

2-5AS is known to play an important role, but most of the details of the mechanisms by which it identifies and acts against infecting viruses remain obscure. Four forms of 2-5AS have been identified by Western and Northern blotting analysis: these measure 40, 46, 69 and 100 kDa (Saunders et al. 1985; Ilson et al. 1986; Chebath et al. 1987a; Rosenblum et al. 1988; Marie et al. 1990). The antiviral functions of the two larger forms have not yet been elucidated. Since the 40- and 46-kDa forms are transcribed in a differential manner of splicing from the same gene on the 12th chromosome, the only differences between them are in their C-terminal regions (Benech et al. 1985; Saunders et al. 1985; Williams et al. 1986). Both the 40- and 46-kDa forms probably have similar functions. The enzymes of the 40- and 69-kDa forms exhibit 65% homology at the amino acid level, and probably represent different gene products. The 69- and 100-kDa forms require not only different conditions for induction, but also appear to differ in their site of action and in many other respects as well (Marie et al. 1990). Each of the four enzymes seems to appear in response to different dsRNA species, act at different sites in the cell and have unique optimum pH values for enzyme activation (Saunders et al. 1985; Rosenblum et al. 1988). Since the activator is supplied by the infected virus as described above, the dsRNA (or infecting virus) is an especially important factor for 2-5AS activation. It has not yet been resolved how each form of 2-5AS is associated with the suppression of various virus infections.

When 2-5AS induction was suppressed by antisense RNA, IFN treatment blocked IFV replication, but not EMCV and VSV multiplication (Benedetti et al. 1987). This result implies that this system is associated with suppression of multiplication of the EMCV and VSV. While this appears to be true for EMCV, no definite evidence is available linking inhibition of VSV growth to the 2-5A/RNase L system. Indeed, there are data suggesting that induction of 2-5AS is not necessary for the development of resistance to VSV infection. K/Balb cells showing no signs of 2-5AS induction have been induced to develop resistance to VSV by induction of the P1 kinase (Hovanessian et al. 1981). HEC-1 cells that lack 2-5AS and P1 kinase were found to have no resistance to infection by VSV or sindbis virus (Verhaegen et al. 1980).

Introduction of cDNA encoding a 40-kDa form of 2-5AS into T98G or CHO cells revealed that the constitutive expression of the 40-kDa form appears to inhibit the multiplication of MGV, EMCV and vaccinia virus, but not VSV or HSV-2 (Chebath et al. 1987b; Grun et al. 1987). The results for VSV infection are thought to differ from those obtained from the experiment using the antisense RNA described above. This may be due to the introduction of the 40-kDa form alone. This means, in short, activator dsRNA for the 40-kDa form might be produced in EMCV-infected cells (Gribaudo et al. 1991), but is not produced by VSV or HSV-2 infection at an appropriate intracellular site. This suggestion is supported by reports that no 2-5A oligomers are found in IFN-treated and VSV-infected cells (De Ferra and Baglioni 1981), and that the antiviral state to VSV was not established in mouse embryonal carcinoma cells in spite of the induction of 2-5AS because of the lack of P1 kinase (Wood and Hovanessian 1979). On the contrary, HSV-2 infection stimulated synthesis of 2-5A molecules after IFN treatment (Cayley et al. 1984), indicating the possible synthesis of 2-5A oligomers by the 69- or 100-kDa form of 2-5AS. However, no 2-5A oligomer-dependent degradation of rRNA was observed in HSV-2-infected cells. Thus, the synthesized 2-5A may have been somewhat different from native 2-5A associated with RNase L activation.

On the other hand, experiments using antisense RNA or MRC-5 cells have implicated defence mechanisms of IFN in VSV infection; resistance to VSV in MRC-5 cells develops without 2-5AS and P1 kinase (Meurs et al. 1981). It is likely that the contribution of some other mechanism is associated with resistance to VSV. However, there is evidence, described later, suggesting that the 2-5A/RNase L system is significantly engaged in VSV resistance.

This limited and complex reaction to a particular virus may result from the localization of the 2-5AS and P1 kinase, activator dsRNA and synthesized 2-5A oligomers.

Differences between the mechanisms of EMCV and VSV replication have been well characterized. In picornaviruses, including EMCV and poliovirus, the virion genome RNA serves as mRNA (positive, single-stranded RNA) for viral protein synthesis. Replication of the virus RNA occurs in association with smooth membranes in the cytoplasm. On the other hand, genomic RNA of VSV is a negative, single-stranded RNA. Viral protein synthesis requires mRNA transcription from the negative genomic RNA. The resulting N protein works to initiate the process of RNA replication. VSV may replicate its RNA in association with the cytoskeletal system (microtubules). The difference in replication sites may play a significant role in the supplementation of activator dsRNA. For instance, the variation in antiviral effects in the 2-5A/RNase L system may reflect differences in the amount of induced 2-5AS, the site of localization of RNase L, the site of virus replication and 2′,5′-phosphodiesterase activity.

In the 2-5A/RNase L system, the breakdown of virus RNA (mRNA) by RNase L is ultimately connected with the antiviral activity. Consequently, attempts have been made to activate RNase L by introducing chemically synthesized 2-5A molecules and its analogues directly into the intact cells.

However, some viruses, such as EMCV and HIV, are able to suppress RNase L production or inhibit the activation of the enzyme (Cayley et al. 1982; Silverman et al. 1982; Wu et al. 1986). In addition, it has been demonstrated that some kinds of cells show low RNase L activity (Taira et al. 1985).

3.2 Antiviral Activity of 2-5A Molecules

Studies of the pathway of the 2-5A/RNase L system have shown that cells in which IFN fails to induce 2-5AS are unable to establish the antiviral state because of the lack of 2-5A oligomers. Similar results may be expected when the infecting virus is a type which does not supply effective activator dsRNA. However, in this situation, it is still possible to produce an antiviral state based on the 2-5A/RNase L system by directly introducing 2-5A molecules into the intact cells from outside.

In initial experiments, 2-5A is added to an extracted fraction of IFN-treated cells (a cell-free system) to evaluate the degradation of virus mRNA and ribosomal RNA (Sen et al. 1976; Baglioni et al. 1978; Clemens and Williams 1978; Williams and Kerr 1978; Wreschner et al. 1981). Chemically synthesized analogues of 2-5A and 2-5A core were investigated for stability against 2′,5′-phosphodiesterase degradation and their ability to activate RNase L and to inhibit cell growth. These experiments resulted in a significant advance in knowledge about the 2-5A molecules. 2-5A is a 2′,5′-linked oligomer. Activation of RNase L requires the presence of an oligomer of this molecule of at least trimer size, with a di- or triphosphate group at the 5′-terminus of the oligomer and at least two free hydroxy groups at the 3′ position of the terminal or internal ribose (Kerr and Brown 1978; Baglioni et al. 1981). Analogues lacking terminal 2′-OH, 3′-OH or internal 3′-OH groups show decreased biological activity, but their stability increased significantly. The 2-5A trimer is the most potent activator of RNase L. This molecule, lacking the triphosphate of the 5′-terminus (core molecule), shows neither binding nor activating capacity against RNase L (Martin et al. 1979). Concentrations of approximately 20–200 nM of 2-5A are synthesized in the cells treated with IFN in response to viral infection (Williams et al. 1979a). These 2-5A oligomers consist of di- and triphosphorylated dimer, trimer, tetramer and higher oligomers. In addition, 5–50 nM of core 2-5A trimer was also detected in IFN-treated and virus-infected cells (Knight et al. 1980). The concentrations of these 2-5A oligomers would appear to be adequate to activate 2-5A-dependent RNase L. Its half-life is short, only 50 min, because of degradation by 2′,5′-phosphodiesterase or esterase, so that a long-lasting antiviral effect is hard to obtain (Williams et al. 1978).

It is well known that neither IFN nor 2-5AS cDNA works effectively against VSV multiplication in particular cells. However, degradation of VSV mRNA was demonstrated after the addition of 2-5A oligomers to a cell-free system (Sen et al. 1976; Baglioni et al. 1978; Clemens and Williams 1978; Wreschner et al. 1981; Cayley et al. 1982; Silverman et al. 1982). Therefore, RNase L is effectively activated by 2-5A, even though VSV infection is apparently inhibited by the 2-5A/RNase L system.

Thus, if 2-5A is to be used for treatment of viral infection, two significant problems remain to be resolved: (1) 2-5A trimer is unable to penetrate the intact cell membrane because of its highly negative polarity; (2) 2-5A trimer has a short half-life in the cells.

Various 2-5A analogues with higher efficiencies of RNase L activation, higher stability and lower cytotoxicity have been chemically synthesized (Cayley and Kerr 1982; Eppstein et al. 1982; Goswami et al. 1982; Imai et al. 1982; Lee and Suhadolnik 1983; Sawai et al. 1983; Kariko et al. 1987). Furthermore, various methods have been developed to introduce 2-5A molecules directly into intact cells. At the time of this writing, 2-5A has been successfully inserted into cells by hypertonic shock, calcium phosphate coprecipitation, microinjection, encapsulation into liposomes, permeabilization with lysolecithin or conjugation with poly (L-lysine) (Williams et al. 1979b; Hovanessian and Wood 1980; Higashi and Sokawa 1982; Bayard et al. 1985, 1986; Federico et al. 1986) and much significant data has been recovered.

These studies have introduced both natural 2-5A and 2-5A derivatives, such as the 2′-O-phosphoglyceryl analogue, the tailed 2-5A analogue, the xyloadenosine analogue, and the cordycepin and phosphorothioate analogue. Both direct and indirect involvement of these 2-5A molecules in the inhibition of viral replication has been observed. These data from direct introduction of 2-5A into intact cells have been agreement with the experiments where the 2-5AS gene was inserted; function of the 2-5A was inferred from the observed inhibition of EMCV multiplication, and somewhat surprisingly, of VSV, HSV-1 and HSV-2

(Williams et al. 1979b; Hovanessian and Wood 1980; Bayard et al. 1985; Fujihara et al. 1989).

Introduction of 2-5A into EMCV-infected BHK cells permeabilized by hypertonic medium causes about a 100-fold reduction in virus yields at a concentration of 10 μm 2-5A (Williams et al. 1979b). This reduction is dependent, however, on the multiplicity of infection of the virus. 2-5A is often overpowered at high multiplicities of infection (m.o.i. = 20) of a virus.

Exposure of VSV-infected cells to calcium-phosphate coprecipitated 2-5A resulted in the inhibition of VSV replication; yield was reduced by 90–100% after treatment with 10–100 nM of 2-5A (Hovanessian and Wood 1980). Conjugation of 2-5A tetramer with poly (L-lysine) of the polypeptide carrier enables it to penetrate intact cells. These hybrid molecules are able to bind and activate RNase L. Incubation of the molecular hybrids with VSV-infected L1210 cells resulted in a dose-dependent inhibition of virus growth by 99% at a conjugate concentration of 180 nM (Bayard et al. 1986). The response to 2-5A in these experiments is similar to that observed in the cell-free system. Pretreatment of 2-5A molecules before viral infection is necessary for maximum activity.

2-5A trimer has been introduced into HSV-infected BHK cells in the form of a calcium phosphate precipitate; here, it inhibited over 90% of HSV-1 syncytium formation at a concentration of 800 nM and over 50% of HSV-2 plaque formation at a concentration of 1 μM. Application of 2-5A trimer as an ointment of polyethylene glycol into guinea pig vagina almost completely prevented the lethal effect of HSV-2 infection (Fujihara et al. 1989).

External application of 2-5A requires treatment within 3 h postinfection for a maximum inhibitory effect. 2-5A molecules must be prevented from acting on the early events of mRNA transcription. The antiviral action of the 2-5A/RNase L system works insufficiently in VSV- or HSV-infected cells as described in Section 3.1. However, the results in the above study seemed to indicate that a significant supply of 2-5A can bring about the reduction in virus yield. Perhaps most noteworthy is the time of the 2-5A formation in IFN-treated and virus-infected cells. Limitation on the antiviral function might be due to an inadequate or delayed supply of 2-5A oligomers. Therefore, the yields of VSV or HSV can be reduced by 2-5A oligomers supplied in sufficient quantity by deliberate superinfection (inoculation) with other viruses such as EMCV.

Various chemical derivatives of 2-5A with triphosphate at the 5′-terminus inhibit VSV or HSV multiplication at a concentration lower than that for natural 2-5A because of their increased stability. One of these analogues, the 2′-O-phosphoglyceryl derivative [(2′-5′)An-PGro] exhibits almost complete resistance to 2′,5′-phosphodiesterase, but remains sensitive to phosphesterase activity. After the introduction of this compound into HeLa cells through microinjection, it showed a powerful suppression of VSV, more than 90% at a concentration of approximately 100 nM (Bayard et al. 1984).

This derivative can also be encapsulated in protein-A-bearing liposomes.The delivery of this liposome-encapsulated analogue into L1210 cells is achieved with the help of monoclonal antibodies directed against the appropriate

Table 1. Effect of natural 2-5A and its analogues on viral replication

Oligomer	Introduction	Virus	Inhibition (%)	Cells	Reference
Natural 2-5A	Hypertonic shock	EMCV	90	BHK	Williams et al. (1979b)
	Calcium phosphate coprecipitation	VSV	90–100	L929	Hovanessian and Wood (1980)
	Conjugation with poly (L-lysine)	VSV	99	L1210	Bayard et al. (1986)
	Calcium phosphate coprecipitation	HSV-1	90	BHK	
		HSV-2	50	BHK	Fujihara et al. (1989)
(2′-5′) An-PGro	Microinjection	VSV	90	HeLa	Bayard et al. (1984)
	Encapsulation into liposomes	VSV	90	HeLa	Bayard et al. (1985)
2-5A Hexylmorphiline	Microinjection	VSV	100	HeLa	Defilippi et al. (1986)

class I major histocompatibility complex-encoded proteins expressed by the cells. The liposome encapsulation system for resistance to VSV infection provided tenfold more efficiency than microinjection (Bayard et al. 1985). Similarly, the microinjection of 2-5A hexylmorpholine analogues into HeLa cells before infection with VSV or MGV at an m.o.i. of 10 provided complete inhibition of VSV replication, however, it was poorly effective in inhibiting MGV yield (Defilippi et al. 1986). Maximum inhibitions to VSV and MGV multiplications by this analogue were achieved at concentrations of 10 nM and 30 μM, respectively. One explanation for the poor results of 2-5A against MGV infection is that the replication site of this virus may be in a region of the cell which is somehow sheltered by physical or biochemical structures, thereby impeding the diffusion of 2-5A to the replication site (Defilippi et al. 1986; Table 1).

The 3′-O-methylated analogue of 2-5A specifically inhibited methylation of vaccinia virus RNA by a virus methyltransferase at submicromolar concentrations. This analogue is at least ten times more active than natural 2-5A (Goswami et al. 1982).

3.3 Antiviral Activity of Core 2-5A

Nonphosphorylated 2-5A and its chemically synthesized or modified core molecules have been proven to exhibit antiviral activity in spite of their inability to activate RNase L. It is easier for them to cross the cell membrane than for natural 2-5A with its 5′-phosphate group. 2-5A cores were not shown to inhibit protein synthesis, but to be active in the inhibition of cellular DNA synthesis and mitogenetic proliferation (Kimchi et al. 1979). These functions of core molecules may involve pathways different from those utilized by the 2-5A-dependent RNase L. Therefore, elucidation of the inhibition mechanism(s) of virus replication by the cores will be a key step in the development of new antiviral drugs.

Exogenously administrated 2-5A cores hardly induced antiviral activity against SFV and EMCV, and were only slightly effective against HSV infection (Eppstein et al. 1983). Many derivatives of core 2-5A have also been produced and their antiviral activities evaluated. The xyloadenosine analogue of the 2-5A core (xylo 2-5A) turned out to be 120 times more stable than the core 2-5A and the inhibitory effect on HSV-1 and HSV-2 replication in 3T3 cells was more than 100-fold greater (Eppstein et al. 1983).

Xylo 2-5A has little effect on VSV, EMCV and paraIFV type-3 infections; however, it is a potent agent against HSV and vaccinia virus. Xylo 2-5A even works against HSV if imposed upon the cell subsequent to HSV infection. This finding indicates that it acts by a different mechanism than natural 2-5A. It is likely that the xylo 2-5A core exerts its antiviral effect through the inhibition of DNA synthesis. In addition, xyloadenosine showed antiviral activity against HSV at 30 μM, indicating that the effect of the xylo 2-5A core on HSV is dependent on a degradation product of the core (Eppstein et al. 1983). The core type (2′,5′)A4-poly(L-lysine) conjugate exhibited lower activity on VSV than the parental conjugate at the same final concentration (Bayard et al. 1986). (2′,5′)An-PGro and hexylmorpholine 2-5A cores were completely ineffective against VSV infection (Bayard et al. 1985). In contrast, the yield of MGV is inhibited by the hexylmorpholine 2-5A core at a relatively high concentration of 500 μM. Defilippi et al. (1986) has proposed that the effect of this core on MGV could be due to catabolic products liberated by intracellular processing.

Application of the cordycepin (3′-deoxy adenosine) analogue with 2′,5′-linkage of 2-5A [ppp(2-5)dAn)] into L929 cells resulted in dose-dependent inhibitions of protein synthesis and HSV-2 plaque formation at the same level as that observed by natural 2-5A molecules (Lee and Suhadolnik 1983; Fujihara et al. 1989). This compound is incapable of activating RNase L (Sawai et al. 1983). In addition, the inhibitory activity of the natural 2-5A core on DNA synthesis disappeared by the replacement of a single 2′,5′-bond with a 3′,5′-bond. The 2′,5′-phosphorothioate tetramer 5′-monophosphate analogues, the 2′,5′-cordycepin trimer core, and the cordycepin trimer 5′-mono-, -di-, and -triphosphate analogues were potent inhibitors of human immunodeficiency virus type-1 (HIV-1) reverse transcriptase (RT) (Montefiori et al. 1989). The cordycepin core trimer and its 5′-monophosphate derivative were encapsulated in liposome to evaluate the inhibitory activity to HIV-1 in vivo. Exposure of HIV-1-infected H9 cells to the liposome recognized by antibodies specific for CD3 resulted in an almost 100% inhibition of virus production (Müller et al. 1991; Table 2). The inhibition mechanism of these analogues is described in more detail in another chapter.

Table 2. Effect of the 2-5A core and its analogues on viral replication

Oligomer	Virus inhibition (+)	Virus inhibition (−)	Cells	Reference
2-5A Core	HSV	SFV, EMCV	3T3	Eppstein et al. (1983)
Xylo2-5A Core	HSV-1, HSV-2 Vaccinia virus	VSV, EMCV	3T3, Vero	Eppstein et al. (1983)
Hexylmorpholine 2-5A Core	MGV	VSV	HeLa	Defilippi et al. (1986)
Cordycepin core or 5′-monophosphate	HSV-2		HeLa	Fujihara et al. (1989)
	HIV		H9	Müller et al. (1991)
Phosphorothioate core or 5′-monophosphate	VSV		HeLa	Charachon et al. (1990)

2-5A and its analogues are potentially effective therapeutic agents because of their ability to induce antiviral activity in cells insensitive to IFN, and in those lacking 2-5AS or RNase L (Taira et al. 1985; Coccia et al. 1988; Fujii et al. 1988a). In fact, introduction of 2-5A into cells lacking 2-5AS resulted in an antiviral state against VSV infection (Federico et al. 1986).

3.4 Other Utilizations of Natural 2-5A Analogues

Phosphorothioate analogues of the 2-5A dimer and trimer core with stereoconfiguration of the 2′,5′-phosphodiester bonds have also been synthesized to evaluate the binding and activation processes of RNase L. Though the 2-5A core does not itself activate RNase L, some forms of the 2′,5′-phosphorothioate trimer core (RpRp, SpRp and RpSp diastereoisomers), which have the ability to penetrate the cell membrane, were able to activate RNase L; the SpSp diasteromer was not (Kariko et al. 1987). The SpSp, pSpSp and pSpSpSp analogues were effective inhibitors of RNase L (Kariko et al. 1987; Charachon et al. 1990). These analogues, therefore, promise to be useful in studies of the 2-5A/RNase L system and, eventually, in antiviral chemotherapy. The nucleotide binding site(s) of RNase L and/or other 2-5A binding proteins are currently under investigation using photoaffinity labelling of the 2- and 8-azido trimer 5′-triphosphate photoprobes of 2-5A (Suhadolnik et al. 1988; Li et al. 1990).

4 The Fluctuation of the 2-5A/RNase L System

Some viruses seem to be capable of fluctuating the antiviral state in IFN-treated cells as a result of the action of their replication. The mechanism which inhibits

Table 3. Modulation of the 2-5A/RNase L system by viral infection

Viruses	Activities	Reference
DNA viruses		
Adenovirus	Suppression of IFN-inducible gene expression by ElA gene product	Anderson and Fennie (1987) Gutch and Reich (1991)
EB virus	Inhibition of 2-5AS activity by EBERI (?) Suppression of 2-5AS induction	Fujii and Oguma (1986)
HBV	Suppression of IFN-inducible gene expression by TP of DNA polymerase	Foster et al. (1991)
RNA viruses		
MSV	Constitutive production of 2-5AS	David et al. (1989)
Mo-MLV	Suppression of IFN-induced 2-5AS induction (decrease in 2-5AS mRNA) Inhibitor of 2-5AS	Fujii et al. (1988c)
HTLV-I	Suppression of IFN-induced 2-5AS induction	Fujii et al. (1992)
HIV-1	Change in distribution of RNase L and 2-5A	Schröder et al. (1989)
Sendai virus	Suppression of IFN-induced 2-5AS induction	Crespi et al. (1988)
Measles and SSPE virus	Suppression of IFN-induced 2-5AS induction (translation ?)	Fujii et al. (1990b)
Mumps virus	Suppression of IFN-induced 2-5AS induction (transcription ?)	Fujii et al. (1988b, 1990b)
Rotavirus	Inhibitor of 2-5AS	Fujii et al. (unpubl. data)

the P1 kinase system is well documented. The 2-5A/RNase L system is also known to be affected by some viral infections. The modification occurs in one of three patterns: (1) increase in 2-5AS; (2) inhibition or suppression of 2-5AS; and (3) changes in 2-5A metabolism and the mode of RNase L distribution. In general, inhibitory and blocking effects on the P1 kinase system are frequently noted in acute viral infections (Table 3). Changes in the 2-5A/RNase L system, in contrast, are frequently found in persistent viral infections. Variations in antiviral function caused by viral infection may correlate with the pathogenesis of virus infections. Most is known about the latter, so they will be discussed first.

4.1 Persistent Infection

Mouse sarcoma virus (MSV: retroviridea)-transformed NIH 3T3 cells express significant amounts of 2-5AS compared to uninfected NIH 3T3 cells without IFN treatment; subsequent antiviral activity for MGV is established in the transformed cells (David et al. 1989). This augmentation may be regulated by the *cis*-element within the 5′-terminus untranslated region located upstream of the 2-5AS gene, in other words, it is important that the site of integration of the MSV genome is in the host cell DNA.

On the other hand, leukemia virus, which belongs to the same retroviridae as MSV, often induces modifications differing from those wrought by MSV. In Balb 3T3 cells persistently infected with Moloney murin leukemia virus (Mo-MLV), the induction of 2-5AS by IFN is markedly suppressed; other antiviral activities for VSV are also observed to decrease at the same time (Fujii et al. 1988c). In other words, not only are cells hampered in establishing the 2-5A/RNase L pathway due to the reduced 2-5AS induction, but other antiviral systems for VSV (probably including the P1 kinase system) may also be suppressed. Furthermore, in Balb 3T3cells with persistent infections of both Mo-MLV and MSV, markedly less 2-5AS activity by IFN occurs than in Balb 3T3 cells persistently infected with MSV. This implies that the effect of Mo-MLV on 2-5AS reduction probably counteracts the stimulation by MSV in expressing 2-5AS.

The suppression of the 2-5A/RNase L system described above may be due to the decrease in enzyme production at the level of transcription and/or translation of 2-5AS mRNA, or the presence of some inhibitor of the enzyme. IFN affects the maturation or assembly process of retrovirus replication in persistently infected cells. Inhibition of virus release results in an accumulation of virus structure proteins (Aboud and Hassan 1983). These components might engage in the inhibition of 2-5AS activity. Northern and dot-blot analyses indicate that the decrease in 2-5AS mRNA is the result of the suppression of transcription. In addition, cell lysate of Mo-MLV-infected cells inhibits 2-5AS activity, indicating the degradation of 2-5A formed by 2-5AS, or the presence of RNA similar to TAR RNA produced by HIV-1.

Cell lines persistently infected with human T-lymphotropic virus type-1 (HTLV-I) exhibit differences in the spontaneous production of IFN-γ, sensitivity to IFN (antiviral activity and cell proliferation inhibition), and IFN-induced 2-5AS induction (Fujii et al. 1990a, 1991). Among cells with HTLV-I, those producing IFN-γ show extremely poor induction of 2-5AS by endogenous IFN-γ or exogenous IFN-α (Fujii et al. 1992). In these cells, no resistance to VSV infection develops in spite of the constitutive production of IFN-γ, or even after IFN-α treatment. Northern blot analysis shows a decrease in 2-5AS mRNA, indicating that the reduction in 2-5AS is the result of suppression at the transcription level. These results are in agreement with those from Balb 3T3 cells persistently infected with Mo-MLV, with the exception of IFN production. In contrast, in an IFN-γ nonproducer cell line such as OKM-2, significant activity of the enzyme is induced by IFN-γ. An inverse relationship between IFN production and 2-5AS induction is demonstrated in these cell lines with HTLV-1. However, poor induction is improved after long-term cultivation of cells with IFN-α. Therefore, it is reasonable to infer that the suppression of IFN-induced 2-5AS induction is the result of downregulation of IFN receptors by IFN-γ produced spontaneously. Cells infected with Mo-MLV or HTLV-I may also produce other factors which tend to suppress ISRE (IFN-stimulated response element). These may have been implicated in the inhibition of the complex formation of F1 or Fg participating in the control of transcription (Coccia et al.

1991). The suppression mechanism(s) of 2-5AS mRNA transcription and the identification of intracellular factors during retrovirus infection remain to be clarified.

Human immunodeficiency virus (HIV-1) infection also influences the 2-5A/RNase L system, causing a modulation of the nuclear matrix-associated 2-5A metabolism (Schröder et al. 1989). The correlation between the significant levels of 2-5A molecules and RNase L and the failure of the HIV-1-infected cells to release progeny virus is convincing evidence that RNase L degrades HIV-1 transcripts. However, alterations of the amount of 2-5A and of the activity of RNase L have been demonstrated in H9 cells infected with HIV-1. The amount of 2-5A formed by 2-5AS increases up to 5.5-fold during the first 3 days after infection; subsequently, the amount drops to the initial values because of the reduced 2-5AS activity. Extracellular release of HIV-1 accompanying a fall in both 2-5AS activity and 2-5A amount is also reported in H9 cells with HIV-1. In addition, decreased RNase L and disturbed induction of 2-5AS have been confirmed to occur in peripheral blood mononuclear leukocytes (PBML) from patients with AIDS (Preble et al. 1985; Wu et al. 1986). Furthermore, endogenous production of IFN-α prevents the HSV-1-mediated induction of the latent HIV-1 provirus. This is the result of the alteration of binding affinity of the NF-kB-specific protein induced by HSV-1 (Popik and Pitha 1991). However, the tat protein is able to inhibit the IFN-mediated restriction for HIV-1 replication (Popik and Pitha 1992). Therefore, it is likely that the tat protein is also a potent inhibitor of IFN functions through the suppression of one two dsRNA-dependent enzymes.

A decrease in antiviral activity increases the sensitivity to superinfection by various other viruses. Many retroviruses produce *trans*-acting factors concerning transcriptional regulation (Nevins 1989), or factors having the ability to inhibit antiviral activity. Some of these are the IE110 and IE175 proteins produced by HSV-1 or HSV-2, and the NF-kB binding protein supplied by HSV or HTLV-I infection (Popik and Pitha 1991; Vlach and Pitha 1992). These factors are able to stimulate HIV-1 replication. TAR-RNA derived from HIV-1 augments HSV multiplication by inhibiting the P1 kinase system (Gunnery et al. 1990). Similarly, failure of 2-5AS induction permits vaccinia virus infection and replication; this then produces a factor(s) activating LTR of HIV-1 (Stellrecht et al. 1992).

In cases of dual infection with HTLV-I and HIV-1, HTLV-I provides a steady supply of NF-kB binding protein through its production of the Tax-P40 protein; subsequently, the binding protein augments HIV-1 replication (Leung and Nabel 1988). HIV-1 multiplication is inhibited by endogenously expressed IFN-α or IFN-γ (Bednarik et al. 1989; Popik and Pitha 1991). These results suggest that the first step in suppressing HIV-1 replication would be to decrease production of NF-kB binding protein and to increase the effectiveness of IFN.

Inhibition and blocking of the P1 kinase and 2-5A/RNase L systems in persistent infection with retroviruses may not only attenuate the resistance of the host against external infection by other viruses, but it may also contribute to the expression of endogenous latent retroviruses. For instance, 2-5A compounds are effective inhibitors of DNA topoisomerase relating to DNA replication and gene expression (Castora et al. 1991). 2-5A may be capable of regulating the expression of certain genes (Endogenous retrovirus genome) by virtue of a direct inhibition of DNA topoisomerase I. Such changes in the infected cells and organisms may represent the transition from the latent to the active state of the virus or the onset of illness.

Several studies of the suppression of IFN-induced 2-5AS induction have been reported in persistent infection by viruses other than retroviruses. Vero cells persistently infected with Sendai virus or SSPE virus show a decrease in antiviral activity for the sindbis virus without any reduction in IFN receptors (Crespi et al. 1986, 1988). This decrease is correlated with the suppression of 2-5AS induction. These persistently infected cells have an approximately tenfold lower basal 2-5AS activity than uninfected Vero cells. In addition, P1 kinase activity is not induced in these cells.

Persistent measles or SSPE virus infection also suppresses IFN-induced 2-5AS induction in various cell lines, indicating the reduction in IFN-related antiviral states (Fujii et al. 1988b, 1990b). This suppression by measles virus, however, is not seen in all persistently infected cell lines. The dependence on the combination of cells and virus strain implies the participation of cellular factors induced by viral infection. 2-5AS mRNA induced by IFN treatment in both persistently infected and uninfected cells shows no great differences, thus the site or sites of suppression may be at any of several stages of post-transcription. Several factors relating to the inhibition of macromolecular synthesis have been reported in measles virus infection (Fujii et al. 1978; Haase et al. 1985). Furthermore, it has been reported that measles virus transcription may be dependent on the action of cell-type-specific factors, the state of differentiation or a cellular double-stranded RNA-specific unwinding modification activity (Miller and Carrigan 1982; Baczko et al. 1988; Schneider-Schaulies et al. 1990; Rataul et al. 1992). Some cellular factor plays a key role in transcription and/or translation of 2-5AS mRNA. Nevertheless, the mechanism(s) contributing to this poor translation of 2-5AS mRNA are still unknown. On the other hand, these persistently infected cells can be classified into two types: the fusion clone and the nonfusion clone. The pattern of induction of 2-5AS mRNA differs in nonfusion-type and fusion-type clones. Less 2-5AS mRNA (about 1.8 kb) is found in the fusion-type clone. Reduced 2-5AS activity of the fusion clone may be in part due to suppression at the transcription level. It is, however, unclear what influence the fusion clone (protein) of measles virus has over transcription activity.

Mumps virus, paramyxoviridae, is also capable of establishing a persistent infection in various cell lines. In all cells with persistent infection, 2-5AS induction is markedly inhibited, along with the antiviral activity for VSV (Fujii et al. 1988b, 1990b). Even after IFN treatment, only infinitesimal amounts of the 1.8-kb mRNA are detectable in cells with persistent mumps virus infection. Such an inhibition of 2-5AS induction must be due to action of the mumps virus at the transcription level; this inhibition is observed regardless of the type of host cell. A functional protein (s), derived originally from the virus or the cellular genome after infection, might influence the consensus sequence concerning the regulation of 2-5AS gene expression. AGAANNGAAA is known as the consensus sequence of the regulating *cis*-element of the 2-5AS promoter (Blanar et al. 1989). A sequence similar to the 2-5AS promoter was found in the promoter region of the HSP 70 (70-kDa heat shock protein) gene. Consequently, induction of HSP 70 is markedly reduced by mumps virus infection (Fujii et al. unpubl. data). It is, therefore, proposed that this concensus sequence might act as one of the key targets for suppression.

The Epstein-Barr virus (EBV) is known to cause latent or persistent infection, Burkitt's lymphoma, nasopharyngeal carcinoma and lymphoproliferative disease. In general, the EBV-transformed B-lymphocytes do not permit productive infection. These persistently infected cells express nine latent proteins and small, non-polyadenylated RNAs known as EBER-1 and EBER-2 (Bhat and Thimmappaya 1985; Lotz et al. 1985; Kitajewski et al. 1986; Longnecker and Kieff 1990). The small RNAs are similar to the adenovirus VAI RNA in size and predicted secondary structure, and they can bind P1 kinase (Rosa et al. 1981; Clark et al. 1991). Since adenovirus VAI RNA prevents P1 kinase activation as an antagonist (Schneider et al. 1985; Kitajewski et al. 1986; Davies et al. 1989), it could be expected that EBER-1 also has an inhibitory effect on the enzyme. Indeed, in vitro experiments have improved inhibition of protein synthesis by P1 kinase in the presence of EBER-1 RNA (Clark et al. 1990). However, EBERs do not decrease the inhibitory effect of IFN on VSV replication in EBV-transformed lymphocytes, nor does EBER deletion render EBV-transformed B-lymphocytes susceptible to an IFN effect on cell proliferation or EBV replication (Swaminathan 1992). These effects might be due to the limited location of EBER RNAs in nuclei (Howe and Steitz 1986). In contrast, EBV-producing cells show poorer induction of 2-5AS than other EBV persistently infected cells (Fujii and Oguma 1986). This is in disagreement with the finding that significant activity of 2-5AS is found in PBMLs obtained from patients with EBV infection (Morag et al. 1982). These PBMLs consist of both EBV-infected B-lymphocytes and EBV-negative lymphocytes. Therefore, failure of 2-5AS induction is likely to occur in EBV-producing B-lymphocytes.

Suppression of 2-5AS induction in these persistently infected cells might be associated with some implicated pathogenesis caused by dual infection by external viruses or the expression of endogenous retroviruses.

4.2 Acute Infection

In acute infections, adenoviruses and reoviruses are known to influence the induction or the activity of 2-5AS. Reovirus serotype-1 and bovine rotavirus produce sigma-3 protein and VP2, respectively. These proteins can bind to the activator dsRNA (Boyle and Holmes 1986; Imani and Jacobs 1988; Schiff et al.

1988). Consequently, these proteins are also expected to influence the 2-5AS activity through removal of activator dsRNA. Extracts from MA 104 cells infected with human rotavirus, reoviridae, inhibited 2-5AS activity by 50% in reaction mixtures obtained from IFN-treated MA 104 cells. The inhibitory effect of the extracts could be overcome by adding excess dsRNA (Fujii et al. unpubl. data). Therefore, the substance in the extracts most probably has a function similar to that of the sigma-3 protein or VP2.

Adenovirus synthesizes a small RNA called VAI RNA, which plays an important role in the translation of virus mRNA in a later stage of virus infection. VAI RNA transcribed by RNA polymerase III was observed to be in competition with activator dsRNA for the binding site of P1 kinase, then inhibiting the activation of the enzyme (Schneider et al. 1985; Kitajewski et al. 1986; Davies et al. 1989; Mellits et al. 1990). Apart from this, adenoviruses produce some proteins encoded by the E1A and E1B regions of the virus genome. These proteins can control the expression of host cellular and viral genes (Nevins 1989). One of the proteins from the E1A gene, the E1A 289-amino acid protein (289 protein) inhibited the expression of various IFN-inducible genes at the transcription level (Anderson and Fennie 1987; Gutch and Reich 1991; Kalvakolanu et al. 1991). Such inhibition has been shown to be based on insufficient IFN-stimulated gene factor-3 (ISGF-3) transcriptional complex. Transcriptional repression was mediated through the deficiency of both ISGF-3 and ISGF-3 subunits in nuclear extracts from IFN-treated E1A-producing cells (Gutch and Reich 1991; Kalvakolanu et al. 1991).

In cells infected with adenovirus, induction of 2-5AS and P1 kinase is probably inhibited. In fact, no antiviral effect of IFN-α on VSV or EMCV is noted in HeLa cells and A549 cells infected with adenovirus type 5. Furthermore, induction of 561, 1–8 and 2-5AS mRNAs by IFN is blocked in adenovirus E1A-expressing cells at the transcription level. In human fetal kidney cells transformed by the E1A gene of adenovirus type 5 (293 cells) IFN-induced 2-5AS induction was poor, and the antiviral activity for VSV was also small in spite of significant levels of P1 kinase (Fujii et al. unpubl. data). Naturally, under these conditions, adenovirus type 5 itself is not inhibited by IFN-α. However, A549 cells are enable to commence antiviral activity for adenovirus type 2 by pretreatment with IFN-γ (Mayer et al. 1992). At the time of this writing, it remains to be settled whether this difference is due to a distinction between IFN types or the virus strains.

Human hepatitis B virus (HBV), like adenoviruses, is thought to inhibit the expression of IFN-inducible genes. Constitutive expression of the terminal protein region (TP) of HBV DNA polymerase resulted in a decrease in cellular responses to IFN and dsRNA (Foster et al. 1991). IFN therapy for patients with hepatitis leads to increased 2-5AS activity and amounts of β_2-macroglobulin in their serum (Quiroga et al. 1988). Among patients with chronic HBV infection, those with IFN-α-resistant virus show a low level of β_2-macroglobulin in their hepatocytes, indicating that TP production may prevent a response to therapy through the suppression of IFN-inducible gene expression (Foster et al. 1991).

HBV infection probably suppresses the induction of 2-5AS and P1 kinase. Under these conditions, attention should be paid to superinfection of hepatitis C and D viruses. The inhibition of IFN-inducible gene expression leads to a decrease in the level of proteins relating to immune response systems, such as HLA expression. Therefore, the immune surveillance system might be disturbed by adenovirus or HBV infection because of insufficient expression of virus antigens. These phenomena might contribute to the establishment of persistent or chronic infection.

Vaccinia virus infection influences both the 2-5A/RNase L system and the P1 kinase system. P1 kinase is inhibited by the E3L and K3L proteins induced by the virus (Watson et al. 1991; Davies et al. 1992). The 2-5A/RNase L system also fluctuates due to the induction of a phosphatase in connection with 2-5A inactivation (Paez and Esteban 1984b). However, as the vaccinia virus replication is suppressed by this system, inhibitory effects of this enzyme might not be crucial.

5 Conclusions

IFN induces a multifaceted response to viral infection. 2-5AS, Mx proteins and P1 kinase manifest, on their own, rather restricted antiviral properties, but the presently available evidence depicts what is sometimes a superbly concerted response. Cell lines lacking these enzymes are unable to muster this response, and are thus rather defenseless against viral infection. However, introduction of natural 2-5A or 2-5A analogues into these defective cells has proven to enable them to commence antiviral activity through the 2-5A/RNase L system or other pathway.

Many viruses can establish persistent or latent infections in human or in culture systems. The establishment of these persistent infections is, in part, the result of the disturbance of the IFN-induced antiviral state. Apparently, 2-5A/RNase L and P1 kinase systems are suppressed by various viral infections; the resulting vulnerable host cells then become subject to unceasing attack by external viruses as well as increased replication of endogenous viruses. Though the multiplication of VSV or EMCV is effectively reduced by IFN, superinfection of vaccinia virus or adenovirus results in the augmentation of VSV or EMCV replication. Especially, in the infection of HSV and retrovirus with a ready establishment of persistent infection, this inhibitory function influences the defense mechanisms and the host sensitivity to other viruses. Superinfection of HSV to HIV-1-infected cells results in the application of factors able to activate HIV-1 replication in *trans*. Under these conditions, the multiplication of HSV is not inhibited by IFN because of TAR RNA and tat protein production by HIV-1. In this context, some factors (IE110, IE175, NF-kB and LBP) participating in HIV-1 replication are supplied by HSV. By introducing 2-5A derivatives into the cells, it is possible to inhibit the replication of HSV and the augmentation of HIV-1. Thus, systems able to restore or elevate antiviral

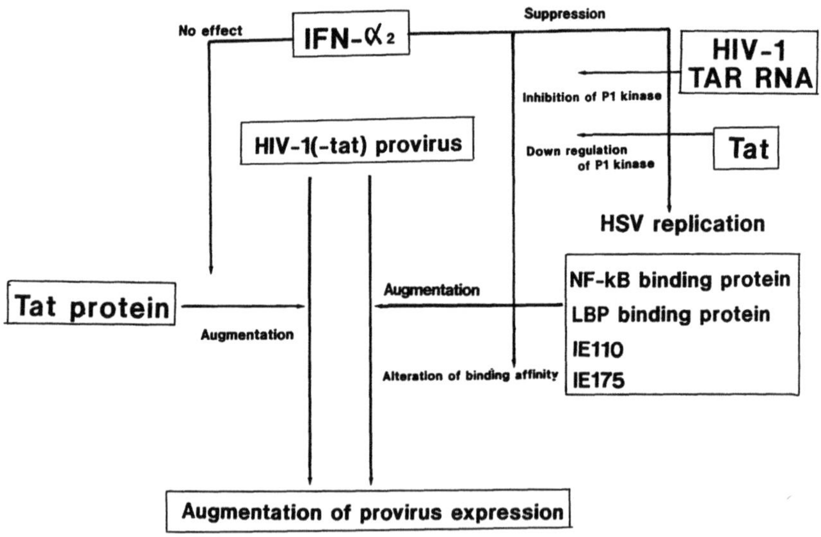

Fig. 1. Antiviral state and viral infection. Augmentation of HIV-1 provirus expression is dependent on some factors regulating the *cis*-element of HIV-1 LTR. These factors are supplied by several viruses. Superinfected HSV is able to replicate and to supply effectively these factors (IE110, IE175, NF-kB and LBP) due to suppression of the antiviral function of IFN by TAR RNA and tat protein. Endogenous expression of IFN results in the suppression of HIV-1 multiplication through modulations of binding affinity and/or production of NF-kB. Introduction of 2-5A oligomers into cells might produce an inhibiting effect on the augmentation of HIV-1 because of the suppression of HSV replication by the oligomers

activity will be required for the successful treatment of these viral infections (Fig. 1).

The role of these inhibitory functions on the pathogenesis of disease has not been defined. However, the appearance of various diseases may be dependent on a background of superinfection or dual infection. Blockage of the inhibitory effects of viruses on the antiviral state is one of the physician's most useful therapeutic weapons. 2-5A and its derivatives hold exciting promise as therapeutic agents in the future.

References

Aboud M, Hassan Y (1983) Accumulation and breakdown of RNA deficient intracellular virus particles in interferon-treated NIH3T3 cells chronically producing Molony murine leukemia virus. J Virol 45:489–495

Aguet M (1990) The interferon-γ receptor: a comparison with other cytokine receptors. J Interferon Res 10:551–558

Anderson KP, Fennie EH (1987) Adenovirus early region 1 modulation of interferon antiviral activity. J Virol 61:787–795

Baczko K, Liebert UG, Cattaneo R, Billeter MA, Roos RP, ter Meulen V (1988) Restriction of measles virus gene expression in measles virus inclusion body encephalitis. J Infect Dis 158:144–150

Baglioni C, Minks MA, Maroney PA (1978) Interferon action may be mediated by activation of a nuclease by pppA2′p5′A2′p5′A. Nature 273:684–687

Baglioni C, D'Alessandro SB, Nilsen TW, den Hartog JAJ, Crea R, van Boom JH (1981) Analogs of (2′-5′)oligo (A): endonuclease activation and inhibition of protein synthesis in intact cells. J Biol Chem 256:3253–3257

Ball LA (1979) Induction of 2′5′-oligoadenylate synthetase activity and a new protein by chick interferon. Virology 94:282–296

Bayard B, Bisbal C, Silhol M, Cnockaert J, Huez G, Lebleu B (1984) Increased stability and antiviral activity of 2′-O-phosphoglyceryl derivatives of (2′-5′) oligo (adenylate): role of phosphodiesterases and phosphatases in (2′-5′) oligo (adenylate) catabolism. Eur J Biochem 142:291–298

Bayard B, Leserman LD, Bisbal C, Lebleu B (1985) Antiviral activity in L1210 cells of liposome-encapsulated (2′-5′) oligo (adenylate) analogues. Eur J Biochem 151:319–325

Bayard B, Bisbal C, Lebleu B (1986) Activation of ribonuclease L by (2′-5′)(A)4-poly(L-lysine) conjugates in intact cells. Biochemistry 25:3730–3736

Bednarik DP, Mosca JD, Raj NBK, Pitha PM (1989) Inhibition of human immunodeficiency virus (HIV) replication by HIV-*trans*-activated α_2-interferon. Proc Natl Acad Sci USA 86:4958–4962

Benech P, Merlin G, Revel M, Chebath J (1985) The 3′ end structure of the human (2′-5′) oligo A synthetase gene: prediction of two distinct proteins with cell type-specific expression. Nucleic Acids Res 13:1267–1281

Benedetti AD, Pytel BA, Baglioni C (1987) Loss of (2′-5′) oligoadenylate synthetase activity by production of anti-sense RNA results in lack of protection by interferon from viral infections. Proc Natl Acad Sci USA 84:658–662

Bhat RA, Thimmappaya B (1985) Construction and analysis of additional adenovirus substitution mutants confirm the complementation of VAI RNA function by two small RNAs encoded by Epstein-Barr virus. J Virol 56:750–756

Bischoff JR, Samuel CE (1989) Mechanism of interferon action: activation of the human P1/eIF-2 protein kinase by individual reovirus s-class mRNAs: s1 mRNA is a potent activator relative to s4 mRNA. Virology 172:106–115

Black TL, Safer B, Hovanessian A, Katze MG (1989) The cellular 68,000-Mr protein kinase is highly autophosphorylated and activated yet significantly degraded during poliovirus infection: implications for translational regulation. J Virol 63:2244–2251

Blanar MA, Baldwin Jr AS, Flavell RA, Sharp PA (1989) A gamma-interferon-induced factor that binds the interferon response of the MHC class I gene, H-2Kb. EMBO J 8:1139–1144

Boyle JF, Holmes KV (1986) RNA-binding proteins of bovine rotavirus. J Virol 58:561–568

Capon DJ, Shepard HM, Goeddel DV (1985) Two distinct families of human and bovine interferon-α genes are coordinately expressed and encode functional polypeptides. Mol Cell Biol 5:768–779

Castora FJ, Erickson CE, Kovacs T, Lesiak K, Torrence PF (1991) 2′,5′-Oligoadenylates inhibit relaxation of supercoiled DNA by calf thymus DNA topoisomerase I.J Interferon Res 11:143–149

Cayley PJ, Kerr IM (1982) Synthesis, characterisation and biological significance of (2′-5′) oligoadenylate derivatives of NAD^+, ADP-ribose and adenosine (5′) tetraphospho-(5′)adenosine. Eur J Biochem 122:601–608

Cayley PJ, Knight M, Kerr IM (1982) Virus-mediated inhibition of the ppp (A2′p) A system and its prevention by interferon. Biochem Biophys Res Commun 104:376–382

Cayley PJ, Davies JA, Mccullagh KG, Kerr IM (1984) activation of the ppp (A2′p) nA system in interferon-treated, herpes simplex virus-infected cells and evidence for novel inhibitors of the ppp (A2′p) nA-dependent RNase. Eur J Biochem 143:165–174

Charachon G, Sobol RW, Bisbal C, Salehzada T, Silhol M, Charubala R, Pfleiderer W, Lebleu B, Suhadolnik RJ (1990) Phosphorothioate analogues of (2′-5′) (A) 4: agonist and antagonist activities in intact cells. Biochemistry 29:2550–2556

Chebath J, Benech P, Hovanessian A, Galabru J, Revel M (1987a) Four different forms of interferon-induced 2′,5′-oligo (A) synthetase identified by immunoblotting in human cells. J Biol Chem 262:3852–3857

Chebath J, Benech P, Revel M, Vigneron M (1987b) Constitutive expression of (2′-5′) oligo A synthetase confers resistance to picornavirus infection. Nature 330:587–588

Clark PA, Sharp NA, Clemens MJ (1990) Translational control by the Epstein-Barr virus small RNA EBER-1. Eur J Biochem 193:635–641

Clark PA, Schwemmle M, Schikinger J, Hilse K, Clemens MJ (1991) Binding of Epstein-Barr virus small RNA EBER-1 to the double-stranded RNA-activated protein kinase DAI. Nucleic Acids Res 19:243–248

Clemens MJ, Williams BRG (1978) Inhibition of cell-free protein synthesis by pppA2′p5′A2′p5′A: a novel oligonucleotide synthesized by interferon-treated L cell extracts. Cell 13:565–572

Coccia EM, Ferderico M, Romeo G, Affabris E, Cofano F, Rossi GB (1988) Interferons-α,β-and-γ-resistant friend cell variants exhibiting receptor sites for interferons but no induction of 2-5A synthetase and 67K protein kinase. J Interferon Res 8:113–127

Coccia EM, Vaiman D, Raber J, Marziali G, Fiorucci G, Orsatti R, Cohen B, Nissim N, Romeo G, Affabris E, Chebath J, Battistini A (1991) Protein binding to the interferon response enhancer correlates with interferon induction of 2′-5′-oligoadenylate synthetase in normal and interferon-resistant friend cells. J Virol 65:2081–2087

Crespi M, Chiu M, Schoub BD, Lyons SF (1986) Effect of interferon on Vero cells persistently infected with SSPE virus and lytically infected measles virus. Arch Virol 90:87–96

Crespi M, Chiu M, Struthers JK, Schoub BD, Lyons SF (1988) Effect of interferon on Vero cells persistently infected with Sendai virus compared to Vero cells persistently infected with SSPE virus. Arch Virol 98:235–251

David S, Nissim A, Chebath J, Salzberg S (1989) 2′-5′-Oligoadenylate synthetase gene expression in normal and murine sarcoma virus-transformed NIH 3T3 cells. J Virol 63:1116–1122

Davies MV, Furtado M, Hershey JWB, Thimmappaya B, Kaufman RJ (1989) Complementation of adenovirus virus-associated RNA I gene deletion by expression of a mutant eukaryotic translation initiation factor. proc Natl Acad Sci USA 86:9163–9167

Davies MV, Elroy-Stein O, Jagus R, Moss B, Kaufman RJ (1992) The vaccinia virus K3L gene product potentiates translation by inhibiting double-stranded-RNA-activated protein kinase and phosphorylation of the alpha subunit of eukaryotic initiation factor 2. J Virol 66:1943–1950

De Ferra F, Baglioni C (1981) Viral messenger RNA unmethylated in the 5′-terminal guanosine in interferon-treated HeLa cells infected with vesicular stomatitis virus. Virology 112:426–435

Defilippi P, Huez G, Verhaegen-Lewalle M, Clercq ED, Imai J, Torrence P, Content J (1986) Antiviral activity of a chemically stabilized 2-5A analog upon microinjection into HeLa cells. FEBS Lett 198:326–332

Eppstein DA, Marsh YV, Schryver BB, Larsen MA, Barnett JW, Verheyden JPH, Prisbe EJ (1982) Analogs of (A2′p) nA: correlation of structure of analogs of ppp (A2′p) 2A and (A2′p) 2A with stability and biological activity. J Biol Chem 257:13390–13397

Eppstein DA, Barnatt JW, Marsh YV, Gosselin G, Imbach JL (1983) Xyloadenosine analogue of (A2′p) 2A inhibits replication of herpes simplex viruses 1 and 2. Nature 302:723–724

Federico M, Romeo G, Affabris E, Coccia EM, Rossi GB (1986) 2′,5′-Oligoadenylate synthetase-uninducible alpha/beta-interferon-resistant friend cells develop an antiviral state when permeabilized with lysolecithin and treated with 2′,5′-oligoadenylate oligomers. J Interferon Res 6:233–240

Foster GR, Ackrill AM, Goldin RD, Kerr IM, Thomas HC, Stark GR (1991) Expression of the terminal protein region of hepatitis B virus inhibits cellular responses to interferons α and γ and double-stranded RNA. Proc Natl Acad Sci USA 88:2888–2892

Fujihara M, Milligan JR, Kaji A (1989) Effect of 2′,5′-oligoadenylate on herpes simplex virus-infected cells and preventive action of 2′,5′-oligoadenylate on the lethal effect of HSV-2. J Interferon Res 9:691–707

Fujii N, Oguma K (1986) Induction of oligo-2′,5′-adenylate synthetase in human lymphoid cells treated with 5-azacytosine and 5-iododeoxyuridine. J Gen Virol 67:2521–2526

Fujii N, Minagawa T, Kato T, Iida H (1978) Thymidine metabolism in cells treated with DNA-Suppressing factor (DSF). Microbiol Immunol 22:133–141

Fujii N, Indoh T, Murakami T, Kimura K, Oguma K, Tsukada Y (1988a) Suppression of interferon-induced oligo-2′,5′-adenylate synthetase induction in human hepatoma cell line, Li-7, Tumor Res 23:103–109

Fujii N, Oguma K, Kimura K, Yamashita T, Ishida S, Fujinaga K, Yashiki T (1988b) Oligo-2′,5′-adenylate synthetase activity in K562 cell lines persistently infected with measles or mumps virus. J Gen Virol 69:2085–2091

Fujii N, Oguma K, Yamashita T, Fujinaga K, Kakinuma M, Yashiki T (1988c) Decrease of oligo-2′,5′-adenylate synthetase activity in balb3T3 cell persistently infected with Moloney murine leukemia virus. Virus Res 10:303–314

Fujii N, Kwon K, Isogai E, Isogai H, Indoh T, Murakami T, Kimura K, Sekiguchi S, Oguma K (1990a) Effect of interferon on cells persistently infected with human T cell leukemia virus (HTLV-I). Tumor Res 25:1–6

Fujii N, Kimura K, Murakami T, Indoh T, Ishida S, Fujinaga K, Oguma K (1990b) Suppression of interferon-induced oligo-2′,5′-adenylate synthetase induction in persistent infection. J Gen Virol 71:3071–3074

Fujii N, Kwon K, Yashiki T, Sekiguchi S, Isogai E, Isogai H, Oguma K (1991) Correlation between oligo-2′,5′-adenylate synthetase and expression of human T-lymphotropic virus type-I specific gag protein. Tumor Res 26:11–15

Fujii N, Kwon K, Yashiki T, Kimura K, Isogai E, Isogai H, Sekiguchi S, Oguma K (1992) Oligo-2′,5′-adenylate synthetase activity in cells persistently infected with human T-lymphotropic virus type-I (HTLV-I). Microbiol Immunol 36:425–429

Garber EA, Chute HT, Condra JH, Gotlib L, Colonno RJ, Smith RG (1991) Avian cells expressing murine Mxl protein are resistant to influenza virus infection. Virology 180:754–762

Gastl G, Huber C (1988) The biology of interferon actions. Blut 56:193–199

Goswami BB, Crea R, Van Boom JH, Sharma OK (1982) 2′-5′-Linked oligo (adenylic acid) and its analogs: a new class of inhibitors of mRNA methylation. J Biol Chem 257:6867–6870

Gribaudo G, Lembo D, Cavallo G, Landolfo S, Lengyel P (1991) Interferon action: binding of viral RNA to the 40-kilodalton 2′-5′-oligoadenylate synthetase in interferon-treated HeLa cells infected with encephalomyocarditis virus. J Virol 65:1748–1757

Grun J, Kroon E, Zoller B, Krempien U, Jungwirth C (1987) Reduced steady-state levels of vaccinia virus-specific early mRNAs in interferon-treated chick embryo fibroblasts. Virology 158:28–33

Gunnery S, Rice AP, Robertson HD, Mathews MB (1990) Tat-responsive region RNA of human immunodeficiency virus 1 can prevent activation of the double-stranded-RNA-activated protein kinase. Proc Natl Acad Sci USA 87:8687–8691

Gutch MJ, Reich NC (1991) Repression of the interferon signal transduction pathway by the adenovirus E1A oncogene. Proc Natl Acad Sci USA 88:7913–7917

Haase AT, Gantz D, Eble B, Walker D, Stowring L, Ventura P, Blum H, Wietgrefe S, Zupancic M, Tourtellotte W, Gibbs CJ, Norrby E, Rozenblatt S (1985) Natural history of restricted synthesis and expression of measles virus genes in subacute sclerosing panencephalitis. Proc Natl Acad Sci USA 82:3020–3024

Hannigan GE, Williams BRG (1991) Signal transduction by interferon alpha through arachidonic acid metabolism. Science 251:204–207

Hersh CL, Brown RE, Roberts WK, Swyryd EA, Kerr IM, Stark GR (1984) Simian virus 40-infected, interferon-treated cells contain 2′,5′-oligoadenylates which do not activate cleavage of RNA. J Biol Chem 259:1731–1737

Higashi Y, Sokawa Y (1982) Microinjection of interferon and 2′,5′-oligoadenylate into mouse L cells and their effects on virus growth. J Biochem 91:2021–2028

Hovanessian AG, Wood JN (1980) Anticellular and antiviral effects of pppA (2′p5′A) n. Virology 101:81–90

Hovanessian AG, Meurs E, Montagnier L (1981) Lack of systematic correlation between the interferon mediated antiviral state and the levels of 2-5A synthetase and protein kinase in three different types of murine cells. J Interferon Res 1:179–190

Howe G, Steitz JA (1986) Localization of Epstein-Barr virus-encoded small RNA by in situ hybridization. Proc Natl Acad Sci USA 87:8790–8794

Ilson DH, Torrence PF, Vilcek J (1986) Two molecular weight forms of human 2′,5′-oligoadenylate synthetase have different activation requirements. J Interferon Res 6:5–12

Imai J, Johnston MI, Torrence PF (1982) Chemical modification potentiates the biological activities of 2-5A and its congeners. J Biol Chem 257:12739–12745

Imani F, Jacobs BL (1988) Inhibitory activity for the interferon-induced protein kinase is associated with the reovirus serotype 1 σ3 protein. Proc Natl Acad Sci USA 85:7887–7891

Kalvakolanu DVR, Bandyopadhyay SK, Harter ML, Sen GC (1991) Inhibition of interferon-inducible gene expression by adenovirus E1A proteins: block in transcriptional complex formation. Proc Natl Acad Sci USA 88:7459–7463

Kariko K, Li SW, Sobol RW, Suhadolnik RJ, Charubala R, Pfleiderer W (1987) Phosphorothioate analogues of 2′,5′-oligoadenylate. Activation of 2′,5′-oligoadenylate-dependent endoribonuclease by 2′,5′-phosphorothioate cores and 5′-monophosphates. Biochemistry 26: 7136–7142

Kerr IM, Brown RE (1978) pppA2′p5′A2′p5′A: an inhibitor of protein synthesis synthesized with an enzyme fraction from interferon-treated cells. Proc Natl Acad Sci USA 75:256–260

Kerr IM, Stark GR (1991) The control of interferon-inducible gene expression. FEBS Lett 285:194–198

Kimchi A, Shure H, Revel M (1979) Regulation of lymphocyte mitogenesis by (2′-5′) oligoadenylate. Nature 282:849–851

Kitajewski J, Schneider RJ, Safer B, Munemitsu SM, Samuel CE, Thimmappaya B, Shenk T (1986) Adenovirus VAI RNA antogonizes the antiviral action of interferon by preventing activation of the interferon-induced eIF-2α kinase. Cell 45:195–200

Klein JB, Schepers TM, Dean WL, Sonnenfeld G, McLeish KR (1990) Role of intracellular calcium concentration and protein kinase C activation in IFN-stimulation of U937 cells. J Immunol 144:4305–4311

Knight M, Cayley PJ, Silverman RH, Wreschner DH, Gilbert CS, Brown RE, Kerr IM (1980) Radioimmune, radiobinding and HPLC analysis of 2-5A and related oligonucleotides from intact cells. Nature 288:189–192

Kumar R, Tiwari RK, Kusari J, Sen GC (1987) Clonal derivatives of the RD-114 cell line differ in their antiviral and gene-inducing responses to interferons. J Virol 61:2727–2732

Lee C, Suhadolnik RJ (1983) Inhibition of protein synthesis by the cordycepin analog of (2′-5′) ppp (Ap) nA, (2′-5′) ppp (3′dAp) n3′dA, in intact mammalian cells. FEBS Lett 157:205–209

Lee TG, Tomita J, Hovanessian AG, Katze MG (1990) Purification and partial characterization of a cellular inhibitor of the interferon-induced protein kinase of Mr 68,000 from influenza virus-infected cells. Proc Natl Acad Sci USA 87:6208–6212

Leung K, Nabel GJ (1988) HTLV-1 transactivator induces interleukin-2 receptor expression through an NF-kB-like factor. Nature 333:776–778

Lewis JA (1988) Induction of an antiviral state by interferon in the absence of elevated levels of 2,5-oligo (A) synthetase and eIF-2 kinase. Virology 162:118–127

Li SW, Moskow JJ, Suhadolnik RJ (1990) 8-Azido double-stranded RNA photoaffinity probes. J Biol Chem 265:5470–5474

Longnecker R, Kieff E (1990) A second Epstein-Barr virus membrane protein (LMP2) is expressed in latent infection and colocalizes with LMP1. J Virol 64:2319–2326

Lotz M, Tsoukas CD, Fong S, Carson DA, Vaughan JH (1985) Regulation of Epstein-Barr virus infection by recombinant interferons. Selected sensitivity to interferon-gamma. Eur J Immunol 15:520–525

Lutfalla GL, Gardiner K, Proudhon D, Vielh E, Uze G (1992) The structure of the human interferon α/β receptor gene. J Biol Chem 267:2802–2809

Marie I, Svab J, Robert N, Galabru J, Hovanessian AG (1990) Differential expression and distinct structure of 69- and 100-kDa forms of 2-5A synthetase in human cells treated with interferon. J Biol Chem 265:18601–18607
Martin EM, Birdsall NJM, Brown RE, Kerr IM (1979) Enzymic synthesis, characterisation and nuclear-magnetic-resonance spectra of pppA2′p5′A2′p5′A and related oligonucleotides: comparison with chemically synthesised material. Eur J Biochem 95:295–307
Mayer A, Gelderblom H, Kumel G, Jungwirth C (1992) Interferon-γ-induced assembly block in the replication cycle of adenovirus 2: augmentation by tumour necrosis factor-α. Virology 187:372–376
Mellits KH, Kostura M, Mathews MB (1990) Interaction of adenovirus VA RNA1 with the protein kinase DAI: nonequivalence of binding and function. Cell 61:843–852
Meurs E, Hovanessian AG, Montagnier L (1981) Interferon-mediated antiviral state in human MRC5 cells in the absence of detectable levels of 2-5A synthetase and protein kinase. J Interferon Res 1:219–232
Meurs E, Chong K, Galabru J, Thomas NSB, Kerr IM, Williams BRG, Hovanessian AG (1990) Molecular cloning and characterization of the human double-stranded RNA-activated protein kinase induced by interferon. Cell 62:379–390
Miller CA, Carrigan DR (1982) Reversible repression and activation of measles virus infection in neural cells. Proc Natl Acad Sci USA 79:1629–1633
Montefiori DC, Sobol RW, Li SW, Reichenbach NL, Suhadolnik RJ, Charubala R, Pfleiderer W, Modliszewski A, Robinson WE, Mitchell WM (1989) Phosphorothioate and cordycepin analogues of 2′,5′-oligoadenylate: inhibition of human immunodeficiency virus type 1 reverse transcriptase and infection in vitro. Proc Natl Acad Sci USA 86:7191–7194
Morag A, Tobi M, Ravid Z, Revel M, Schattner A (1982) Increased (2′-5′)-oligo-A synthetase activity in patients with prolonged illness associated with serological evidence of persistent Epstein-Barr virus infection. Lancet 27:744
Müller WEG, Weiler BE, Charubala R, Pfleiderer W, Leserman L, Sobol RW, Suhadolnik RJ, Schröder HC (1991) Cordycepin analogues of 2′,5′-oligoadenylate inhibit human immunodeficiency virus infection via inhibition of reverse transcriptase. Biochemistry 30:2027–2033
Nevins JR (1989) Mechanisms of viral mediated *trans*-activation. In: Maramorosch K, Murphy FA, Shatkin AJ (eds) Advances in virus research vol 37. Academic Press, San Diego, pp 35–83
Nilsen TW, Maroney PA, Baglioni C (1982) Synthesis of (2′,-5′) oligoadenylate and activation of an endoribounuclease in interferon-treated HeLa cells infected with reovirus. J Virol 42:1039–1045
Paez E, Esteban M (1984a) Resistance of vaccinia virus to interferon is related to an interference phenomenon between the virus and the interferon system. Virology 134:12–28
Paez E, Esteban M (1984b) Nature and mode of action of vaccinia virus products that block activation of the interferon-mediated ppp (A2′p) nA-synthetase. Virology 134:29–39
Pain VM (1986) Inhibition of protein synthesis in mammalian cells. Biochem J 235:625–637
Pavlovic J, Zurcher T, Haller O, Staeheli P (1990) Resistance to influenza virus and vesicular stomatitis virus conferred by expression of human MxA protein. J Virol 64:3370–3375
Pavlovic J, Haller O, Staeheli P (1992) Human and Mouse Mx proteins inhibit different steps of the influenza virus multiplication cycle. J Virol 66:2564–2569
Penn LJZ, Williams BRG (1984) Interferon-induced 2-5A synthetase activity in human peripheral blood mononuclear cells after immunization with influenza virus and rubella virus vaccines. J Virol 49:748–753
Pestka S, Langer JA, Zoon KC, Samuel CE (1987) Interferons and their actions. Annu Rev Biochem 56:727–777
Pfeffer LM, Eisenkraft BL, Reich NC, Improta T, Baxter G, Daniel-Issakani S, Strulovici B (1991) Transmembrane signaling by interferon involves diacylglycerol production and activation of the isoform of protein kinase C in Daudi cells. Proc Natl Acad Sci USA 88:7988–7992
Popik W, Pitha PM (1991) Inhibition by interferon of herpes simplex virus type 1-activated transcription of tat-defective provirus. Proc Natl Acad Sci USA 88:9573–9577
Popik W, Pitha PM (1992) Transcriptional activation of the tat-defective human immunodeficiency virus type-1 provirus: effect of interferon. Virology 189:435–447

Preble OT, Rook AH, Steis R, Silverman RH, Krause D, Quinnan GV, Masur H, Jacob J, Longo D, Gelmann EP (1985) Interferon-induced 2′-5′ oligoadenylate synthetase during interferon therapy in homosexual men with Kaposi's sarcoma: marked deficiency in biochemical response to interferon in patients with acquired immunodeficiency syndrome. J Infect Dis 152:457–465

Quiroga JA, Mora I, Porres JC, Carreno V (1988) Elevation of 2′,5′-oligoadenylate synthetase activity and HLA-I associated β_2-microglobulin in response to recombinant interferon-α administration in chronic HBeAg-positive hepatitis. J Interferon Res 8:755–763

Rataul SM, Hirano A, Wong TC (1992) Irreversible modification of measles virus RNA in vitro by nuclear RNA-unwinding activity in human neuroblastoma cells. J Virol 66:1769–1773

Reich NC, Pfeffer LM (1990) Evidence for involvement of protein kinase C in the cellular response to interferon alpha. Proc Natl Acad Sci USA 87:8761–8765

Reid LE, Brasnett AH, Gilbert CS, Porter ACG, Gewert DR, Stark GR, Kerr IM (1989) A single DNA response element can confer inducibility by both α- and γ-interferons. Proc Natl Acad Sci USA 86:840–844

Rice AP, Kerr SM, Roberts WK, Brown RE, Kerr IM (1985) Novel 2′,5′-oligoadenylates synthesized in interferon-treated, vaccinia virus-infected cells. J Virol 56:1041–1044

Rosa MD, Gottlieb E, Lerner MR, Steitz JA (1981) Striking similarities are exhibited by two small Epstein-Barr virus encoded ribonucleic acids and the adenovirus-associated ribonucleic acids VAI and VAII. Mol Cell Biol 1:785–796

Rosenblum MG, Cheung L, Kessler D (1988) Differential activity of the 30-kD and the 100-kD forms of 2′-5′ An synthetase induced by recombinant human interferon-α and interferon-γ. Interferon Res 8:275–282

Roy S, Katze MG, Parkin N, Edery I, Hovanessian AG, Sonenberg N (1990) Control of the interferon-induced 68-kilodalton protein kinase by the HIV-1 tat gene product. Science 247:1216–1219

Roy S, Agy M, Hovanessian AG, Sonenberg N, Katze MG (1991) The integrity of the stem structure of human immunodeficiency virus type 1 tat-responsive sequence RNA is required for interaction with the interferon-induced protein kinase. J Virol 65:632–640

Rysiecki G, Gewert DR, Williams BRG (1989) Constitutive expression of a 2′,5′-oligoadenylate synthetase cDNA results in increased antiviral activity and growth suppression. J Interferon Res 9:649–657

Samuel CE (1991) Antiviral actions of interferon: interferon-regulated cellular proteins and their surprisingly selective antiviral activities. Virology 183:1–11

Saunders ME, Gewert DR, Tugwell ME, McMahon M, Williams BRG (1985) Human 2-5A synthetase: characterization of a novel cDNA and corresponding gene structure. EMBO J 4:1761–1768

Sawai H, Imai J, Lesiak K, Johnston MI, Torrence PF (1983) Cordycepin analogues of 2-5A and its derivatives: chemical synthesis and biological activity. J Biol Chem 258:1671–1677

Schiff LA, Nibert ML, Co MS, Brown EG, Fields BN (1988) Distinct binding sites for zinc and double-stranded RNA in the reovirus outer capsid protein 3. Mol Cell Biol 8:273–283

Schmidt A, Chernajovsky Y, Shulman L, Federman P, Berissi H, Revel M (1979) An interferon-induced phosphodiesterase degrading (2′-5′) oligoadenylate and C-C-A terminus of tRNA. Proc Natl Acad Sci USA 76:4788–4792

Schneider RJ, Safer B, Munemitsu SM, Samuel CE, Shenk T (1985) Adenovirus VAI RNA prevents phosphorylation of the eukaryotic initiation factor 2α subunit subsequent to infection. Proc Natl Acad Sci USA 82:4321–4325

Schneider-Schaulies S, Liebert UG, Baczko K, ter Meulen V (1990) Restricted expression of measles virus in primary rat astroglial cells. Virology 177:525–534

Schröder HC, Wenger R, Kuchino Y, Müller WEG (1989) Modulation of nuclear matrix-associated 2′,5′-oligoadenylate metabolism and ribonuclease L activity in H9 cells by human immunodeficiency virus. J Biol Chem 264:5669–5673

Sen GC, Lebleu B, Brown GE, Kawakita M, Slattery E, Lengyel P (1976) Interferon, double-stranded RNA and mRNA degradation, Nature 264:370–373

Silverman RH, Cayley PJ, Knight M, Gilbert CS, Kerr IM (1982) Control of the ppp (A2′p) nA system in HeLa cells: effect of interferon and virus infection. Eur J Biochem 124:131–138

Staeheli P (1990) Interferon-induced proteins and the antiviral state. Adv Virus Res 38:147–200
Stellrecht KA, Sperber K, Pogo BGT (1992) Activation of the human immunodeficiency virus type 1 long terminal repeat by vaccinia virus. J Virol 66:2051–2056
Suhadolnik RJ, Kariko K, Sobol RW, Li SW, Reichenbach NL (1988) 2- and 8-Azido photoaffinity probes. 1. Enzymatic synthesis, characterization, and biological properties of 2- and 8-azido photoprobes of 2-5A and photolabeling of 2-5A binding proteins. Biochemistry 27:8840–8846
Swaminathan S, Tomkinson B, Kieff E (1991) Recombinant Epstein-Barr virus with small RNA (EBER) genes deleted transforms lymphocytes and replicates in vitro. Proc Natl Acad Sci USA 88:1546–1550
Taira H, Yamamoto F, Furusawa M, Sawai H, Kawakita M (1985) Comparative studies on (2′-5′) oligoadenylate-related enzyme systems and the antiviral effect of interferon in two mouse cell lines which differ in (2′-5′) oligoadenylate sensitivity of their protein synthesizing system. J Interferon Res 5:583–596
Taniguchi T (1988) Regulation of cytokine gene expression. Annu Rev Immunol 6:439–464
Thomis DC, Doohan JP, Samuel CE (1992) Mechanism of interferon action: cDNA structure, expression, and regulation of the interferon-induced, RNA-dependent P1/eIF-2 protein kinase from human cells. Virology 188:33–46
Uze G, Lutfalla G, Gresser I (1990) Genetic transfer of a functional human interferon receptor into mouse cells: cloning and expression of its cDNA. Cell 60:225–234
Vaiman D, Pietrokovsky S, Cohen B, Benech P, Chebath J (1990) Synergism of type I and type II interferons in stimulating the activity of the same DNA enhancer. FEBS Lett 265:12–16
Verhaegen M, Divizia M, Vandenbussche P, Kuwata T, Content J (1980) Abnormal behavior of interferon-induced enzymatic activities in an interferon-resistant cell line. Proc Natl Acad Sci USA 77:4479–4483
Vlach J, Pitha PM (1992) Herpes simplex virus type 1-mediated induction of human immunodeficiency virus type 1 provirus correlates with binding of nuclear proteins to the NF-KB enhancer and leader sequence. J Virol 66:3616–3623
Watson JC, Chang H, Jacobs BL (1991) Characterization of a vaccinia virus-encoded double-stranded RNA-binding protein that may be involved in inhibition of the double-stranded RNA-dependent protein kinase. Virology 185:206–216
Weissmann C, Weber H (1986) The interferon genes. Prog Nucleic Acid Res Mol Biol 33:251–300
Whitaker-Dowling P, Youngner J (1983) Vaccinia rescue of VSV from interferon-induced resistance: reversal of translation block and inhibition of protein kinase activity. Virology 131:128–136
Whitaker-Dowling P, Youngner J (1986) Vaccinia-mediate rescue of encephalomyocarditis virus from the inhibitory effects of interferon. Virology 152:50–57
Williams BRG, Kerr IM (1978) Inhibition of protein synthesis by 2′-5′ linked adenine oligonucleotides in intact cells. Nature 276:88–90
Williams BRG, Kerr IM, Gilbert CS, White CN, Ball LA (1978) Synthesis and breakdown of pppA2′p5′A2′p5′A and transient inhibition of protein synthesis in extracts from interferon-treated and control cells. Eur J Biochem 92:455–462
Williams BRG, Golgher RR, Brown RE, Gilbert CS, Kerr IM (1979a) Natural occurrence of 2-5A in interferon-treated EMC virus-infected L cells. Nature 282:582–586
Williams BRG, Golgher RR, Kerr IM (1979b) Activation of a nuclease by pppA2′p5′A2′p5′A in intact cells. FEBS Lett 105:47–52
Williams BRG, Saunders ME, Willard HF (1986) Interferon-regulated human 2-5A synthetase gene maps to chromosome 12. Somatic Cell Mol Genet 12:403–408
Wood JN, Hovanessian AG (1979) Interferon enhances 2-5A synthetase in embryonal carcinoma cells. Nature 282:74–76
Wreschner DH, McCauley JW, Skehel JJ, Kerr IM (1981) Interferon action-sequence specificity of the ppp (A2′p) nA-dependent ribonuclease. Nature 289:414–417
Wu JM, Chiao JW, Maayan S (1986) Diagnostic value of the determination of an interferon-induced enzyme activity: decreased 2′,5′-oligoadenylate dependent binding protein activity in AIDS patient lymphocytes. AIDS Res 2:127–131

The 2-5A System and HIV Infection

H.C. Schröder[1], M. Kelve[2], and W.E.G. Müller[1]

1 Immunodeficient State in HIV Infection

The human immunodeficiency virus type 1 (HIV-1) is the etiologic agent of acquired immunodeficiency syndrome (AIDS). The progression of this retroviral disease is associated with various clinical manifestations, including the acquisition of an immunodeficient state, the frequent presence of neurological disorders, and some malignancies (reviewed in Barré-Sinoussi et al. 1983; Wong-Staal and Gallo 1985; Fauci 1988). Immunologic dysfunctions caused by HIV-1 infection include disorders in the production of cytokines (Murray et al. 1984; Abb et al. 1986). For example, a significant decrease in the production of interferon-α (IFN-α) by cultured peripheral blood mononuclear cells (PBMC) from patients infected with HIV-1 has been reported (Rossol et al. 1989; Voth et al. 1990). In addition, the production of IFN-γ is decreased in patients at late CDC stage III and CDC IV (Rossol et al. 1989). However, at late stages of the disease, the IFN-γ levels and, in particular, IFN-α levels increase (Buimovici-Klein et al. 1986; Skidmore and Mawson 1987; Rossol et al. 1989). Recent results suggest that HIV is capable of inducing in vitro production of conventional acid-stabile IFN-α, while only low levels of acid-labile IFN-α are present in sera of patients with AIDS-related complex (ARC) and AIDS (Capobianchi et al. 1988). There are also some hints that one enzyme of the 2′,5′-oligoadenylate (2-5A) pathway, the ribonuclease L (RNase L), is inactive in blood lymphocytes of AIDS patients (Carter et al. 1987), despite a high level of its activator 2-5A, indicating the presence of an inhibitor in HIV-infected cells which interferes with the proper functioning of this antiviral pathway. Therefore restoration of the natural 2-5A synthetase (2-5OAS)/RNase L and dsRNA-dependent protein kinase pathways may help to overcome the depressed immune status of ARC and AIDS patients.

2 The 2-5A Pathway

The 2-5OAS/RNase L pathway is part of the cellular antiviral mechanism that is activated by IFN (Lengyel 1982; Pestka et al. 1987; Fig. 1). The 2-5OAS

[1]Institut für Physiologische Chemie, Abteilung Angewandte Molekularbiologie, Johannes Gutenberg-Universität, Duesbergweg 6, D-55099 Mainz, Germany
[2]Institute of Chemical Physics and Biophysics, Akadeemia tee 23, EE0026 Tallinn, Estonia

Progress in Molecular and Subcellular Biology, Vol. 14
W.E.G. Müller/H.C. Schröder (Eds.)
© Springer-Verlag Berlin Heidelberg 1994

isoenzymes (EC 2.7.7.19) (Laurent et al. 1983; Chebath et al. 1987a) convert ATP into 2-5A, i.e., a series of 2′,5′-linked oligoadenylates mostly consisting of two to four adenylate residues with a 5′-terminal triphosphate, and pyrophosphate. However, at least one 2-5OAS is also capable of synthesizing oligomers containing more than ten adenylate residues (Yang et al. 1981). Figure 2a shows the structural formula of the 2-5A trimer core (= A_3), which is formed from the 2-5A trimer (= p_3A_3) by dephosphorylation. The 2-5OAS isoenzymes are activated by synthetic and naturally occurring dsRNA or RNA stem-loop structures, which are present in, e.g., heterogeneous nuclear RNA (Hubbell et al. 1991) or the *trans*-acting response element (TAR)-RNA sequence of HIV-1 (Schröder et al. 1990e). The small isoenzyme forms of 2-5OAS (40 and 46 kDa) are derived from the same gene by differential splicing (Benech et al. 1985); the 67/69-kDa and 100-kDa forms are probably encoded by different genes(s) (Marié and Hovanessian 1992). The formation of these enzymes is induced by IFN (Cohen et al. 1988; Rutherford et al. 1988) but they are also detectable at lower levels in cells not treated with IFN (Nilsen et al. 1982b, c).

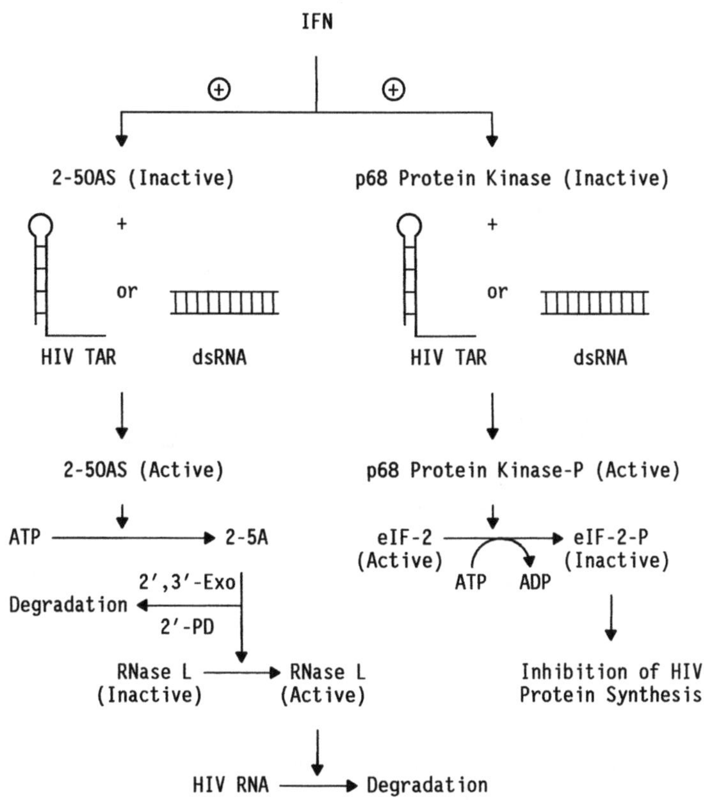

Fig. 1. 2-5OAS/RNase L and dsRNA-dependent protein kinase pathway

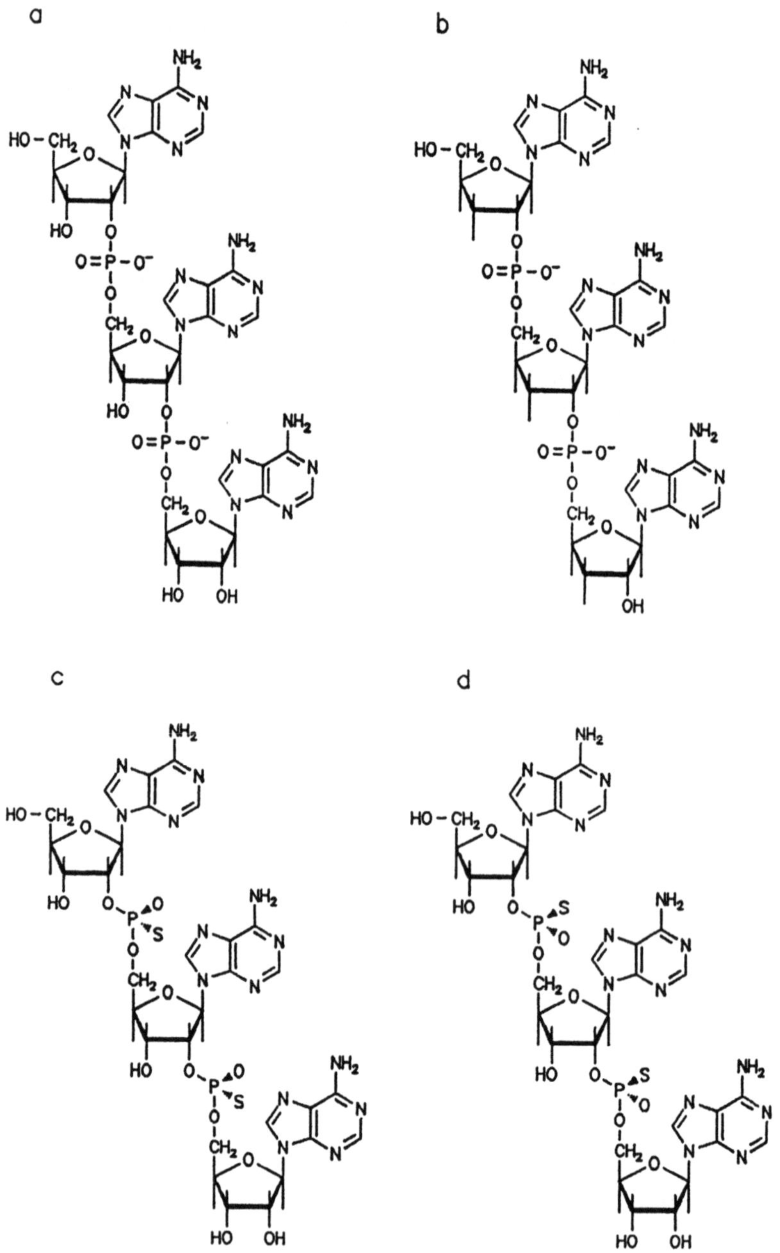

Fig. 2. Structural formulae of **a** A_3; **b** Co_3; **c** $A_{Rp}A_{Rp}A$; **d** $A_{Sp}A_{Sp}A$

2-5A containing a 5′-terminal triphosphate acts as allosteric activator of an endoribonuclease, the RNase L (EC 3.1.2.7) (Silverman et al. 1988; Bisbal et al. 1989; Fig. 1). The RNase L degrades preferentially viral and cellular, single-stranded RNA (Baglioni et al. 1979; Nilsen et al. 1982a, b; Schröder et al. 1989), primarily after U-A, U-G, and U-U sequences (Floyd-Smith et al. 1981; Wreschner et al. 1981). The consequence is an inhibition of protein synthesis (Kerr and Brown 1978). However, 2-5A is rapidly inactivated by two nucleases: a relatively unspecific phosphodiesterase (EC 3.1.1.3) (Schmidt et al. 1979; Johnston and Hearl 1987) and the 2′,3′-exoribonuclease (EC 3.1.13.4) (Müller et al. 1980; Schröder et al. 1980, 1984), which is specific for oligoadenylate.

The efficiency of the antiviral activity of the 2-5A system is demonstrated by the following findings: (1) production of antisense RNA to 2-5OAS in transfected cells has been shown to result in a loss of protection by IFN from viral infections (De Benedetti et al. 1987). (2) After transfection with a 2-5OAS-encoding cDNA CHO cells were found resistant to infection with picornaviruses also in the absence of IFN treatment (Chebath et al. 1987b).

The 2-5A metabolic enzymes are present in both the nucleus and the cytoplasm (Nilsen et al. 1982c; Laurent et al. 1983). The nuclear enzymes are associated with the nuclear matrix (Schröder et al. 1988, 1989); the same structure also harbors the cellular and viral mRNA (e.g., HIV mRNA) precursors (Verheijen et al. 1988; Müller et al. 1989, 1990b). Therefore, the production of 2-5A and the 2-5A-mediated degradation of HIV-1 mRNA proceed closely associated with the same nuclear substructure.

3 2-5A Metabolism in HIV-1-Infected Cells

After infection of cells in vitro, the level of 2-5A was found to be inversely correlated with virus production. In human T-cells H9 infected with HIV-1, there is a strong increase in the level of 2-5A during the initial stage of infection until onset of virus production (Schröder et al. 1988, 1989). This increase is due to a transient activation of nuclear matrix-associated 2-5OAS. On the other hand, the activity of 2′,3′-exoribonuclease, which degrades 2-5A, changes only slightly. The activity of nuclear matrix-associated RNase L, which is activated by 2-5A, also increases after infection of cells with HIV-1. Maximal enzyme activities are reached at days 2 to 3 postinfection; during this period of time, no virus particles are released from the cells. Possibly viral RNA molecules occurring at this time point are rapidly degraded; RNase L would be suitable for that purpose (Schröder et al. 1989). Later during infection, the 2-5OAS activity and the activity of RNase L strongly decrease, simultaneous with the rise in viral protein synthesis and virus release.

The increase in 2-5OAS activity at days 2 and 3 after infection was found to be paralleled by an enhancement of the 2-5OAS mRNA level (Ushijima et al. 1993). Later, the amount of 2-5OAS mRNA decreases, simultaneous with the appearance of HIV-*tat* transcripts. This result suggests that after HIV infection either an induction of 2-5OAS gene expression or an alteration in 2-5OAS mRNA half-life occurs. Since the half-life, e.g., of the 69-kDa 2-5OAS is ≈ 8 h (Hovanessian et al. 1988), it is quite possible that the HIV-caused alterations in nuclear matrix-associated 2-5OAS activity are regulated at the transcriptional and/or posttranscriptional level.

The inverse correlation of the 2-5A level and HIV production suggests that agents extending the time period during which high levels of 2-5A are present in the cell may retard or even prevent the release of HIV-1. Indeed, it was found that the release of HIV-1 can be suppressed by application of agents stimulating the dsRNA-dependent antiviral response of cells, e.g., IFN (Ho et al. 1985), certain lectins (Schröder et al. 1990a), or dsRNA, e.g. mismatched dsRNA (ampligen) (Carter et al. 1987; Montefiori and Mitchell 1987). Hence, modulation of the dsRNA-dependent 2-5OAS/RNase L system by application of these agents appears to be a promising strategy in the treatment of AIDS.

The antiviral activity of dsRNA does not seem to require IFN production. For example, Gendelman et al. (1990) demonstrated that treatment of HIV-infected monocytes with poly(I)·poly(C) results in high levels of 2-5OAS and complete inhibition of HIV gene expression but no detectable IFN-α activity.

Two further strategies to control HIV infection are the enhancement of the efficiency of the intracellular 2-5A system (1) by the development of 2-5A analogues which are better activators of RNase L or more resistant against the 2-5A degrading nuclease, and (2) by a gene technological approach, called "intracellular immunization" (see Sect. 9). Table 1 summarizes different strategies using the 2-5A system to suppress HIV replication.

Table 1. Strategies to suppress HIV replication using the 2-5A system

Strategy	Compound[a]
Induction of 2-5OAS	IFN[1], dsRNA[2,3], lectins[4]
Activation of 2-5OAS	Mismatched dsRNA[5-7]
Stabilization of 2-5A	Cordycepin analogues[8], phosphorothioate analogues[9]
Activation of RNase L	Phosphorothioate analogues[10-12]
2-5A-caused inhibition of RT	Cordycepin analogues[13,14] phosphorothioate analogues[10,13]
2-5A-caused inhibition of topoisomerase I	Long-chain 2-5A[15,16]
Blockage of Tat-caused inhibition of activation of 2-5OAS by TAR	Penicillamine[17]
Intracellular immunization	HIV-LTR – 2-5OAS hybrid gene[18,19] HIV-LTR – IFN hybrid gene[20]

[a]1, Ho et al. (1985); 2, Strayer et al. (1991); 3, Carter et al. (1991); 4, Schröder et al. (1990a); 5, Carter et al. (1987); 6, Montefiori and Mitchell (1987); 7, Ushijima et al. (1993); 8, Suhadolnik et al. (1983); 9, Kariko et al. (1987c); 10, Suhadolnik et al. (1989); 11, Kariko et al. (1987a); 12, Kariko et al. (1987b); 13, Montefiori et al. (1989); 14, Müller et al. (1991); 15, Castora et al. (1991); 16, Schröder et al. (1993); 17, Schröder et al. (1990e); 18, Schröder et al. (1990c); 19, Schröder et al. (1990d); 20, Bednarik et al. (1989).

Besides HIV-1-infected cells, alterations in the 2-5A system have been detected also in cells infected by other viruses, e.g., herpes simplex virus (Cayley et al. 1984). However, in the latter case the inhibitory effect of 2-5A, synthesized after treatment of the infected cells with IFN, on virus growth was found to be impaired; most likely due to the appearance of 2-5A derivatives inhibiting the activation of RNase L by authentic 2-5A after virus infection. The occurrence of novel 2-5A derivatives has also been reported in vaccinia virus-infected cells (Rice et al. 1985).

4 Modulation of 2-5OAS/RNase L Activity by HIV-1 RNA and Protein

4.1 Tat-TAR Interaction

The *trans*-activator protein Tat of HIV-1 is known to cause a strong enhancement of HIV-1 long-terminal repeat (LTR)-directed gene expression, either (or both) at the transcriptional or posttranscriptional level (Rosen 1991). These effects are thought to be mediated by the binding of Tat to the *trans*-acting response element (TAR) (Dingwall et al. 1989; Müller et al. 1990b). The TAR sequence, which is located within the HIV-1 LTR, is also present in the 5′ leader of all HIV-1 transcripts (Rosen and Pavlakis 1990). This sequence forms a stable stem-loop structure due to a large inverted repeat, which is required for Tat activation (Karn 1991). Moreover, it was shown that in addition to the viral Tat cellular proteins bind to TAR-RNA (Gaynor et al. 1989). Wu et al. (1988) and Gatignol et al. (1989) reported that three host cell-encoded, nuclear polypeptides of 100, 62 (61–63), and 46 kDa bind to TAR-RNA, very likely in a direct way. Tat binds also strongly to the nuclear matrix (Müller et al. 1989, 1990b), which is the nuclear binding site for both host cell and viral mRNA (Schröder et al. 1987a,b, 1989; Cook 1989). In this way, Tat may function as a linker in HIV RNA-matrix attachment (Müller et al. 1990b).

4.2 Activation of 2-5OAS by HIV TAR

Usually the activity of the dsRNA-dependent 2-5OAS is measured in the presence of a synthetic dsRNA such as poly(I)·poly(C). In in vitro studies Schröder et al. (1990e) demonstrated that the TAR sequence, which is present in the 5′ leader of HIV-1 mRNA, is also able to bind and activate purified 2-5OAS. Interestingly, binding of TAR to the synthetase in vitro was prevented by addition of Tat protein (Schröder et al. 1990e), which binds to this sequence with a high affinity (Müller et al. 1990b). Displacement of the synthetase at the TAR sequence by Tat results in the abolition of TAR-mediated activation of the enzyme. This, in turn, can be prevented by addition of Zn^{2+} and Cd^{2+} chelators, e.g., *o*-phenanthroline and penicillamine (Schröder et al. 1990e). *o*-Phenanthroline possibly impairs Tat function by complexing Zn^{2+} ions which are present in the native Tat protein (a dimer of two identical subunits, Frankel et al. 1988) (Müller et al. 1990b). D-Penicillamine (Schulof et al. 1986) may act in a similar way; however, this drug can also form disulfide bonds with the cysteine residues at the functional binding domain of Tat protein.

Transient expression assays with the HIV-1 TAR sequence fused to a reporter gene (bacterial chloramphenicol acetyltransferase, CAT) revealed that the levels of both CAT mRNA (linked to HIV-1 LTR) and CAT protein were decreased after treatment of the cells (HeLa) with IFN (Schröder et al. 1990e). However, if the cells were cotransfected with a *tat* sequence containing plasmid, CAT gene expression was insensible to the action of IFN. This result indicates that those mRNAs, which are linked to TAR sequence, are prone to nucleolytic degradation.

The preferential cleavage of viral mRNA seen in infected cells (Baglioni et al. 1979; Nilsen et al. 1982a) might occur through a localized activation of 2-5OAS and RNase L (Fig. 3). Because 2-5A, formed by the synthetase and containing a 5′-triphosphate, is rapidly inactivated by cellular nucleases (2′,3′-exoribonuclease and 2′-phosphodiesterase) and phosphatases (Bayard et al. 1984) (see Fig. 1), its action must necessarily be restricted to its site of formation. 2-5A with no or only one 5′-phosphate, or consisting of less than three adenylate residues, is essentially unable to activate the RNase L. Therefore, it is assumed that 2-5A once formed by a localized activation of 2-5OAS, e.g., by the TAR segment at the 5′-terminus of HIV mRNA, preferentially activates those RNase L molecules which are present in the nearest neighborhood. This should result in preferential degradation of the viral RNA (here: HIV mRNA), and might be responsible for the low abundance of HIV-1 mRNAs during the initial phase of infection or during prolonged latency periods. A localized activation of the 2-5A-dependent RNase L has also been demonstrated in in vitro studies using other systems, e.g., by hybridizing oligo(U) to the poly(A) tail of mRNA or poly(I) to the poly(C) stretch of EMC virus RNA; in both cases a preferential degradation of the partially double-stranded RNAs by extracts from IFN-treated cells was found (Nilsen and Baglioni 1979; Baglioni et al. 1984).

This specifically antiviral response, triggered by the HIV-1 TAR sequence, is thought to be antagonized by Tat (see above). Binding of Tat to TAR prevents binding of TAR to 2-5OAS and, hence, the activation of the 2-5A pathway. The consequence would be an increase in stability of HIV mRNAs during later, productive stages of infection. Therefore, compounds interfering with the TAR-Tat interaction, e.g., agents complexing metal ions required for Tat function (see above), should help to restore the antiviral state in the cell by rendering the TAR segment capable of activating 2-5OAS.

4.3 Interaction of HIV TAR with p68 Protein Kinase

The TAR segment of the HIV-1 leader is also able to modulate the second pathway involved in the IFN-mediated antiviral response of cells, the dsRNA-

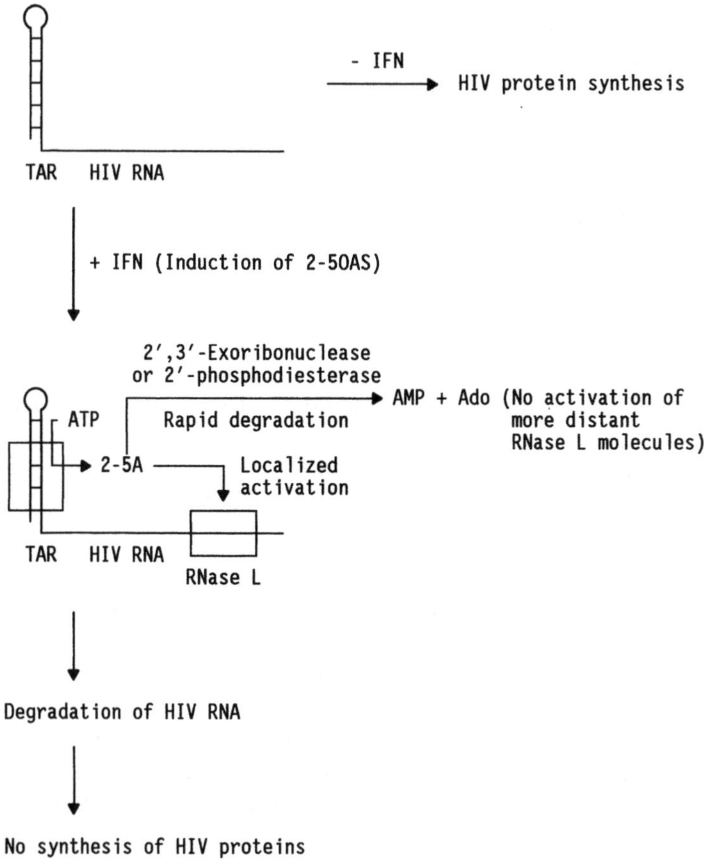

Fig. 3. Localized activation of dsRNA-dependent 2-5OAS and RNase L by HIV TAR RNA

dependent protein kinase/eIF-2 (eukaryotic initiation factor-2) system (Edery et al. 1989; SenGupta and Silverman 1989; Ushijima et al. 1993).

Binding of dsRNA to the dsRNA-dependent kinase (p68 kinase) results in autophosphorylation of the enzyme. The phosphorylated kinase then catalyzes the phosphorylation of the α-subunit of eIF-2. After phosphorylation eIF-2 is incapable of recycling and, consequently, initiation of mRNA translation is impaired (Farrell et al. 1977; Fig. 1).

The p68 kinase is able to bind to immobilized TAR RNA. Binding of the kinase to TAR RNA has been shown to result in an activation (Edery et al. 1989) or inhibition (Gunnery et al. 1990) of the enzyme. In our hands, the bound enzyme was capable of autophosphorylation (Ushijima et al. 1993). Interestingly, binding and activation of dsRNA-dependent kinase by the TAR segment are abolished by addition of Tat protein (Ushijima et al. 1993). As mentioned above,

Tat also suppresses the binding of purified 2-5OAS to TAR RNA-cellulose. In addition, SenGupta et al. (1990) presented evidence that Tat protein specifically enhances translation of mRNA containing 5′ leader of HIV-1.

On the other hand, Roy et al. (1990) demonstrated that productive infection of cells by HIV-1 results in a decrease in the amount of cellular p68 kinase. The amount of the enzyme was also found to be reduced in IFN-treated HeLa cells stably expressing *tat*. The authors concluded that the translational *trans*-inhibitory effect of the TAR RNA region of HIV-1 mediated by activation of the dsRNA-dependent kinase may be downregulated by *tat* during productive HIV-1 infection.

5 Modulation of Intracellular Antiviral Mechanisms by 2-5A Analogues

The cellular antiviral 2-5A system can be used as a means to suppress HIV replication. Possible strategies to enhance the efficiency of the 2-5A system are (see also Table 1):

- the development of compounds [particularly dsRNA analogues such as mismatched dsRNA, poly(I)·poly($C_{12}U$)], which are inducers of 2-5OAS and thus increase the production of 2-5A;
- the development of 2-5A analogues, which display a higher stability against nucleolytic degradation;
- the development of 2-5A analogues, which are phosphatase-resistant or do not require phosphorylation for activation of RNase L; and
- the development of 2-5A analogues, which are more potent activators of RNase L.

Different structural analogues of 2-5A with modifications in the sugar moiety, the nucleobase, or the internucleotide linkages were synthesized to enhance the stability towards 2-5A nucleases. Indeed, it was found that two classes of 2-5A analogues, the cordycepin (3′-deoxyadenosine; Co) and phosphorothioate analogues, display a pronounced anti-HIV activity in vitro (Doetsch et al. 1981; Montefiori et al. 1989; Suhadolnik et al. 1989; Müller et al. 1991). The structural formulae of some 2-5A analogues are shown in Fig. 2b–d.

5.1 *Cordycepin Analogues*

A promising compound in this respect is the 2′,5′-cordycepin core trimer (Co_3) (Doetsch et al. 1981; Fig. 2b). Müller et al. (1991) found that the metabolically stable cordycepin analogues of the 2-5A trimer carrying a 5′-monophosphate (pA_3) as well as its dephosphorylated core (A_3) potently inhibit HIV replication, when delivered into target cells (H9) via liposomes. The liposomes were directed to the T-cell receptor molecules CD3 by specific antibodies (Renneisen et al. 1990). Treatment of HIV-1-infected H9 cells with as little as 1 μM Co_3 or its

5′-monophosphate derivative (pCo_3) resulted in an almost total inhibition of virus production (Müller et al. 1991). The authentic 2-5A core and its 5′-monophosphate had only a very small inhibitory effect. Under the experimental conditions used, Co_3 did not serve as prodrug of cordycepin (Montefiori et al. 1989). Similar results have recently been obtained by using nanoparticles as a carrier system (Müller et al. unpubl.).

5.2 Phosphorothioate Analogues

The metabolic stability of the 2′,5′-internucleotide linkage was also enhanced by the introduction of phosphorothioate functions (Kariko et al. 1987a,b,c). The dimers and trimers of the diastereomeric phosphorothioate derivatives of 2-5A were first synthesized enzymatically (Kariko et al. 1987c). Later, the chemical syntheses of the two diastereomeric phosphorothioate dimers, four diastereomeric trimers (Pfleiderer and Charubala 1987) (the structural formulae of $A_{Rp}A_{Rp}A$ and $A_{Sp}A_{Sp}A$ are shown in Fig. 2c, d), and eight diastereomeric tetramers (see Charubala and Pfleiderer, this Vol.) have been achieved. The metabolic stability of the 2′,5′-phosphorothioate derivatives is much higher than that of authentic 2-5A, especially in the S_p configuration (Kariko et al. 1987c). These chiral molecules exhibit remarkable biological activities. The 2′,5′-phosphorothioate dimer 5′-triphosphate can bind to RNase L and activate the enzyme; this property has also been found with three of the four 2′,5′-phosphorothioate trimer cores ($A_{Rp}A_{Rp}A$, $A_{Sp}A_{Rp}A$, and $A_{Rp}A_{Sp}A$); however, the $A_{Sp}A_{Sp}A$ trimer diastereomer inhibits the activation process (Kariko et al. 1987a). The $A_{Sp}A_{Sp}A$ trimer diastereomer and its 5′-monophosphate ($pA_{Sp}A_{Sp}A$) are the most effective known inhibitors of RNase L (Kariko et al. 1987a,b).

After microinjection of the 2′,5′-phosphorothioate tetramer 5′-monophosphates into the cytoplasm of HeLa cells and infection of the cells with vesicular stomatitis virus (VSV), an antiviral effect was observed with the following compounds: $pA_{Rp}A_{Rp}A_{Rp}A$, $pA_{Sp}A_{Rp}A_{Rp}A$, and $pA_{Rp}A_{Sp}A_{Sp}A$. The $pA_{Sp}A_{Sp}A_{Sp}A$ isomer (at a concentration of 10^{-6} M) displayed no effect (Charachon et al. 1990). By microinjecting the $pA_{Sp}A_{Sp}A_{Sp}A$ isomer together with authentic 2-5A, a reduction or elimination of the protection against VSV infection was found (Charachon et al. 1990). Therefore, the $pA_{Sp}A_{Sp}A_{Sp}A$ phosphorothioate tetramer is a suitable antagonist to selectively inhibit the RNase L system and thereby probe the functional significance of RNase L.

The anti-HIV activity of a series of diastereomeric pairs of 2′,5′-phosphorothioates was determined in HIV-infected MT2 cells (Montefiori et al. 1989). The $pA_{Rp}A_{Sp}A_{Sp}A$ tetramer diastereomer displayed the strongest anti-HIV activity with optimal inhibition at 0.04 μM. At this concentration $pA_{Rp}A_{Sp}A_{Sp}A$ does not inhibit cell growth. Authentic 2-5A had no anti-HIV activity up to a concentration of 250 μM.

5.3 Cellular Uptake of 2-5A Analogues

One problem in the application of 2-5A analogues in HIV-infected cell systems could be their inability to be taken up by cells across cell membranes. Current efforts are directed toward developing oligonucleotides, which can permeate the cell membrane (see Charubala and Pfleiderer, this Vol.). One possible way is the conjugation of the 2-5A derivatives with poly(L-lysine). The uptake of 2-5A analogues into cells can also be achieved by encapsulation of the compounds in liposomes (Bayard et al. 1985; see also above). However, there is evidence that some 2-5A analogues are taken up in lymphoblasts and HIV-infected cells without the need for a vehicle (Montefiori et al. 1989. Suhadolnik et al. 1989). For example, it was shown that – although oligonucleotides are in general not capable of transmembrane passage – the ^{32}P-2-5A trimer core and the ^{3}H-2′,5′-cordycepin trimer core are taken up by intact lymphoblasts in culture (Suhadolnik et al. 1983).

6 Inhibition of Reverse Transcriptase by 2-5A Analogues

Surprisingly, the cordycepin analogues do not stimulate 2-5A-dependent RNase L activity (Sawai et al. 1983; Müller et al. 1991), although they inhibit HIV production. Hence, the cordycepin analogues also display no effect on the amount of cellular RNA and protein (Müller et al. 1991), in contrast to A_3 and pA_3 which activate RNase L, thereby causing degradation of viral and cellular mRNA and a reduction in viral and cellular protein synthesis (Nilsen and Baglioni 1979; Baglioni et al. 1984). The target of Co_3 and pCo_3 was found to be the HIV-1 reverse transcriptase (RT) (Müller et al. 1991). Using the viral RNA genome and the cellular $tRNA^{Lys.3}$ as template/primer, a 90% inhibition of enzyme activity was detectable by these compounds at a concentration of 10 µM. The enzyme was not inhibited by A_3 or p_3A_3 (concentration: 0.25–256 µM).

The mode of action of the cordycepin analogues might be as follows (Fig. 4). The HIV-1 RT most likely binds to the anticodon domain of $tRNA^{Lys.3}$ (Barat et al. 1989), which serves as primer for RT in the virus-infected cells. This binding region is comprised of the sequence 5′-A-C-U-S [modified uridine ester]-U-U-R[modified adenosine]-A-U*[pseudouridine]-C-U-G-3′ (the anticodon S-U-U is underlined). Complex formation between the RT and this sequence, which contains four uridines (one of them a modified one) in a row, could be weakened by pCo_3.

In an independent approach, using a competition assay, Sobol et al. (1991) found that the binding of oligo$(dT)_8$ as a primer to the recombinant RT is inhibited by the 5′-triphosphate of 2-5A as well as $tRNA^{Lys.3}$. This result also suggests a direct interaction of 2-5A or – with a higher efficiency – analogues of 2-5A with the binding of RT to the primer required to start viral DNA synthesis.

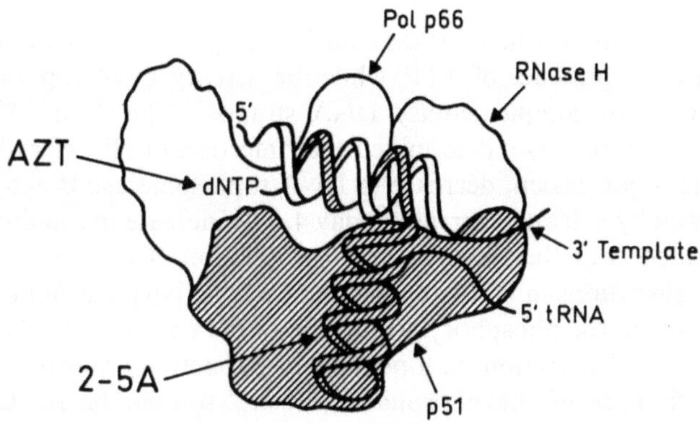

Fig. 4. Inhibition of HIV-1 RT by 2-5A analogues. The different target sites of 3′-azido-3′deoxythymidine (*AZT*) and 2-5A analogues at the hypothetical complex of tRNA, RNA template, and RT are shown. The HIV RT is a heterodimer of a p51 and a p66 subunit (Kohlstaedt et al. 1992). The anticodon and dihydrouridine stems and loops of the tRNA presumably bind to the p51 subunit. 2-5A analogues prevent binding of RT to the tRNA

Even more effective as inhibitors of HIV-1 RT are the diastereomeric *R*p and *S*p phosphorothioate tetramer 5′-monophosphate analogues of 2-5A (Montefiori et al. 1989). The most effective inhibitor of HIV-1 RT is the 2′,5′-phosphorothioate trimer 5′-triphosphate ($p_3A_3\alpha S$); a 50% inhibition was found at 0.5 μM. The presence of the 5′-monophosphorothioate function in the 2-5A derivative confers this molecule with an enhanced metabolic stability, because the phosphorothioate group is not hydrolyzed by cellular phosphatases. Adenosine 5′-*O*-phosphorothioate (AMPS) does not inhibit RT activity up to 200 μM, suggesting that the observed inhibition is not caused by degradation products of the phosphorothioate derivatives.

Moreover, the 2′,5′-phosphorothioate cores and their 5′-monophosphates are able to bind to RNase L and to activate the enzyme (Kariko et al. 1987a). Therefore, these 2-5A derivatives might have a dual mechanism of action: the same molecules are able to activate the cellular antiviral system and to suppress virus replication via a direct inhibition of RT.

7 Inhibition of DNA Topoisomerase I by 2-5A

7.1 Alterations of DNA Topoisomerase Activities in HIV-Infected Cells

DNA topoisomerases are thought to be involved in the regulation of important nuclear events such as DNA replication, transcription, and chromatin assembly (reviewed in Wang 1985; Liu 1989). It is likely that DNA topoisomerases also affect virus replication; e.g., both provirus integration and expression might require topological changes. Two types of DNA topoisomerases exist:

topoisomerase I, which makes transient single-strand breaks in DNA, does not require the presence of ATP, while the activity DNA topoisomerase II, which cleaves both complementary DNA strands, depends on ATP (Wang 1985). Matthes et al. (1990) determined that infection of H9 cells with HIV-1 in vitro results in a transient decrease in DNA topoisomerase II activity (at days 1–3), followed by a drastic increase at day 4. The increase in topoisomerase II activity coincided with the onset of virus production. The underlying mechanism of these alterations in DNA topoisomerase II activity was found most likely to be a change in the phosphorylation state of the enzyme (Matthes et al. 1990). The HIV-caused alterations in topoisomerase II activity were found to be associated with changes in the phosphatidylinositol/protein kinase C pathway in the course of virus infection (Müller et al. 1990a). Inhibition of protein kinase C-mediated phosphorylation of DNA topoisomerase II results in an inhibition of HIV-1 production (Matthes et al. 1990). Phosphorylation (Pommier et al. 1990) and poly(ADP-ribosyl)ation (Jongstra-Bilen et al. 1983) also seem to be involved in posttranslational regulation of activity of the type I enzyme. We found that nuclear matrix-associated topoisomerase II activity and production of HIV-1 can be inhibited by natural lignans (Schröder et al. 1990b).

7.2 Cellular Topoisomerase I

Castora et al. (1991) reported that the activity of calf thymus DNA topoisomerase I is inhibited by 2-5A. Later, Schröder et al. (1993) demonstrated that 2-5A is an efficient inhibitor of the type I topoisomerase activity present both in uninfected and HIV-infected cells. Inhibition of the enzyme activity by 2-5A depends on the chain length of the oligomer and the presence of 5′-phosphate. The 5′-triphosphorylated 2-5A oligomers consisting of more than four adenylate residues are most active (inhibition of DNA relaxation by $\geq$ 50% at 1 μM concentration); the 5′-triphosphorylated 2-5A hexamer exhibits the strongest effect (Schröder et al. 1993). The 2-5A core dimer and trimer inhibit the enzyme only at very high concentrations (1 mM). It should be noted that the 2-5A trimers are also potent activators of 2-5A-dependent RNase L, whereas the 2-5A dimer is unable to activate this enzyme (Baglioni et al. 1978).

Usually the 2-5OAS isoenzymes catalyze the synthesis of 2-5A consisting of two to four 2′,5′-linked adenylate residues. Treatment of cells with the mismatched dsRNA, poly(I)·poly($C_{12}U$), causes the production of longer 2-5A oligomers (Schröder et al. 1993; see also Müller et al., this Vol.). This also results in an inhibition of DNA relaxing activity of nuclear DNA topoisomerase I (Schröder et al. 1993).

7.3 HIV-Associated Topoisomerase I

A type I topoisomerase activity has been reported to be associated with HIV-1 particles (Priel et al. 1990). The virus-associated enzyme which can be inhibited by the topoisomerase I inhibitor, camptothecin (Priel et al. 1991), was found to be antigenetically related to cellular topoisomerase I, but differed from this enzyme in its response to ATP. It is unclear whether the HIV-1-associated topoisomerase I activity is due to cellular enzyme molecules packaged in virus particles during viral assembly or whether it represents a novel class of the enzyme. A type I topoisomerase activity has been reported also to be associated with other virus particles, e.g., vaccinia virus (Shuman and Moss 1987), Rous sarcoma virus (Weis and Faras 1981), and equine infectious anemia virus (Priel et al. 1990). 2-5A is also capable of inhibiting topoisomerase I activity in isolated HIV-1 particles. At a concentration of 10 μM p_3A_3 significantly inhibits relaxing activity of DNA topoisomerase I in HIV lysates, in contrast to the A_3 core and the 5′-monophosphate, pA_3 (Schröder et al. 1993).

7.4 Mechanism of Action

The mechanism of action through which 2-5A oligomers inhibit DNA relaxing activity of DNA topoisomerase I is not known. It might be possible that 2-5A interacts with the ATP-binding site found in eukaryotic type I topoisomerases (Castora and Kelly 1986). Binding of ATP at this site may be involved in the modulation of enzyme activity (Trask et al. 1984). Interestingly, the HIV-associated topoisomerase I was found to be more strongly inhibited by ATP than the cellular enzyme (Priel et al. 1990).

The finding that 2-5A may inhibit DNA topoisomerase I activity may have implications for the understanding of the antiviral mechanism of 2-5A. DNA topoisomerase I represents a novel target for 2-5A analogues and the development of longer 2-5A analogues which are better inhibitors of this enzyme may be a promising strategy for anti-HIV chemotherapy. It should be noted that also inhibitors of DNA topoisomerase II such as coumermycin A1 (Baba et al. 1989) and certain lignans (Schröder et al. 1990b) are able to block HIV replication in vitro. The latter compounds (arctigenin and trachelogenin) also strongly suppress the HIV integration reaction (Pfeifer et al. 1992).

8 Stimulation of 2-5A Metabolism by Lectins

Retardation of HIV-1 release due to a delay in the decrease in the intracellular level of 2-5A has also been found after treatment of cells, infected with HIV-1, with a D-galactose-specific lectin from a sponge (*Chondrilla nucula*) (Schröder et al. 1990a). This lectin causes an induction of 2-5OAS activity both in noninfected and HIV-infected H9 cells (Schröder et al. 1990a). The mechanism of action of the lectin is still unclear. It appears likely that the alteration in 2-5OAS activity is correlated with the influence of the lectin on cell proliferation (Wells and Mallucci 1985; Smekens et al. 1986). The opposite effect has been observed with another lectin, concanavalin A. The inhibition of the

IFN-α-induced increase in 2-5OAS level caused by this lectin is assumed to be due to an inhibition of the internalization of IFN-α after binding to its specific cell surface receptor (Faltynek et al. 1988).

9 "Intracellular Immunization" of Cells with HIV-LTR – 2-5OAS Hybrid DNA

The selectivity of the dsRNA-dependent 2-5OAS/RNase L system is limited by the fact that the RNase L, which is activated by 2-5A, is not able to discriminate between viral and cellular RNA, and hence, also degrades cellular RNA (Nilsen et al. 1982b; Schröder et al. 1989). Therefore, Schröder et al. (1990c, d) put the synthesis of 2-5A under the control of the HIV-1-encoded Tat protein. The aim of this "intracellular immunization" approach was to increase the efficiency of the 2-5A-mediated antiviral response selectively in the infected cells expressing Tat protein. Tat protein is known to cause a strong increase in the steady-state level of HIV-1 LTR-specific mRNAs (Rosen and Pavlakis 1990), mediated by the interaction of Tat with the TAR sequence located within the HIV-1 LTR (Müller et al. 1990b). Tat present in the infected cell was indeed capable of activating in *trans* the expression of 2-5OAS, linked to an HIV 3′-LTR, while in its absence the enzyme is expressed only slightly.

The expression vector used (pU3R-III/2-5AS) contained a human 2-5OAS cDNA encoding an active form of the enzyme (9-21 cDNA, Benech et al. 1985), located 3′ to an HIV 3′-LTR. This vector was used for stable transfection of CD4 receptor bearing HeLa-T4$^+$ cells, which can be infected with HIV-1. It was found that the LTR-directed expression of the hybrid DNA can indeed be activated in *trans* by infection with HIV-1 or by the HIV-1 *tat* gene product (Schröder et al. 1990c, d). HIV replication after infection of the stably transfected HeLa-T4$^+$ cells with HIV-1 was strongly inhibited. In nontransfected cultures or after transfection with the selectable marker plasmid only, the percentage of cells expressing p17 and p24 *gag* protein of HIV-1 was about 60% 5 days after infection. However, after stable transfection with pU3R-III/2-5AS the number of p17- and p24-positive cells was reduced to about 2%. The RT activity as a measure of virus production in the culture fluid of transfected cells was also markedly decreased. This result shows that transfection of cells with a plasmid containing 2-5OAS under control of HIV-1 LTR is able to efficiently confer these cells with the capacity to suppress HIV replication in vitro.

In a similar "intracellular immunization" approach, it was shown that the introduction of a hybrid IFN gene behind the HIV-1 LTR also efficiently inhibited HIV-1 replication in vitro (Bednarik et al. 1989). However, in this case, it cannot be excluded that IFN, synthesized by the transfected cells in the presence of Tat, is released from the cells and then acts on other cells, irrespective of whether these cells are infected or not. An enhanced resistance to HIV-1 replication has also been achieved in cells of the human promonocytic U937 line after stable transfection with a human IFN-β gene under the control of an MHC promoter fragment (Mace et al. 1991); the level of 2-5OAS was found to be increased and the effect on HIV-1 replication could be abolished by addition of anti-IFN-β antibodies to the culture medium.

A series of further strategies for "intracellular immunization" has been developed: application of antisense RNA, ribozymes, or dominant-negative mutants (Malim et al. 1989; Trono et al. 1989; Sarver et al. 1990). One problem in these studies was that the stably transfected cells were not able to synthesize sufficient amounts of the respective gene products when their synthesis was performed by RNA polymerase II. In case of intracellular immunization of cells by the 2-5OAS gene described above, a high level of expression (and selective expression in the infected cells) was achieved by putting this gene under the control of HIV-LTR. Another possibility to synthesize higher amounts of antisense RNA (but without selectivity for infected cells), e.g., is the linkage of the DNA to be expressed with a transcription unit used by RNA polymerase III. It is known that tRNAs transcribed by RNA polymerase III are 100 times more abundant than poly(A)-containing mRNAs. This strategy has been employed to inhibit replication of Moloney murine leukemia virus by expression of higher amounts of tRNA- and virus-specific antisense RNA-hybrid transcripts and to suppress HIV-1 replication by overexpression of tRNA-TAR sequences (which "catch" the Tat protein) (Sullenger et al. 1990).

10 Summary

2′,5′-Oligoadenylates (2-5A) have an essential role in the establishment of the antiviral state of a cell exposed to virus infection. The key enzymes of the 2-5A system are the 2-5A forming 2′,5′-oligoadenylate synthetase (2-5OAS), the activity of which depends on the presence of viral or cellular double-stranded RNA (dsRNA), and the 2-5A-activated ribonuclease (RNase L). Basic research in recent years has shown that the 2-5A system is a promising target for anti-HIV chemotherapy, particularly due to its interaction with double-stranded segments within HIV RNA. Two new strategies have been developed which yield a selective antiviral effect of 2-5A against HIV-1 infection: (1) development of 2-5A analogues displaying a dual mode of action (activation of RNase L and inhibition of HIV-1 RT) and (2) intracellular immunization of cells against HIV-1 infection by application of the HIV-1-LTR – 2-5OAS hybrid gene. A further strategy is the inhibition of DNA topoisomerase I by longer 2-5A oligomers.

Acknowledgements. This work was supported by grants from the Bundesgesundheitsamt (FVP 5/88; A2 and A3).

References

Abb J, Piechowiak H, Zachoval R, Zachoval V, Deinhardt F (1986) Infection with human T-lymphotropic virus type III and leukocyte interferon production of homosexual men. Eur J Clin Microbiol 5:365–368

Baba M, Pauwels R, Balzarini J, Schols D, De Clercq E (1989) Coumermycin A1 is a potent inhibitor of human immunodeficiency virus (HIV) replication in vitro. Int J Exp Clin Chemother 2:15–20

Baglioni C, Minks MA, Maroney PA (1978) Interferon action may be mediated by activation of a nuclease by pppA2′p5′A2′p5′A. Nature 273:684–687

Baglioni C, Maroney PA, West DK (1979) 2′5′Oligo(A) polymerase activity and inhibition of viral RNA synthesis in interferon-treated HeLa cells. Biochemistry 18:1765–1770

Baglioni C, De Benedetti A, Williams GJ (1984) Cleavage of nascent reovirus mRNA by localized activation of the 2′-5′-oligoadenylate-dependent endoribonuclease. J Virol 52:865–871

Barat C, Lullien V, Schatz O, Keith G, Nugeyre MT, Grüninger-Leitch F, Barré-Sinoussi F, LeGrice SFJ, Darlix JL (1989) HIV-1 reverse transcriptase specifically interacts with the anticodon domain of its cognate primer tRNA. EMBO J 8:3279–3285

Barré-Sinoussi F, Chermann JC, Rey F, Nugeyre MT, Chamaret S, Gruest J, Dauguet C, Axler-Blin C, Vezinet-Brun F, Rouzioux C, Rozenbaum W, Montagnier L (1983) Isolation of a T-lymphotropic retrovirus from a patient at risk for acquired immune deficiency syndrome (AIDS). Science 220:868–871

Bayard B, Bisbal C, Silhol M, Cnockaert J, Huez G, Lebleu B (1984) Increased stability and antiviral activity of 2′-O-phosphoglyceryl derivatives of (2′-5′)oligo(adenylate). Roles of phosphodiesterases and phosphatases in (2′-5′)oligo(adenylate) catabolism. Eur J Biochem 142:291–298

Bayard B, Leserman LD, Bisbal C, Lebleu B (1985) Antiviral activity in L1210 cells of liposome-encapsulated (2′-5′)oligo(adenylate) analogues. Eur J Biochem 151:319–325

Bednarik DP, Mosca JD, Raj NBK, Pitha PM (1989) Inhibition of human immunodeficiency virus (HIV) replication by HIV-trans-activated α2-interferon. Proc Natl Acad Sci USA 86:4958–4962

Benech P, Mory Y, Revel M, Chebath J (1985) Structure of two forms of the interferon-induced (2′-5′)oligo A synthetase of human cells based on cDNAs and gene sequences. EMBO J 4:2249–2256

Bisbal C, Salehzada T, Lebleu B, Bayard B (1989) Characterization of two murine (2′-5′)$(A)_n$-dependent endonucleases of different molecular mass. Eur J Biochem 179:595–602

Buimovici-Kleim E, Lange M, Klein RJ, Grieco MH, Cooper LZ (1986) Long-term follow-up of serum interferon and its acid stability in a group of homosexual men. AIDS Res 2:99–108

Capobianchi MR, De Marco F, Di Marco P, Dianzani F (1988) Acid-labile human interferon alpha production by peripheral blood mononuclear cells stimulated by HIV-infected cells. Arch Virol 99:1–19

Carter WA, Strayer DR, Brodsky I, Lewin M, Pellegrino MG, Einck L, Henriques HF, Simon GL, Parenti DM, Scheib RG, Schulof RS, Montefiori DC, Robinson WE, Mitchell WM, Volsky DJ, Paul D, Paxton H, Meyer III WA, Karikó K, Reichenbach N, Suhadolnik RJ, Gillespie DH (1987) Clinical, immunological, and virological effects of ampligen, a mismatched double-stranded RNA, in patients with AIDS or AIDS-related complex. Lancet i:1286–1292

Carter WA, Ventura D, Shapiro DE, Strayer DR, Gillespie DH, Hubbell HR (1991) Mismatched double-stranded RNA, ampligen (poly(I)·poly(C_{12}U)), demonstrates antiviral and immunostimulatory activities in HIV disease. Int J Immunopharmacol 13 (Suppl 1):69–76

Castora FJ, Kelly WG (1986) ATP inhibits nuclear and mitochondrial type I topoisomerases from human leukemia cells. Proc Natl Acad Sci USA 83:1680–1684

Castora FJ, Erickson CE, Kovacs T, Lesiak K, Torrence PF (1991) 2′,5′-Oligoadenylate inhibits relaxation of supercoiled DNA by calf thymus DNA topoisomerase I. J Interferon Res 11:143–149

Cayley PJ, Davies JA, McCullagh KG, Kerr IM (1984) Activation of the ppp$(A2'p)_n$A system in interferon-treated, herpes simplex virus-infected cells and evidence for novel inhibitors of the ppp$(A2'p)_n$A-dependent RNase. Eur J Biochem 143:165–174

Charachon G, Sobol RW, Bisbal C, Salehzada T, Silhol M, Charubala R, Pfleiderer W, Lebleu B, Suhadolnik RJ (1990) Phosphorothioate analogues of (2′-5′)$(A)_4$: agonist and antagonist activities in intact cells. Biochemistry 29:2550–2556

Chebath J, Benech P, Hovanessian A, Galabru J, Revel M (1987a) Four different forms of interferon-induced 2′,5′-oligo(A) synthetase identified by immunoblotting in human cells. J Biol Chem 262:3852–3857

Chebath J, Benech P, Revel M, Vigneron M (1987b) Constitutive expression of (2′-5′) oligo A synthetase confers resistance to picornavirus infection. Nature 330:587–588
Cohen B, Peretz D, Vaiman D, Benech P, Chebath J (1988) Enhancer-like interferon responsive sequences of the human and murine (2′-5′)oligoadenylate synthetase gene promotors. EMBO J 7:1411–1419
Cook PR (1989) The nucleoskeleton and the topology of transcription. Eur J Biochem 185:487–501
De Benedetti A, Pytel BA, Baglioni C (1987) Loss of (2′-5′)oligoadenylate synthetase activity by production of antisense RNA results in lack of protection by interferon from viral infections. Proc Natl Acad Sci USA 84:658–662
Dingwall C, Ernberg I, Gait MJ, Green SM, Heaphy S, Karn J, Lowe AD, Singh M, Skinner MA, Valerio R (1989) Human immunodeficiency virus-1 tat protein binds TAR RNA in vitro. Proc Natl Acad Sci USA 86:6925–6929
Doetsch PW, Suhadolnik RJ, Sawada Y, Mosca JD, Flick MB, Reichenbach NJ, Dang AQ, Wu JM, Charubala R, Pfleiderer W, Henderson EE (1981) Core (2′-5′)oligoadenylate and the cordycepin analog: inhibitors of Epstein-Barr virus-induced transformation of human lymphocytes in the absence of interferon. Proc Natl Acad Sci USA 78:6699–6703
Edery I, Petryshyn R, Sonenberg N (1989) Activation of double-stranded RNA-dependent kinase (dsI) by the TAR region of HIV-1 mRNA: a novel translational control mechanism. Cell 56:303–312
Faltynek CR, Princler GL, Ruscetti FW, Birchenall-Sparks M (1988) Lectins modulate the internalization of recombinant interferon-alpha A and the induction of 2′,5′-oligo(A) synthetase. J Biol Chem 263:7112–7117
Farrell PJ, Balkow K, Hunt T, Jackson RJ, Trachsel H (1977) Phosphorylation of initiation factor eIF-2 and the control of reticulocyte protein synthesis. Cell 11:187–200
Fauci AS (1988) The human immunodeficiency virus: infectivity and mechanisms of pathogenesis. Science 239:617–622
Floyd-Smith G, Slattery E, Lengyel P (1981) Interferon action: RNA cleavage pattern of a (2′-5′)oligoadenylate-dependent endonuclease. Science 212:1030–1032
Frankel AD, Bredt DS, Pabo CO (1988) Tat protein from human immunodeficiency virus forms a metal-linked dimer. Science 240:70–73
Gatignol A, Kumar A, Rabson A, Jeang KT (1989) Identification of cellular proteins that bind to the human immunodeficiency virus type 1 trans-activation-responsive TAR element RNA. Proc Natl Acad Sci USA 86:7828–7832
Gaynor R, Soultanakis E, Kuwabara M, Garcia J, Sigman DS (1989) Specific binding of HeLa cell nuclear protein to RNA sequences in the human immunodeficiency virus transactivating region. Proc Natl Acad Sci USA 86:4845–4862
Gendelman HE, Friedman RM, Joe S, Baca LM, Turpin JA, Dveksler G, Meltzer MS, Dieffenbach C (1990) A selective defect of interferon alpha production in human immunodeficiency virus-infected monocytes. J Exp Med 172:1433–1442
Gunnery S, Rice AP, Robertson HD, Mathews MB (1990) Tat-responsive region RNA of human immunodeficiency virus 1 can prevent activation of the double-stranded-RNA-activated protein kinase. Proc Natl Acad Sci USA 87:8687–8691
Ho DD, Hartshorn KL, Rota TR, Andrews CA, Kaplan JC, Schooley RT, Hirsch MS (1985) Recombinant human interferon alpha-A suppresses HTLV-III replication in vitro. Lancet i:602–604
Hovanessian AG, Svab J, Marie I, Robert N, Chamaret S, Laurent AG (1988) Characterization of 69- and 100-kDa forms of 2-5A-synthetase from interferon-treated human cells. J Biol Chem 263:4945–4949
Hubbell HR, Sheetz PC, Iogal SS, Brodsky I, Kariko K, Li S-W, Suhadolnik RJ, Sobol RW (1991) Heterogeneous nuclear RNA from hairy cell leukemia patients activates 2′,5′-oligoadenylate synthetase. Anticancer Res 11:1927–1932
Johnston MI, Hearl WG (1987) Purification and characterization of a 2′-phosphodiesterase from bovine spleen. J Biol Chem 262:8377–8382

Jongstra-Bilen J, Ittel M-E, Niedergang C, Vosberg H-P, Mandel P (1983) DNA topoisomerase I from calf thymus is inhibited in vitro by poly(ADP-ribosylation). Eur J Biochem 136:391–396

Kariko K, Li SW, Sobol RW, Suhadolnik RJ, Charubala R, Pfleiderer W (1987a) Phosphorothioate analogues of 2′,5′-oligoadenylate. Activation of 2′,5′-oligoadenylate-dependent endoribonuclease by 2′,5′-phosphorothioate cores and 5′-monophosphates. Biochemistry 26:7136–7142

Kariko K, Li SW, Sobol RW, Suhadolnik L, Reichenbach NL, Suhadolnik RJ, Charubala R, Pfleiderer W (1987b) Phosphorothioate analogs of 2-5A: elucidation of the stereochemical course of the enzymes of the 2-5A synthetase/RNase L system. Nucleosides Nucleotides 6:173–184

Kariko K, Sobol RW, Suhadolnik L, Li SW, Reichenbach NL, Suhadolnik RJ, Charubala R, Pfleiderer W (1987c) Phosphorothioate analogues of 2′,5′-oligoadenylate. Enzymatically synthesized 2′,5′-phosphorothioate dimer and trimer: unequivocal structural assignment and activation of 2′,5′-oligoadenylate-dependent endoribonuclease. Biochemistry 26:7127–7135

Karn J (1991) Control of human immunodeficiency virus replication by the *tat*, *rev*, *nef* and protease genes. Curr Opinion Immunol 3:526–536

Kerr IM, Brown RE (1978) pppA2′p5′A2′p5′A: an inhibitor of protein synthesis synthesized with an enzyme fraction from interferon-treated cells. Proc Natl Acad Sci USA 75:256–260

Kohlstaedt LA, Wang J, Friedman JM, Rice PA, Steitz TA (1992) Crystal structure at 3.5 Å resolution of HIV-1 reverse transcriptase complexed with an inhibitor. Science 256:1783–1790

Laurent G St, Yoshie O, Floyd-Smith G, Samanta H, Sehgal PB (1983) Interferon action: two (2′-5′)$(A)_n$ synthetases specified by distinct mRNAs in Ehrlich ascites tumor cells treated with interferon. Cell 33:95–102

Lengyel P (1982) Biochemistry of interferons and their actions. Annu Rev Biochem 51:251–282

Liu LF (1989) DNA topoisomerase poisons as antitumor drugs. Annu Rev Biochem 58:351–375

Mace K, Seif I, Anjard C, De Maeyer-Guignard J, Dodon MD, Gazzolo L, De Maeyer E (1991) Enhanced resistance to HIV-1 replication in U937 cells stably transfected with the human IFN-β gene behind an MHC promoter fragment. J Immunol 147:3553–3559

Malim MH, Böhnlein S, Hauber J, Cullen BR (1989) Functional dissection of the HIV-1 Rev *trans*-activator – Derivation of a *trans*-dominant repressor of Rev function. Cell 58:205–214

Marié I, Hovanessian AG (1992) The 69-kDa 2-5A synthetase is composed of two homologous and adjacent functional domains. J Biol Chem 267:9933–9939

Matthes E, Langen P, Brachwitz H, Schröder HC, Maidhof A, Weiler BE, Renneisen K, Müller WEG (1990) Alteration of DNA topoisomerase II activity during infection of H9 cells by human immunodeficiency virus type I in vitro: a target for potential therapeutic agents. Antiviral Res 13:273–286

Montefiori DC, Mitchell WM (1987) Antiviral activity of mismatched double-stranded RNA against human immunodeficiency virus in vitro. Proc Natl Acad Sci USA 84:2985–2989

Montefiori DC, Sobol RW, Li SW, Reichenbach NL, Suhadolnik RJ, Charubala R, Pfleiderer W, Modliszewski A, Robinson WE, Mitchell WM (1989) Phosphorothioate and cordycepin analogues of 2′,5′-oligoadenylate: inhibition of human immunodeficiency virus type 1 reverse transcriptase and infection in vitro. Proc Natl Acad Sci USA 86:7191–7194

Müller WEG, Schröder HC, Zahn RK, Dose K (1980) Degradation of 2′,5′-linked oligoriboadenylates by 3′-exoribonuclease and 5′-nucleotidase. Hoppe-Seyler's Z Physiol Chem 361:469–472

Müller WEG, Wenger R, Reuter P, Renneisen K, Schröder HC (1989) Association of tat protein and viral mRNA with nuclear matrix from HIV-1 infected cells. Biochim Biophys Acta 1008:208–212

Müller WEG, Matthes E, Reuter P, Wenger R, Friese K, Kuchino Y, Schröder HC (1990a) Effect of nonviable preparations from human immunodeficiency virus type 1 on nuclear matrix-associated DNA polymerase α and DNA topoisomerase II activities. J AIDS 3:1–10

Müller WEG, Okamoto T, Reuter P, Ugarkovic D, Schröder HC (1990b) Functional characterization of Tat protein from human immunodeficiency virus; evidence that Tat links viral RNAs to nuclear matrix. J Biol Chem 265:3803–3808

Müller WEG, Weiler BE, Charubala R, Pfleiderer W, Leserman L, Sobol RW, Suhadolnik RJ, Schröder HC (1991) Cordycepin analogues of 2′,5′-oligoadenylate inhibit human immunodeficiency virus infection via inhibition of reverse transcriptase. Biochemistry 30:2027–2033

Murray HW, Rutin BY, Masur H, Roberts R (1984) Impaired production of lymphokines and immune (gamma) interferon in the acquired immunodeficiency syndrome. N Engl J Med 310:883–889

Nilsen TW, Baglioni C (1979) Mechanism for discrimination between viral and host mRNA in interferon-treated cells. Proc Natl Acad Sci USA 76:2600–2604

Nilsen TW, Maroney PA, Baglioni C (1982a) Synthesis of (2′-5′)oligoadenylate and activation of an endoribonuclease in interferon-treated HeLa cells infected with reovirus. J Virol 42:1039–1045

Nilsen TW, Maroney PA, Robertson HD, Baglioni C (1982b) Heterogeneous nuclear RNA promotes synthesis of (2′,5′)oligoadenylate and is cleaved by the (2′,5′)oligoadenylate-activated endoribonuclease. Mol Cell Biol 2:154–160

Nilsen TW, Wood DL, Baglioni C (1982c) Presence of 2′,5′-oligo(A) and of enzymes that synthesize, bind, and degrade 2′,5′-oligo(A) in HeLa cell nuclei. J Biol Chem 257:1602–1605

Pestka S, Langer JA, Zoon KC, Samuel CE (1987) Interferons and their actions. Annu Rev Biochem 56:727–777

Pfleiderer W, Charubala R (1987) Synthesis of P-thioadenylyl(2′-5′)adenosine and P-thioadenylyl(2′-5′)-P-thioadenylyl(2′-5′)adenosine. Nucleosides Nucleotides 6:513–516

Pfeifer K, Merz H, Steffen R, Müller WEG, Trumm S, Schulz J, Eich E, Schröder HC (1992) In-vitro anti-HIV activity of lignans. Differential inhibition of HIV-1 integrase reaction, topoisomerase activity and cellular microtubules. J Pharm Med 2:75–97

Pommier Y, Kerrigan D, Hartman KD, Glazer RI (1990) Phosphorylation of mammalian DNA topoisomerase I and activation by protein kinase C. J Biol Chem 265:9418–9422

Priel E, Showalter SD, Roberts M, Oroszlan S, Segal S, Aboud M, Blair DG (1990) Topoisomerase I activity associated with human immunodeficiency virus (HIV) particles and equine infectious anemia virus core. EMBO J 9:4167–4172

Priel E, Showalter SD, Blair DG (1991) Inhibition of human immunodeficiency virus (HIV-1) replication in vitro by nontoxic doses of camptothecin, a topoisomerase I inhibitor. AIDS Res Hum Retrovir 7:65–72

Renneisen K, Leserman L, Matthes E, Schröder HC, Müller WEG (1990) Inhibition of expression of human immunodeficiency virus-1 in vitro by antibody-targeted liposomes containing antisense RNA to the env region. J Biol Chem 265:16337–16342

Rice AP, Kerr SM, Roberts WK, Brown RE, Kerr IM (1985) Novel 2′,5′-oligoadenylates synthesized in interferon-treated, vaccinia virus-infected cells. J Virol 56:1041–1044

Rosen CA (1991) Regulation of HIV gene expression by RNA-protein interactions. Trends Genet 7:9–14

Rosen CA, Pavlakis GN (1990) Tat and Rev: positive regulators of HIV gene expression. AIDS 4:499–509

Rossol S, Voth R, Laubenstein HP, Müller WEG, Schröder HC, Meyer zum Büschenfelde KH, Hess G (1989) Interferon production in patients infected with HIV-1. J Infect Dis 159:815–821

Roy S, Katze MG, Parkin NT, Edery I, Hovanessian AG, Sonenberg N (1990) Control of the interferon-induced 68-kilodalton protein kinase by the HIV-1 *tat* gene product. Science 247:1216–1219

Rutherford MN, Hannigan GE, Williams BRG (1988) Interferon-induced binding of nuclear factors to promoter elements of the 2-5A synthetase gene. EMBO J 7:751–759

Sarver N, Cantin EM, Chang PS, Zaia JA, Ladne PA, Stephens DA, Rossi JJ (1990) Ribozymes as potential anti-HIV-1 therapeutic agents. Science 247:1222–1225

Sawai H, Imai J, Lesiak K, Johnston MI, Torrence PF (1983) Cordycepin analogues of 2-5A and its derivatives. Chemical synthesis and biological activity. J Biol Chem 258:1671–1677

Schmidt A, Chernajovsky Y, Shulman L, Federman P, Berissi H, Revel M (1979) An interferon-induced phosphodiesterase degrading (2′-5′)oligoadenylate and the C-C-A terminus of tRNA. Proc Natl Acad Sci USA 76:4788–4792

Schröder HC, Zahn RK, Dose K, Müller WEG (1980) Purification and characterization of a poly(A)-specific exoribonuclease from calf thymus. J Biol Chem 255:4535–4538

Schröder HC, Gosselin G, Imbach J-L, Müller WEG (1984) Influence of the xyloadenosine analogue of 2′,5′-oligoriboadenylate on poly(A)-specific, 2′,5′-oligoriboadenylate degrading 2′,3′-exoribonuclease and further enzymes involved in poly(A) (+)mRNA metabolism. Mol Biol Rep 10:83–89

Schröder HC, Bachmann M, Diehl-Seifert B, Müller WEG (1987a) Transport of mRNA from nucleus to cytoplasm. Prog Nucleic Acid Res Mol Biol 34:89–142

Schröder HC, Trölltsch D, Friese U, Bachmann M, Müller WEG (1987b) Mature mRNA is selectively released from the nuclear matrix by an ATP/dATP-dependent mechanism sensitive to topoisomerase inhibitors. J Biol Chem 262:8917–8925

Schröder HC, Wenger R, Rottmann M, Müller WEG (1988) Alteration of nuclear (2′-5′)oligoriboadenylate synthetase and nuclease activities preceding replication of human immunodeficiency virus in H9 cells. Biol Chem Hoppe-Seyler 369:985–995

Schröder HC, Wenger R, Kuchino Y, Müller WEG (1989) Modulation of nuclear matrix-associated (2′-5′)oligoadenylate metabolism and ribonuclease L activity in H9 cells by human immunodeficiency virus. J Biol Chem 264:5669–5673

Schröder HC, Kljajic Z, Weiler BE, Gasic M, Uhlenbruck G, Kurelec B, Müller WEG (1990a) The galactose-specific lectin from the sponge *Chondrilla nucula* displays anti-human immunodeficiency virus activity in vitro via stimulation of the (2′-5′)oligoadenylate metabolism. Antivir Chem Chemother 1:99–105

Schröder HC, Merz H, Steffen R, Müller WEG, Sarin PS, Trumm S, Schulz J, Eich E (1990b) Differential in vitro anti-HIV activity of natural lignans. Z Naturforsch 45c:1215–1221

Schröder HC, Ugarkovic D, Merz H, Kuchino Y, Okamoto T, Müller WEG (1990c) Protection of HeLa-T4$^+$ cells against human immunodeficiency virus (HIV) infection after stable transfection with HIV LTR – 2′,5′-oligoadenylate synthetase hybrid gene. FASEB J 4:3124–3130

Schröder HC, Ugarkovic D, Wenger R, Merz H, Müller WEG (1990d) Suppression of HIV-1 replication by a HIV-*trans*-activated enzyme of the cellular antiviral response. AIFO 5: 533–534

Schröder HC, Ugarkovic D, Wenger R, Okamoto T, Müller WEG (1990e) Binding of Tat protein to TAR region of human immunodeficiency virus type 1 blocks TAR-mediated activation of (2′-5′)oligoadenylate synthetase. AIDS Res Hum Retrovir 6:659–672

Schröder HC, Suhadolnik RJ, Pfleiderer W, Charubala R, Müller WEG (1992) (2′-5′)Oligoadenylate and intracellular immunity against retrovirus infection. Int J Biochem 24:55–63

Schröder HC, Kelve M, Schäcke H, Pfleiderer W, Charubala R, Suhadolnik RJ, Müller WEG (1993) Inhibition of DNA topoisomerase I activity by 2′,5′-oligoadenylates and mismatched double-stranded RNA (ampligen) in uninfected and HIV-1-infected H9 cells. Chem-Biol Interactions (in press)

Schulof RS, Scheib RG, Parenti DM, Simon GL, DiGioia RA, Paxton HM, Chandra P, Courtless C, Taguchi TY, Sun KD, Goldstein IA, Sarin PS (1986) Treatment of HTLV-III/LAV patients with D-penicillamine. Arzneim Forsch (Drug Research) 36:1531–1534

Sen GC (1982) Mechanism of interferon action: progress towards its understanding. Prog Nucleic Acid Res Mol Biol 27:105–156

SenGupta DN, Silverman RH (1989) Activation of interferon-regulated, dsRNA-dependent enzymes by human immunodeficiency virus-1 leader RNA. Nucleic Acids Res 17:969–978

SenGupta DN, Berkhout B, Gatignol A, Zhou A, Silverman RH (1990) Direct evidence for translational regulation by leader RNA and Tat protein of human immunodeficiency virus type 1. Proc Natl Acad Sci USA 87:7492–7496

Shuman S, Moss B (1987) Identification of a vaccinia virus gene encoding a type I DNA topoisomerase. Proc Natl Acad Sci USA 84:7478–7482

Silverman RH, Jung DD, Nolan-Sorden NL, Dieffenbach CW, Kedar VP, SenGupta DN (1988) Purification and analysis of murine 2-5A-dependent RNase. J Biol Chem 263:7336–7341

Skidmore SJ, Mawson SJ (1987) Alpha-interferon in anti-HIV positive patients. Lancet ii:520

Smekens M, Dumont JE, Degeyter A, Galand P (1986) Effect of estrogen administration on rat liver 2-5A synthetase activity. Biochim Biophys Acta 887:341–344

Sobol RW, Suhadolnik RJ, Kumar A, Lee BJ, Hatfield DL, Wilson SH (1991) Localization of a polynucleotide binding region in the HIV-1 reverse transcriptase: implications for primer binding. Biochemistry 30:10623–10631

Strayer DR, Carter WA, Pequignot E, Topolsky D, Brodsky I, Suhadolnik RJ, Reichenbach N, Paul D, Einck L, Hubbell ER, Pinto A, Strauss K, Gillespie D (1991) Activity of synthetic dsRNA – ampligen – in HIV disease. Clin Biotechnol 3:169–175

Suhadolnik RJ, Doetsch PW, Devash Y, Henderson EE, Charubala R, Pfleiderer W (1983) 2′,5′-Adenylate and cordycepin trimer cores: metabolic stability and evidence for antimitogenesis without 5′-rephosphorylation. Nucleosides Nucleotides 2:351–366

Suhadolnik RJ, Lebleu B, Pfleiderer W, Charubala R, Montefiori DC, Mitchell WM, Sobol RW, Li SW, Kariko K, Reichenbach NL (1989) Phosphorothioate analogs of 2-5A: activation/inhibition of RNase L and inhibition of HIV-1 reverse transcriptase. Nucleosides Nucleotides 8:987–990

Sullenger BA, Gallardo HF,Ungers GE, Gilboa E (1990) Overexpression of TAR sequences renders cells resistant to human immunodeficiency virus replication. Cell 63:601–608

Trask DK, DiDonato JA, Muller MT (1984) Rapid detection and isolation of covalent DNA/protein complexes: application to topoisomerase I and II. EMBO J 3:671–676

Trono D, Feinberg MB, Baltimore D (1989) HIV-1 gag mutants can dominantly interfere with the replication of the wild-type virus. Cell 59:113–120

Ushijima H, Rytik PG, Schäcke H, Scheffer U, Müller WEG, Schröder HC (1993) Mode of action of the anti-AIDS compound poly(I)·poly($C_{12}U$) (ampligen): activator of 2′,5′-oligoadenylate synthetase and double-stranded RNA-dependent kinase. J Interferon Res 13:161–171

Verheijen R, van Venrooij W, Ramaekers F (1988) The nuclear matrix: structure and composition. J Cell Sci 90:11–36

Voth R, Rossol S, Hess G, Schütt KH, Schröder HC, Meyer zum Büschenfelde K-H, Müller WEG (1990) Differential gene expression of interferon-alpha and tumor necrosis factor-alpha in peripheral blood mononuclear cells from patients with AIDS related complex and AIDS. J Immunol 144:970–975

Wang JC (1985) DNA topoisomerases. Annu Rev Biochem 54:665–697

Weis JH, Faras AJ (1981) DNA topoisomerase activity associated with Rous sarcoma virus. Virology 114:563–566

Wells V, Mallucci L (1985) Expression of the 2-5A system during the cell cycle. Exp Cell Res 159:27–36

Wong-Staal F, Gallo RC (1985) Human T-lymphotropic retroviruses. Nature 317:395–403

Wreschner DH, McCauley JW, Skehel JJ, Kerr IM (1981) Interferon action – sequence specificity of the ppp(A2′p)$_n$A-dependent ribonuclease. Nature 289:414–417

Wu FK, Garcia JA, Harrich D, Gaynor RB (1988) Purification of the human immunodeficiency virus type 1 enhancer and TAR binding proteins EBP-1 and UBP-1. EMBO J 7:2117–2129

Yang K, Samanta H, Dougherty T, Jayaram B, Broeze R, Lengyel P (1981) Interferons, double-stranded RNA, and RNA degradation. Isolation and characterization of homogeneous human (2′-5′)(A)$_n$ synthetase. J Biol Chem 256:9324–9328

2′,5′-Oligoadenylate Synthetase in Autoimmune BB Rats

V. Bonnevie-Nielsen[1]

1 Introduction

In 1974, Wistar rats in the colony of Bio Breeding Laboratories in Worcester, Massachusetts, developed signs of insulin-dependent diabetes (IDDM) (Logothetopoulos et al. 1984; Marliss et al. 1982). This syndrome is believed to be a cell-mediated autoimmune process that selectively destroys the insulin producing β-cells in the islets of Langerhans in the pancreas. Throughout the years many ingenious hypotheses have been created in order to explain the destructive process, but the triggering event still remains elusive. In the many attempts to understand the etiopathogenesis of IDDM great help has come from different animal models like the streptozotocin-treated mice, the NOD (nonobese diabetic) mice, and the spontaneously diabetic BB rat.

The syndrome presented by the BB rats is developed on a genetic background (Jackson et al. 1984) and has many similarities to human type I diabetes (Table 1). By selective breeding, the rats can be divided into DP (diabetes-prone) and DR (diabetes-resistant) (Butler et al. 1983). The DP rats will develop diabetes with an incidence of 60–90%, independent of sex, whereas the incidence is only about 1% in the DR rats. Thus, the animals become diabetic without any chemical or surgical intervention and will suffer death unless insulin treatment is initiated. The classical signs of IDDM, glycosuria, weight loss, and increase in blood glucose begin rather abruptly at a mean age of 90 days and within 1–2 weeks the diabetic syndrome is fully developed (Mordes et al. 1987).

Similar to the pathological observations in the human diabetic syndrome, the islets of Langerhans of acutely diabetic BB rats are infiltrated with mononuclear cells (Dean et al. 1985; Lee et al. 1988). This characteristic finding is called insulitis or better isleitis. The specific character of the pathological process directed against the β-cells is underlined by the fact that the other islet cells (glucagon-producing α-cells, somatostatin-producing δ-cells and PP, pancreatic polypeptide, producing cells) are unaffected. In long-standing diabetes the islets are degenerated with no β-cells left, but with the other cell types still present (Seemayer et al. 1983).

[1] Dept. of Medical Microbiology, Institute of Medical Biology, University of Odense, Winsløwparken 19, DK-5000 Odense C, Denmark

Progress in Molecular and Subcellular Biology, Vol. 14
W.E.G. Müller/H.C. Schröder (Eds.)
© Springer-Verlag Berlin Heidelberg 1994

Table 1. Characteristic findings in human and BB rat diabetes

Increase in plasma glucose
Glycosuria
Insulin deficiency
Weight loss
Ketoacidosis
Death, unless insulin treatment initiated
In pancreatic islets: β-cell loss, conservation of α-, δ- and PP-cells

Table 2. Autoimmune phenomena in BB rats and man

Insulitis (mononuclear cell infiltration in the islets of Langerhans)
Thyroid inflammation
ICA (islet cell antibodies, only in man)
ICSA (β-cell-specific surface antibodies)
Cytotoxic islet cell antibodies
GAD, 64-kDa protein (glutamic acid decarboxylase) autoantibodies
Association with HLA-DQ and DR genes (man) or MHC complex
RTI^u (rat)
Circulating antibodies against thyroid, parietal cells, muscle cells and insulin

In addition to the mononuclear cell infiltration in the islets, other signs of immunological abnormalities are present both with the autoantibodies directed against different tissues and with cellular immune abnormalities (Elder et al. 1983; Like and Rossini 1984; Dyrberg et al. 1988); (Table 2). In humans, a 64-kDa protein and islet cell autoantibodies (ICA) can be detected several years before the appearance of diabetes (Baekkeskov et al. 1987) and in the DP rats these autoantibodies are also recovered several weeks before diagnosis (Baekkeskov et al. 1984). After ca. 10-year's search for the specific 64-kDa antigen, it was finally shown that GAD (glutamic acid decarboxylase) is an enzyme present in high amounts in the islets of Langerhans and that it was also the responsible antigen in eliciting the 64-kDa antibodies (Baekkeskov et al. 1990). The cellular localization of the antigen is primarily the plasma membrane of the β-cells. In BB rats, which develop diabetes, the 64-kDa antibodies are present from days 12 to 22.

Many efforts, although with less luck, have attempted to point out the causes of the β-cell destructive process. It is conceivable that β-cell cytotoxic immune cells are responsible, as demonstrated by the infiltration of mononuclear lymphocytes in the pancreatic islets. These cells are activated T-cells bearing Ia^+ surface antigens (Dean et al. 1985). Also, there is some evidence of a major role played by the NK cells (MacKay et al. 1986; Woda and Padden 1987), but other factors and other immune cells, e.g., antigen-presenting cells (macrophages) and T-helper cells, are not excluded as being essential in the specific cytotoxicity directed against the β-cells.

The significance of the antigen-presenting cellular system in the initial events in β-cell destruction has for obvious reasons attracted much interest. When silica particles, which are selectively destructive to macrophages, are administered to BB rats, diabetes and inflammatory infiltration of the pancreatic islets are prevented (Oschilewski et al. 1985; Amano and Yoon 1990). Supposedly, this is due to a decrease in the amount of macrophage-dependent T-lymphocytes including helper, suppressor, and NK cells.

The DPBB rat, but not the DR rat, exhibits a severe T-cell lymphopenia (Like and Rossini 1984) and lacks functional cytotoxic T-cells (Woda and Padden 1987; Prud'homme et al. 1988). Apparently, however, the responsible gene for this phenomenon is inherited independently of the diabetes gene.

The development of diabetes in the BB rat seems to be related to a certain MHC-haplotype (Holowachuk and Greer 1989). This is quite analogous to human IDDM (type 1 diabetes) and is suggestive of a process in which, MHC-restricted autoaggressive T-lymphocytes play a central role.

In the BB rat, the autoimmune process has the $RT1^u$ rat major histocompatibility complex as a prerequisite and apparently, within the RT1 complex, the RT1.D is most important in this connection (Boitard et al. 1985; Holowachuk and Greer 1989). In the DPBB rat the RT6 antigen is absent in peripheral T-cells, but this is not the case in DR animals. Treatment of DR rats with monoclonal antibodies against RT6.1 depletes $RT6^{\alpha +}$ T-cells and more than 50% of the animals become diabetic (Greiner et al. 1987).

2 Development of Diabetes in BB Rats Is Affected by Viruses

Viruses have been implicated as causative agents in the pathogenesis of diabetes and in some instances an association between virus infection and development of hyperglycaemia has been observed. Proposed mechanisms of action include virus-induced modification of β-cell antigens, direct lysis of β-cells by cytotoxic lymphocytes (Dean et al. 1985) or cytokine-mediated changes in regulatory and/or effector lymphocytes or molecular mimicry (Oldstone 1989). Experiments with EMC virus in mice and rats have shown that a virus may precipitate diabetes. In man, the RNA virus, Coxsackie B, has been isolated from the islets of Langerhans, which has led to the suggestion of a virus as a direct cause of the disease (Yoon et al. 1979; Szopa et al. 1990). The diversity in virus' capability of inducing diabetes has been shown for EMC virus where infection of mice with the EMC-B type leads to IFN production but not disease. The opposite is true for the EMC-D type (Kaptur et al. 1989). The latter observation is important because it points to a more or less direct, i.e. not cytokine-mediated, β-cell destruction exerted by virus. On the other hand, the findings do not exclude the action of cytotoxic effector cells (Prud'homme et al. 1988). In DRBB rats inoculated with parvovirus (Kilham rat virus), an earlier occurrence of diabetes was observed with signs of lymphocytic infiltration in the pancreatic islets. However, DPBB rats only developed diabetes provided that the animals had been reconstituted with spleen cells from DRBB rats (Guberski et al. 1991). This suggests that parvovirus acts by stimulating DR autoreactive cells directed against β-cell destruction.

The above considerations have prompted studies of environmental influences on diabetes development. Interestingly, it was demonstrated that lymphocytic choriomeningitis (LCMV) virus actually reduced diabetes incidence in DPBB rats significantly (Dyrberg et al. 1988) and in concordance with these data, more animals (DP) became diabetic earlier when obtained by caesarian section and reared SPF than in animals reared non-SPF (Like et al. 1991). The infection of DPBB rats with LCMV virus was furthermore shown not to involve cytotoxic lymphocytes (CTL) but rather production of neutralizing antibodies, which was in contrast to the DRBB rats where CTL were responsible for elimination of the virus (Oldstone et al. 1990). Other data indicate that viral infection significantly inhibits effector cell function in DPBB rats, nearly abolishes mononuclear cell infiltration in the islets (Dyrberg et al. 1988) and enhances effector cell activity in DRBB rats (Like et al. 1991).

3 Effects of dsRNA in BB Rats

The immunostimulatory effects of double-stranded RNA (dsRNA) through activation of interferon were discovered several years ago and it is now well known that the inhibiting effect of dsRNA on protein synthesis is related to the dsRNA-dependent enzymes, the 2′,5′-oligoadenylate synthetase (2-5AS) and a protein kinase (Lengyel 1987). In discussing the effect of the 2-5AS enzyme, it is notable that the synthetic dsRNA (poly I:C, i.e. poly inosinic:poly cytidylic acid) dependency of this enzyme can be substituted by an infection with a virus which generates viral dsRNA molecules (Field et al. 1972). The use of dsRNA mimicking a viral infection seems relevant and has been most fruitful even if many unsolved questions still arise regarding the effect of this molecule. For example, in rats that have not been challenged by environmental antigens, i.e. SPF animals, the poly I:C effect was strong resulting in a high percentage of diabetic rats (Like et al. 1991; Thomas et al. 1991). This could be due to an exaggerated presentation of MHC class I molecules on "unprotected" β-cells that are subsequently attacked by a non-discriminating immune system. Also, aggressive action of NK cells towards the β-cells, together with the growth inhibiting effect of the 2-5AS system, would make the β-cells highly vulnerable. In antigen-challenged rats, i.e. non-SPF animals, the poly I:C effect is reduced, possibly due to an immune system that has already experienced many different antigens. In this situation the MHC class I presentation could be downregulated, the 2-5AS reactivity could be levelled off to a low but constitutive expression due to downregulation of α-IFN receptors, and finally, even an antisense effect of poly I:C affecting transcription of the relevant genes could play an important role (De Benedetti et al. 1987). In very simple terms, the immunologically "dirty" animals are better protected against the poly I:C effect than their "clean" counterparts. The question then arises, whether the DPBB rats are in fact able to produce antibodies against the poly I:C molecules or whether poly I:C, similar to the LCMV virus, attaches to and stimulates different lymphocyte subtypes. Both possibilities seem unlikely because the LCMV virus elicits immunosuppression and protection against diabetes in the DPBB rat (Oldstone et al. 1990), whereas poly I:C on the contrary precipitates diabetes much earlier (Nakamura et al. 1991; Bonnevie-Nielsen et al. 1992). On the other hand, we have clear evidence that addition of poly I:C in vitro to the insulin-producing β-cell line (RINm-5) highly induces 2-5AS activity (Bonnevie-Nielsen et al. 1991a). The effect of poly I:C is dose-related and seems to induce effector cells most clearly in DPBB rats, but also DRBB rats become diabetic, although at a lower incidence rate (Sobel et al. 1992). In poly I:C treatment of DPBB rats, the presence of viral antigens apparently suppresses the effector cells or, in other words, protects the β-cells against destruction (Like et al. 1991). Since these different effects are not congruous, the concept of poly I:C solely as a virus-like agent should be carefully evaluated and other poly I:C-dependent mechanisms affecting the islets of Langerhans should be considered. In some cases, however, an interaction between poly I:C and viral antigens could be present, leading to the suggestion that previous or present virus infections might block the transcriptional effect on certain genes induced by poly I:C, i.e. among other MHC class I, 2-5AS, and IFN genes. In other cases, a direct effect of poly I:C on these genes, in particular the MHC class I and/or α-IFN genes, would increase the amount of MHC class I molecules on the β-cells and thus the attraction of cytotoxic T-cells.

Since dsRNA molecules are capable of inducing transcription of the 2-5AS, α-IFN and MHC class I genes (Mémet et al. 1991), it is not unlikely that these genes would be activated in the BB rat, particularly in relation to the development of diabetes. Thus, in our experiments, it was clear that α-IFN and/or poly I:C induced MHC class I molecules and 2-5AS activity in both isolated pancreatic rat islets and in the above-mentioned β-cell line, (RINm-5) (Bonnevie–Nielsen et al. 1991a). Treatment of DRBB or DPBB rats with poly I:C leads to a significantly earlier occurrence of diabetes, enforced when the rats are depleted of RT6.1 lymphocytes or are seropositive for environmental viruses (Like et al. 1991; Thomas et al. 1991). Obviously, an effect of α-IFN could be involved, but high doses of this cytokine induce only a very moderate effect in vitro on 2-5AS in DPBB rat lymphocytes (Bonnevie–Nielsen et al. 1992), very much in contrast to that found in human peripheral blood lymphocytes (PBL) (Bonnevie–Nielsen et al. 1991b).

In untreated DPBB rats the constitutively expressed 2-5AS activity apparently only leads to production of dimers of the 2′,5′-oligoadenylates (Bonnevie-Nielsen et al. 1991b). Figure 1 shows the gel electrophoretic pattern of 2′,5′-oligoadenylates generated by the human and the DRBB lymphocytic 2-5AS enzyme. Also, we could not show any differences in the sensitivity of the 2-5AS activity in PBL homogeneous to different poly I:C concentrations in DRBB rats, DPBB rats or diabetic BB rats (Fig. 2). The lack of poly I:C sensitivity is

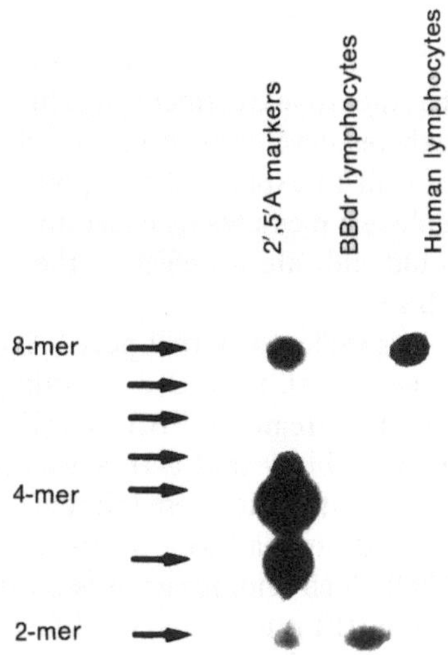

Fig. 1. 2′,5′-oligoadenylates synthesized in PBL from DRBB rat and man. 2′,5′A are molecular markers for 2′,5′-oligoadenylates. In DRBB and human PBL homogenates, dimers or di- and octamers were observed respectively. Samples were run in a gel electrophoresis system as described (Bonnevie-Nielsen et al. 1991b)

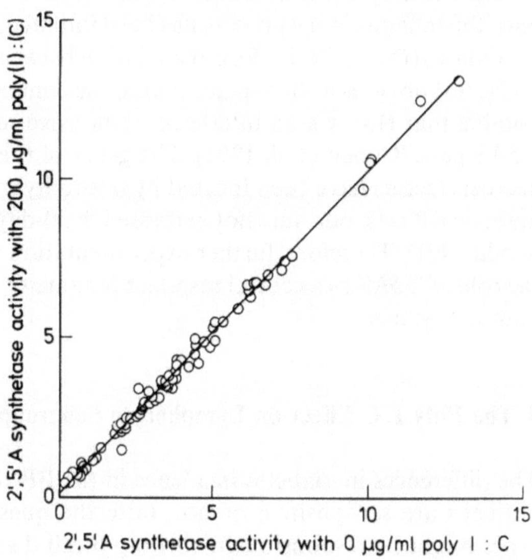

Fig. 2. 2–5AS activity in PBL homogenates from 82 BB rats (n = 24 DRBB, n = 42 DPBB, n = 16 diabetic BB). No effect of a high poly I:C concentration (200 μg/ml) was detected. For details, see Bonnevie-Nielsen et al. (1991b)

characteristic for the 100-kDa isoform of 2-5AS in both man and mice (Chebath et al. 1987). Therefore, the possibility exists that the BB rat mainly expresses the 100-kDa isoform. We found, however, that daily injections with poly I:C for 2 weeks reversed the insensitivity of 2-5AS to poly I:C in vitro and elicited high 2-5AS activities in PBL and isolated lymph nodes (Bonnevie-Nielsen et al. 1992), but we have no evidence of other isoforms being expressed. It should be noted that intensive treatment with high i.v. doses of α-IFN had no detectable effect on diabetes development or 2-5AS activity in PBL from BB rats. These data are difficult to explain at this point, but they seem to underline that the poly I:C molecule mediates its effect directly and not through α-IFN. Alternatively, they could indicate a defect in the α-IFN/2-5AS signal/response pathways in the BB rat.

It is well known that poly I:C treatment affects the expression of α-IFN genes (Mémet et al. 1991) and accordingly higher α-IFN concentrations were found in poly I:C-treated DPBB rats (Ewel et al. 1992). Therefore, in order to evaluate the possibility that α-IFN was responsible in eliciting diabetes, high doses of α-IFN antibodies were injected into the rats with or without poly I:C. However, this did not lead to a postponement of the appearance of diabetes (Ewel et al. 1992). Such findings as ours on the involvement of α-IFN in diabetes precipitation in BB rats, if any, need further clarification (Bonnevie-Nielsen et al. 1992). Thus, experiments on factors affecting α-IFN receptors and binding conditions for α-IFN to rat PBL are still missing. In any case, the different data seem to indicate the existence of different pathways which elicit IFN and poly I:C effects. Possibly, other cytokines like IL-1, IL-2, secreted locally in the pancreas once the inflammatory process has been initiated, are far more relevant cytokine candidates (Dinarello 1989), but a relation between these two cytokines and poly I:C is not obvious at this stage. Within the context of 2-5AS gene expression it is notable that IL-6 via an interferon-responsive sequence enhancer activates the 2-5AS gene (Cohen et al. 1991). The genes of this cytokine and of TNF (tumor necrosis factor) have been located by in situ hybridization in pancreatic islets in diabetic BB rats, but were not expressed in 30-day-old DP or DR rats (Jiang and Woda 1991). Therefore, further experimentation should be aimed at elucidating the role of 2-5AS as a central response parameter to the different cytokines in the immune system.

4 The Poly I:C Effect on Lymphocyte Subgroups

The differences in diabetes incidence in the BB rat, depending on whether viral antigens are seropositive or not, raise the question of what lymphocytic cell types might be responsible in eliciting β-cell destruction. It was found that the MHC antigen, RTI^u class II, was unchanged in DPBB and DRBB rats (Holowachuk and Greer 1989). However, elimination of T-lymphocytes bearing the $RT6.1^+$ antigen (suggested to be protective against diabetes development) by treatment with specific antibodies leads to higher frequency of diabetes when the rats are seropositive for vital antigens (Thomas et al. 1991). A disturbed balance between autoreactive effector cells and regulatory $RT6^+$ cells could play a role in diabetes development, but to what extent cytokines like IFN and the connected activity in the 2-5AS system are involved in these cell types remains to be determined. It was demonstrated that the number of helper-inducer and cytotoxic-suppressor cells did not differ in poly I:C and saline-treated rats. The number of NK cells was higher in the poly I:C group, whereas T-cells expressing the RT6.1 were unchanged (Sobel et al. 1992). Therefore, the increase in NK cell activity in response to poly I:C treatment could be involved in diabetes development (Sobel et al. 1992). It is puzzling that the increased α-IFN activity following poly I:C treatment apparently did not change the antigen presentation in lymphocyte subgroups in PBL. However, it should be noted that the mononuclear cells infiltrating the islets of Langerhans could be a selectively recruited population with specific antigens not detectable in PBL. Thus it has been reported that antigen-presenting macrophages, in fact, precede lymphocytic infiltration in the islets (Walker et al. 1988; Hanenberg et al. 1989). This cell type secretes interleukin I which is essential in activating T-lymphocytes. Furthermore, it was recently demonstrated that dendritic cells may be present earlier than the macrophages and therefore could be even more important in the initial T-lymphocyte stimulation (Ziegler et al. 1992). Further experiments are important because the mononuclear cell infiltration in the pancreatic islets is a main characteristic indication of the autoimmune nature of the disease. It is tempting to speculate that these autoreactive cells, stimulated by locally secreted cytokines, are provoked to express 2-5AS activity which, due to the inhibiting effect on protein synthesis of the activated RNase L, will exert a detrimental effect on the β-cells.

5 Concepts and Hypotheses

The nature and the specificity of virally or poly I:C-induced diabetes is not clear at this stage, neither in the BB rat nor in man. At least two different concepts can be based on the available data. One is the molecular mimicry concept and the other could be called the gene concept. According to the first concept, we suggest that poly I:C serves as a model of a viral infection when injected into the rats. If this molecule or a conversion product thereof is conceived or processed as a foreign viral antigen by the immune system and bears molecular similarities to certain β-cell antigens, the stage is set for an autoimmune attack by cytotoxic T-lymphocytes assisted by an upregulation of MHC class I antigen presentation. Local cytokine production of IL-1 and IL-2 from antigen-presenting macrophages and T-helper cells will enforce the attack against the β-cells. Also, production of α-IFN and, hence, the activation of the 2-5AS system will block cellular growth and protein synthesis. It should be noted that α-IFN does not necessarily need to be involved because poly I:C alone can activate the 2-5AS gene. In this scenario, increased MHC class I presentation will synchronously follow the activity of the macrophages and T-lymphocytes.

The molecular mimicry concept was detailed recently (Kaufman et al. 1992), based on amino acid sequence similarities between the GAD antigen and Coxsackie virus B_4 protein. By infection with Coxsackie virus, viral peptides would be presented in an MHC-restricted manner to T-lymphocytes through the antigen-presenting cells. The presence of GAD epitopes on the β-cells could then be recognized by the T-lymphocytes and an immune attack could be initiated. Consequently, as long as GAD molecules are released from the β-cells, the autoimmune process will proceed.

Following our gene concept, the possibility is considered that poly I:C possesses (among others) an inductive effect on MHC class I and on 2-5AS genes and is capable of activating NK cells. An increased MHC class I presentation on the β-cell surface will accelerate the autoimmune process, assisted by the NK cells. The 2-5AS activity will have the same detrimental effect on protein synthesis as mentioned before. By unifying the components of the two concepts, a sufficient basis for efficient autoimmune destruction of the pancreatic β-cells is created.

The precipitating event for the β-cell destruction in insulin-dependent diabetes in animals and man is the crucial but still unexplained point. However, it seems reasonable to suggest that one possible explanation lies hidden in the precipitating effect of a molecular structure as exemplified by dsRNA.

References

Amano K, Yoon J-W (1990) Studies on automimmunity for initiation of β-cell destruction. Diabetes 39:590–596

Baekkeskov S, Dyrberg T, Lernmark Å (1984) Autoantibodies to a 64-kilodalton islet cell protein precede the onset of spontaneous diabetes in the BB rat. Science 224:1348–1350

Baekkeskov S, Landin M, Kristensen JK, Srikanta S, Bruining GJ, Mandrup-Poulsen T, deBeaufort C, Soeldner JS, Eisenbarth G, Lindgren, Sundquist G, Lernmark Å (1987) Antibodies to a 64.000 M_r human islet cell antigen precede the clinical onset of insulin-dependent diabetes. J Clin Invest 79:926–934

Baekkeskov S, Aanstot H-J, Christgau S, Reetz M, Solimena M, Casalko M, Folli F, Richter-Olesen H, Camilli P (1990) Identification of the 64 K autoantigen in insulin-dependent diabetes as the GABA-synthesizing enzyme glutamic acid decarboxylase. Nature 347:151–156

Banatvala JE (1987) Insulin-dependent (juvenile-onset, type 1) diabetes mellitus Coxsackie B viruses revisited. Prog Med Virol 34:33–54

Boitard C, Michie S, Serrurier P, Butcher GW, Larkins AP, McDevitt HO (1985) In vivo prevention of thyroid and pancreatic autoimmunity in the BB rat by antibody to class II major histocompatibility complex gene products. Proc Natl Acad Sci USA 82:6627–6631

Bonnevie-Nielsen V, Gerdes A-M, Fleckner J, Petersen JS, Michelsen B, Dyrberg T (1991a) Inteferon stimulates the expression of 2′,5′-oligoadenylate synthetase and MHC class I antigens in insulin producing cells. J Interferon Res 11:255–260

Bonnevie-Nielsen V, Husum G, Kristiansen K (1991b) Lymphocytic 2′,5′-oligoadenylate synthetase is insenitive to dsRNA and interferon stimulation in autoimmune BB rats. J Interferon Res 11:351–356

Bonnevie-Nielsen V, Husum G, Markholst H, Dyrberg T (1992) Double stranded RNA, but not α-IFN accelerates onset of diabetes and restores 2′,5′-oligoadenylate synthetase activity in the BB rat. Pediat Adoles Endocinol 23:6–12

Butler L, Guberski DL, Like AA (1983) Genetic analysis of the BB/W diabetic rat. Can J Genet Cytol 25:7–15

Chebath J, Benech P, Hovanessian A, Galabru J, Revel M (1987) Four different forms of interferon-induced 2′,5′ oligo(A) synthetase identified by immunoblotting in human cells. J Biol Chem 262:3852–3857

Cohen B, Gothelf Y, Vaiman D, Chen L, Revel M, Chebath J (1991) Interleukin-6 induces the (2′-5′)oligoadenylate synthetase gene in M1 cells through an effect on the interferon-responsive enhancer. Cytokine 3:83–91

Dean BM, Walker R, Bone AJ, Baird JD, Cooke A (1985) Pre-diabetes in the spontaneously diabetic BB/E rat: lymphocyte subpopulation in the pancreatic infiltrate and expression of rat MHC class II molecules in endocrine cells. Diabetologia 28:464–466

De Benedetti A, Pytel BA, Baglioni C (1987) Loss of (2′-5′)oligoadenylate synthetase activity by production of antisense RNA results in lack of protection by interferon from viral infections. Proc Natl Acad Sci USA 84:658–662

Dinarello CA (1989) Interleukin-1 and its biologically related cytokines. Adv Immunol 44:153–205

Dyrberg T, Schwimmbeck P, Oldstone MBA (1988) Inhibition of diabetes in BB rats by virus infection. J Clin Invest 81:928–931

Elder ME, Maclaren NK (1983) Identification of profound peripheral T lymphocyte immunodeficiencies in the spontaneously diabetic BB rat. J Immunol 130:1723–1731

Ewel CH, Sobel DO, Zeligs BJ, Bellanti JA (1992) Poly I:C accelerates development of diabetes mellitus in diabetes-prone BB rat. Diabetes 41:1016–1021

Field AK, Tytell AA, Lampson GP, DeSomer P (1972) Interferon production linked to the toxicity of polyribosinic-polycytidylic acid. Proc Natl Acad Sci USA 58:1004–1010

Greiner DL, Mordes JP, Handler ES, Angelillo M, Nakamura N, Rossini AA (1987) Depletion of RT6.1$^+$ T lymphocytes induces diabetes in resistant BioBreeding/Worcester (BB/W) rats. J Exp Med 166:461–475

Guberski DL, Thomas VA, Shek WR, Like AA, Handler ES, Rossini AA, Wallace JE, Welsh RM (1991) Induction of type I diabetes by Kilham's rat virus in diabetes-resistant BB/Wor rats. Science 254:1010–1013

Hanenberg H, Kolb-Bachofen V, Kantwerk-Funke G, Kolb H (1989) Macrophage infiltration precedes and is a prerequisite for lymphocytic insulitis in pancreatic islets and pre-diabetic BB rats. Diabetologia 32:126–134

Holowachuk EW, Greer MK (1989) Unaltered class II histocompatibility antigens and pathogenesis of IDDM in BB rats. Diabetes 38:267–271

Jackson RA, Buse JB, Rifai R, Pelletier D, Milford EL, Carpenter CB, Eisenbarth GS, Williams RM (1984) Two genes required for diabetes in BB rats: evidence from cyclical intercrosses and backcrosses. J Exp Med 159:1629–1636

Jiang Z, Woda B (1991) Cytokine gene expression in the islets of the diabetic biobreeding/Worcester rat. J Immunol 146:2990–2994

Kaptur P, Thomas DC, Giron D (1989) Differing attachment of diabetogenic and non-diabetogenic variants of encephalomyocarditis virus to β-cells. Diabetes 38:1103–1108

Kaufman DL, Erlander MG, Clare-Salzler M, Atkinson MA, Marclaren NK, Tobin AJ (1992) Autoimmunity to two forms of glutamate decraboxylase in insulin-dependent diabetes mellitus. J Clin Invest 89:283–292

Lee KU, Lim MK, Amano K, Pak CY, Jarworski MA, Mehta JG, Yoon J-W (1988) Preferential infiltration of macrophages during early stages of insulitis in diabetes-prone BB rats. Diabetes 37:1053–1058

Lengyel P (1987) Double-stranded RNA and interferon action. J Interferon Res 7:511–519

Like AA, Rossini AA (1984) Spontaneous autoimmune diabetes mellitus in the BioBreeding/Worcester rat. Surv Synth Pathol Res 3:131–138

Like AA, Guberski DL, Butler L (1991) Influence of environmental viral agents on frequency and tempo of diabetes mellitus in BB/Wor rats. Diabetes 40:259–262

Logothetopoulos J, Valiquette N, Madura E, Cvet D (1984) The onset and progression of pancreatic insultis in the overt, spontaneously diabetic, young adult BB rat studied by pancreatic biopsy. Diabetes 33:33–36

MacKay P, Jacobson J, Rabinovitch A (1986) Spontaneous diabetes mellitus in the Bio-Breeding/Worcester rat: evidence in vitro for natural cell lysis of islet cells. J Clin Invest 77:916–924

Marliss EB, Nakhooda AF, Poussier P, Sima AA (1982) The diabetic syndrome of the BB Wistar rat: possible relevance to type I (insulin dependent) diabetes in man. Diabetologia 22:225–232

Mémet S, Besancon F, Bourgeade MF, Thang MN (1991) Direct induction of interferon-γ and interferon α/β-inducible genes by double-stranded RNA. J Interferon Res 11:131–141

Mordes JP, Desemone J, Rossini AA (1987) The BB rat. Diabetes Metab Rev 3:725–750

Nakamura N, Tsutsumi Y, Kimata S, Sawada M, Hasegawa G, Kitagawa Y, Nakano K, Kondo M, Nakao H, Makino S (1991) Induction of diabetes by poly I:C and anti-RT6.1 antibody treatment in DR-BB rats. Endocrinol Jpn 38:523–526

Oldstone MBA (1989) Molecular mimicry as a mechanism for the cause and as a probe uncovering etiologic agent(s) of autoimmune disease. Curr Top Microbiol Immunol 145:126–135

Oldstone MBA, Tishon A, Schwimmbeck PL, Shyp S, Lewicki H, Dyrberg T (1990) Cytotoxic T lymphocytes do not control lymphocytic choriomeningitis virus infection of BB diabetes-prone rats. J Gen Virol 71:785–791

Oschilewski U, Kiesel U, Kolb H (1985) Administration of silica prevents diabetes in BB rats. Diabetes 38:197–199

Prud'homme GJ, Lapchak PH, Parfrey NA, Colle E, Guttman RD (1988) Autoimmunity – prone BB rats lack functional cytotoxic T cells. Cell Immunol 114:198–208

Seemayer TA, Colle E, Tannenbaum GS, Oligny LL, Guttman RJ, Goldman H (1983) Spontaneous diabetes mellitus syndrome in the rat. III. Pancreatic alterations in a glycosuric and untreated diabetic BB Wistar-derived rats. Metabolism 32 (Suppl 1):26–32

Sobel D, Newsome J, Ewel CH, Bellanti JA, Abbassi V, Creswell K, Blair O (1992) Poly I:C induces development of diabetes mellitus in BB rats. Diabetes 41:515–520

Szopa TM, Ward T, Dronfield DM, Portwood ND, Taylor KW (1990) Coxsackie B_4 viruses with the potential to damage beta cells of the islets are present in clinical isolates. Diabetologia 33:325–328

Thomas V, Woda BA, Handler ES, Greiner DL, Mordes JP, Rossini AA (1991) Altered expression of diabetes in BB/Wor rats by exposure to viral proteins. Diabetes 40:255–258

Walker R, Bone AJ, Cooke A, Baird J (1988) Distinct macrophage subpopulations in pancreas of prediabetic BB/E rats. Diabetes 37:1301–1304

Woda BA, Padden C (1987) Biobreeding/Worcester (BB/Wor) rats are deficient in the generation of functional cytotoxic T cells. J Immuno 139:1514–1517

Yoon J-W, Austin M, Onodera T, Notkins AL (1979) Virus induced diabetes mellitus. Isolation of a virus from the pancreas of a child with diabetic ketoacidosis. N Engl J Med 300:d1173–1179

Ziegler A-G, Erhard J, Lampeter EF, Nagelkerken LM, Standl E (1992) Involvement of dendritic cells in early insulitis of BB rats. J Autoimmunity 5:571–579

Oligoadenylate and Cyclic AMP: Interrelation and Mutual Regulation

A.V. ITKES[1]

1 Introduction

The problem regarding interconnection and mutual regulation of cyclic adenosine 3′,5′-monophosphate (cAMP) and 2′,5′-olisoadenylate (2-5A) became evident when an involvement of both low molecular weight intracellular regulators in the control of cell proliferation was demonstrated: the action of both cAMP and 2-5A on cell growth processes suggested a coordination of their activity and a regulation of each of these agents by the other.

The role of cAMP in cell growth control has been extensively investigated, and its action as proliferative or antiproliferative agent discussed (see, for example, recent reviews by Chochung et al. 1991a, b; Maenhaut et al. 1991). The partially contradictory data on the effects of cAMP presumably correspond to a multiple mechanism of its action; the expression of different aspects of this complicated, cAMP-dependent mechanism in different cells and tissues can result in the different effects observed. In any case, both positive and negative action of cAMP on cell propagation, as noted above, supports the hypothesis of its coordination with corresponding effects of 2-5A.

Initially, 2-5A was described as a mediator of the antiviral and antiproliferative action of interferons (IFNs) (see the comprehensive reviews in this volume and general review by Pestka et al. 1987). This biological role of 2-5A is now accepted. At the same time, variation in the 2-5A level was found for cells and tissues that were not treated with IFNs: in nongrowing confluent cells and in cells stimulated to divide (Stark et al. 1979; Kimchi et al. 1981; Krishnan and Baglioni 1981; Jacobsen et al. 1983), in the dog liver regenerating after hepatectomy (Etienne-Smekens et al. 1983), etc. These results have shown a strong negative correlation of the 2-5A content and rate of cell proliferation that suggest an involvement of 2-5A in the general mechanism(s) of cell growth regulation. From this point of view, the action of IFNs via 2-5A is only one example of the involvement of this oligonucleotide in the regulation of intracellular processes. This hypothesis was in good agreement with the results on the existence of different forms of 2-5A synthetase with different sensitivity to IFNs,

[1] Institute of Molecular Biology of the Academy of Sciences of Russia, Vavilov str. 32, 117984 Moscow, Russia

Progress in Molecular and Subcellular Biology, Vol. 14
W.E.G. Müller/H.C. Schröder (Eds.)
© Springer-Verlag Berlin Heidelberg 1994

including completely IFN-independent forms (Bonnevienielsen et al. 1991) of the enzyme.

In my opinion, the interregulation of 2-5A and cAMP also shows that 2-5A is a universal intracellular regulator; the data discussed in the present chapter confirm this idea.

2 Interaction of 2-5A and cAMP: Direct Regulation of the Enzymes of cAMP and 2-5A Metabolism

As noted above, 2-5A was described originally as an IFN intermediate; presently, this role of 2-5A is widely accepted. Therefore, it was reasonable to divide the data on the interrelation of the oligoadenylates and cAMP. The first group of results describes the biochemical mechanisms directly connecting 2-5A and cAMP; to the second group belong the data on the interrelationship of cAMP and IFNs, i.e. information on the IFN-dependent regulation of the cAMP system that might be mediated by 2-5A or, in contrast, on the action of IFNs on 2-5A via cAMP.

The first group of results contains data on the regulation of cAMP enzymes and 2-5A metabolism by each of these agents; these will be described in the next three sections.

2.1 2-5A-Dependent Activation of Phosphodiesterase of cAMP

Activation of cAMP phosphodiesterase (cAMP PDE) by 2-5A was described for both the enzyme activity in the cell extracts and for partially purified preparations of cAMP PDE (Itkes and Kochetkova 1981). This effect has been demonstrated in vitro in the presence of chemically synthesized 5′-triphosphorylated 2-5A.

As shown previously, two main forms of cAMP PDE are present in different animal cells and tissues, namely, forms with a high ($Km = n \times 10^{-6}$) or low affinity ($Km = n \times 10^{-5}$ M) to cAMP (Thompson et al. 1984). The sole form of cAMP PDE that was found in NIH 3T3 cell extracts appeared to be the high-affinity form; triphosphorylated 2-5A activates this enzyme ca. two-fold. The kinetics of this activation was found to correspond to a standard noncompetitive mechanism with the constant of activation, Ka, amounting to 5×10^{-8} M (Tunitskaya et al. 1983).

In addition, the activation by cAMP was demonstrated for a partially purified preparation of cAMP PDE from rat liver. This preparation contained the PDE forms with both high and low affinity to cAMP. However, only the high-affinity form was activated by 2-5A; the kinetic mechanism and value of Ka were similar to that for the PDE of the NIH 3T3 cell extract (Tunitskaya et al. 1983).

Thus, 2-5A activates the hydrolysis of cAMP via a high-affinity form of cAMP PDE. This formal conclusion does not mean that this regulatory pathway plays a sufficient role in the actual process of cAMP content control. However, it should be mentioned that the value of the activation constant (5×10^{-8} M) corresponds to the usual intracellular concentration of 2-5A (Knight et al. 1980; Silverman et al. 1982), which makes it possible to affect the cAMP content in vivo via this mechanism.

2.2 cAMP-Dependent Induction of 2-5A Synthetase

Experiments on cAMP-dependent induction of 2-5A synthetase were carried out with cultured cells; Table 1 shows the main results.

As can be seen from Table 1, the cAMP-dependent elevation of 2-5A synthetase activity (at least from the literature cited) is lower than the elevation caused by IFNs that increase the enzyme activity 10- to 100- fold (reviewed by Pestka et al. 1987) or even more (Stark et al. 1979). Moreover, IFN and cAMP demonstrated an additive action, increasing the synthetase activity (Krispin et al. 1984). These data suggest that the induction of 2-5A synthetase by IFNs involves a mechanism independent of cAMP. At the same time, a cAMP-dependent mechanism may also play a role in enzyme induction (for details, see Sect. 3).

2.3 Putative Mechanism of the cAMP-Dependent Induction of 2-5A Synthetase

Preliminary results have shown that cAMP elevates 2-5A synthetase activity via gene activation rather than by enzyme molecule phosphorylation (Itkes et al. 1984b); unfortunately, the mechanism of this activation is still unclear. Nevertheless, new data on cAMP-dependent regulation of gene expression make it possible to propose a hypothetical mechanism for the activation of the 2-5A synthetase gene.

Recently, a few transcription factors were described that mediate cAMP action on corresponding genes. The more universal of these factors is CREB, the

Table 1. Induction of 2-5A synthetase by cAMP in different cell lines

Cell line	Rate of elevation of cAMP content	Rate of activation of 2-5A synthetase[a]	Reference
NIH 3T3	2.8	2.4	Itkes et al. (1984b)
CaOv	Not detected	2.1	Itkes et al. (1982)
L–929	2.0	4.4	Bokhonko et al. (1988)

[a]Data on maximal rates of observed 2-5A synthetase activation; these rates varied depending on conditions of cultivation, mechanism of cAMP elevation, etc.

protein binding to the regulatory site CRE (*c*AMP-*r*esponsive *e*lement) of cAMP-stimulating genes (reviewed by Roesler et al. 1988). The computer analysis carried out in our laboratory demonstrated that in the sequence of the 5′-flanking region of the human 2-5A synthetase gene (Rutherford et al. 1988) there is no site similar to the CRE consensus TGACGTCA. However, the regulatory 5′-region of this gene contains at least three potential transcription factor binding sites that could mediate the action of cAMP.

The first putative site is the sequence similar to the binding region of the C/EBP factor. This protein regulates metabolic reactions including cAMP-dependent processes in liver and fatty tissue; the consensus for the binding factor is ATTGCGCAAT (McKnight et al. 1989). In addition, Metz and Ziff (1991) have found that the transcription factor rNFIL-6 recognizes a fragment of the SRE promoter element of the *c-fos* gene: ATTAGGACAT; this fragment is similar to the C/EBP binding site. It was also shown that binding of the rNFIL-6 to SRE is increased by the phosphorylation of the factor (Metz and Ziff 1991). Hence, both C/EBP and rNFIL-6 (or their analogs in different tissues and other species) could be discussed as possible mediators of gene activation by cAMP.

The corresponding sequence of the 2-5A synthetase gene is shown in Table 2. It is positioned in the gene ca. 350 bp upstream to the multiple IFN-induced RNA start site (Rutherford et al. 1988), i.e. the location is similar to that for the SRE region in the *c-fos* gene. Thus, the sequence of the 2-5A synthetase gene presumably binding C/EBP and/or NFIL-6 factors might act as the transcription regulatory site.

The second sequence found in the synthetase gene is similar to the AP-2 element. This element was originally shown to be a basal transcription enhancer present in the SV-40 virus promoter region and in the human metallothionein gene (Mitchell et al. 1987). Its role as a cAMP-inducible enhancer was demonstrated first for the *b*-globin gene (Imagava et al. 1987) and then for a number of cAMP-dependent genes (Roesler et al. 1988). The identification of the AP-2-like sequence in the human 2-5A synthetase gene is presented in Table 3. It should be mentioned that the sequences of the AP-2 elements vary in different genes; Table 3 shows only the elements most similar to the sequence of the synthetase gene (for more details, see Roesler et al. 1988).

Table 2. The sequences similar to the C/EBP-binding site

Gene	Sequence	Reference
C/EBP consensus	ATTGCG–CAAT	McKnight et al. (1989)
SRE of *c-fos*	ATTAGG–ACAT	Metz and Ziff (1991)
Mutant SRE of *c-fos*	ATTAGC–TCAT	–//–
2-5A synthetase	TTAGAATCAT	Rutherford et al. (1988)

Table 3. AP–2 element-like sequences in the cAMP-dependent genes

Gene	Sequence	Reference
Plasminogen activator pig	− 642 − 635 CCCCACCC	Nagammine et al. (1984)
Tyrosine aminotransferase rat	− 101 − 94 TCCCTCCC	Becker et al. (1987)
Metallothionein II_a human	− 120 − 113 CGCCTGGA	Imagava et al. (1987)
2-5A synthetase human	ca. − 570[a] CCCCTGCA	Rutherford et al. (1988)

[a]Human 2-5A synthetase is transcribed from multiple (not single) IFN-induced RNA start sites (Rutherford et al. 1988), i.e. the distance from the AP-2-like site to the beginning of the starting region is approximated.

Table 4. The NF-$\varkappa$B-binding sequences

Gene	Sequence[a]	Reference
Immunoglobulin $\varkappa$-chain	GGgactttCC	Wirth and Baltimore (1988)
HIV-1 direct repeat	GGgactttCC	Nabel and Baltimore (1987)
$H2K^b$ class I gene	GGgattccCC	Baldwin and Sharp (1987)
SV-40 enchancer	GGaaagtcCC	Baldwin and Sharp (1987)
IL-2 receptor, α-chain	GGgagattCC	Bohnlein et al. (1988)
IFN-β	GGgaaattCC	Goodbourn and Maniatis (1988); Fujita et al. (1989)
2-5A synthetase	GGaaaataCC	Rutherford et al. (1988)

[a]Capital letters show the conserved nucleotides.

The third regulatory element considered as a hypothetical cAMP-dependent regulator is the putative site for transcription factor NF-$\varkappa$B. This factor was first demonstrated to be a lymphoid-specific regulator of immunoglobulin $\varkappa$-chain transcription (Wirth and Baltimore 1988). This element was later registered in other genes including IFN-β (Goodbourn and Maniatis 1988; Fujita et al. 1989). The NF-$\varkappa$B-binding sequences of these genes are shown in Table 4.

The presence of the NF-$\varkappa$B element in the IFN-β gene is more interesting for us because this element was suggested to act cooperatively with the IRF-1 regulatory element (Fujita et al. 1989). A similar element is located also in the 2-5A synthetase gene (reviewed by Porter et al. 1988). This fact allows the assumption that these motifs cooperate in the synthetase gene as well.

The activation of NF-$\varkappa$B by cAMP-dependent protein kinase (Shirakawa and Mizel 1989) is related to an increase in the translocation rate of the factor into cell nuclei; the target of the phosphorylation is presumably I$\varkappa$B, the protein inhibiting NF-$\varkappa$B in cytoplasm (Baeuerle and Baltimore 1988). An analogous mechanism may be involved in the mediation of cAMP action on the 2-5A synthetase gene.

Thus, the data reviewed above demonstrate that the mechanism of stimulation of the 2-5A synthetase gene by cAMP could be multiple, moreover, the action of each of the factors described can vary in different tissues and cell lines, under different kinds of treatment, etc. This complicated situation might account for the absence (at least presently) of clear experimental data on this mechanism(s).

2.4 cAMP-Dependent Phosphorylation of the Inhibitor of 2′-PDE. Inhibition of 2′-PDE

2′-Phosphodiesterase (2′-PDE), the enzyme which hydrolyzes 2-5A to ATP and AMP, is involved in the regulation of biological processes associated with variations in the intracellular 2-5A level including IFN action and alterations in cell proliferation status (Pestka et al. 1987).

The inhibition of 2′-PDE by cAMP was originally shown for NIH 3T3 cell cultures (Itkes et al. 1984b). Studies on the molecular mechanism of the regulation of 2′-PDE by cAMP (Itkes and Severin 1987) have demonstrated that the cAMP-dependent protein kinase phosphorylates the protein inhibitor of 2′-PDE (referred to as I-18); the phosphorylated form of I-18 inhibits the 2′-PDE activity.

The action of phosphorylated I-18 has been demonstrated for highly purified 2′-PDE and the inhibitor preparation from NIH 3T3 cells and for the enzyme/inhibitor complex from rat liver (Itkes et al. 1984a; Itkes and Severin 1987; see Table 5).

The kinetic mechanism for the enzymes and inhibitors from both origins was found to be noncompetitive. The values of the kinetic constants for the enzyme and I-18 from NIH 3T3 cells were $Km = 1.5 \times 10^{-4}$ M (for "core" trimer 2-5A),

Table 5. 2′-Phosphodiesterase-inhibiting activity of I-18 protein from rat liver: activation of the inhibitor by phosphorylation (calculated from data by Itkes and Severin 1987)

Activity of 2′-PDE[a] in the presence of I-18 (cpm)	Phosphorylation of the I-18 molecule (cpm)	Rate of inhibition, (%)
1210	70	0
960	950	21
700	1400	42
420	1700	65
230	2300	81

[a] Purified I-18 was phosphorylated by the catalytic subunit of the cAMP-dependent protein kinase, added to a partially purified preparation of 2′-PDE from rat liver and then incubated with 2-5A.

$Ki = 7 \times 10^{-8}$ M; for the preparations from rat liver: $Km = 3 \times 10^{-4}$ M, $Ki = 10^{-7}$ M.

It should be mentioned that the similar mechanism of cAMP-dependent enzyme regulation, mediated by phosphorylation of protein inhibitor, was found first for protein phosphatase type I (Cohen 1982). This enzyme is regulated by a number of protein inhibitors including the inhibitor type I, which is phosphorylated by cAMP-dependent protein kinase that causes activation of the inhibitor. Thus, the mechanism involved in 2′-PDE regulation by cAMP is not unique; it is also possible that analogous mechanisms are characteristic for other enzymes.

3 Interferons and cAMP

3.1 Involvement of cAMP in the Interferon-Dependent Regulation of the 2-5A System

Early investigations on the involvement of cAMP in the action of IFN have shown (Friedman and Pastan 1969; Weber and Stewart 1975) that, firstly, cAMP potentiates antiviral activity of IFN and, secondly, IFN increases the intracellular cAMP content. The analysis of the multiple action of IFN on macrophage-like cells of wild-type, cAMP-dependent protein kinase-deficient and adenylate cyclase-deficient variants (Schneck et al. 1982) suggests that at least for transformed macrophages, IFN-dependent elevation of intracellular cAMP mediates the stimulation of phagocytosis and growth inhibition; in contrast, the antiviral effect of IFN is cAMP-independent.

However, a number of other works have demonstrated that there is no strict correlation between the effects of IFNs and cAMP elevation and activation of cAMP-dependent processes. Different types of IFN and different cell lines also exhibited differences in the rate of mediation of IFN action by cAMP. The treatment of CaOv cells with IFNs (see below) is an example of IFN action involving the cAMP mediatory mechanism.

Action of native IFN-α on CaOv cell culture (Kalmakhelidze et al. 1987; Balandin et al. 1988) results in an increase in the cAMP content, caused by both activation of adenylate cyclase and inhibition of cAMP PDE (Fig. 1). Similar dynamics of the enzymes of (cAMP metabolism is observed after treatment of the cells with IFN-β (Fig. 2). At the same time, IFN-γ does not affect the content of cAMP in the CaOv cells. Hence, even for the cells elevating cAMP under IFN treatment, the effect depends on the type of IFN used. Nevertheless, CaOv cell culture is a good model for the experiments serving to elucidate the mediation by cAMP of the rise in 2-5A synthetase activity caused by IFNs (α or β).

The involvement of cAMP in the enhancement of 2-5A synthetase activity following IFN treatment of the cells is demonstrated by the data in Table 6.

In these experiments the inhibitor of cAMP PDE, isobutilmethylxantine was used to increase the intracellular concentration of cAMP ca. 2.5-fold. At the

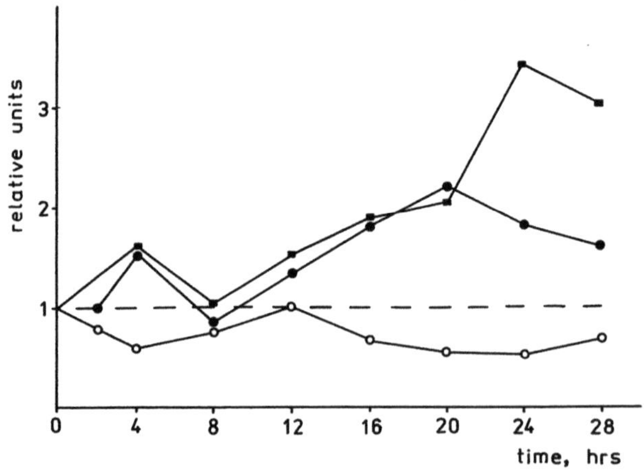

Fig. 1. Dynamics of the components of the cAMP system in the CaOv cell culture treated with native IFN-α, 1200 IU/ml. *Filled circles* activity of the adenylate cyclase; *open circles* activity of cAMP PDE; *filled squares* content of cAMP

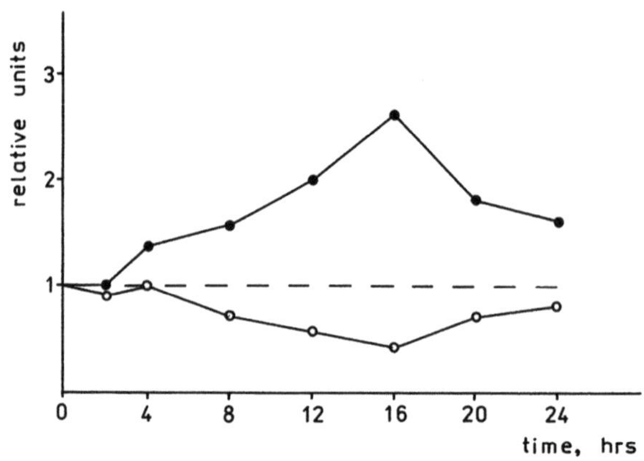

Fig. 2. Dynamics of the components of the cAMP system in the CaOv cell culture treated with native IFN-β, 350 IU/ml. *Filled circles* activity of the adenylate cyclase; *open circles* activity of cAMP PDE

same time, a low dose of IFN-α (25 IU/ml) did not change the level of cyclic nucleotides. Hence, the data show that IFN-α can enhance the 2-5A synthetase level not only via cAMP; but, in addition, higher doses of the IFNs causing the rise in cAMP content also cause the cAMP-mediating contribution to the total elevation of enzyme activity.

This hypothesis could explain the contradictory conclusions of different authors concerning the involvement of cAMP in IFN action, especially in the

Table 6. Activity of 2-5A synthetase in (CaOv cells treated with isobutilmethylxantine and IFN-α. (Krispin et al. 1984)

Isobutilmethylxantine (0.1 mM) in the cell culture media	Native human IFN-α (25 IU/ml) in the cell culture media	2-5A synthetase activity (pmol/ mg cell protein)
−	−	20
+	−	51
−	+	72
+	+	150

Table 7. Antiviral effect of the isobutilmethylxantine-induced elevation of the intracellular cAMP content in the CaOv cell culture. (Krispin et al. 1984)

Concentration of isobutil-methylxantine in the cell culture media (mM)	Activity of 2-5A synthetase (%)	Yield of EMCV, plaque-forming units $\times 10^{-7}$
0	100	1.3
0.1	130	0.7
0.5	230	0.1

modulation of 2-5A synthetase. The multiple mechanism of this modulation makes it possible to develop the experimental models exhibiting IFN-stimulating enzyme activation which can be independent of cAMP or partially dependent, etc. This varies according to the type of IFN, the kind of cell line (which can be partially resistant to IFNs and can have a normal or mutant system of cAMP-dependent phosphorylation and a system of 2-5A-connected proteins), and many other factors.

3.2 The Cyclic AMP/2-5A System Mimics Partially the Antiviral Activity of IFNs

The data discussed above suggest that if the experimental model corresponds to all conditions regarding the involvement of cAMP in the mechanism of 2-5A synthetase activation, the final effect of the rise in cAMP content is the development of an antiviral state of the cells even in the absence of IFN. This antiviral state could be observed only for viruses sensitive to 2-5A, because other IFN-dependent antiviral mechanisms (Pestka et al. 1987) evidently are not activated as a consequence of the rise in cAMP. When the encephalomyocarditis virus (EMCV), which is highly sensitive to 2-5A, was used (Krispin et al. 1984), both an antiviral effect of cAMP (the content of which was increased by isobutilmethylxantine, the inhibitor of cAMP PDE) and a correlation of this effect with enhancement of 2-5A synthetase activity were found (Table 7).

4 Cyclic AMP-Dependent Phosphorylation Causes the Elevation of the 2-5A Level Correlating with Antiproliferative Effects

The partial mediation of IFN-induced 2-5A enhancement by cAMP also means the mediation of the antiproliferative action of IFN by cAMP because 2-5A is involved in this action (see Sect. 1; review by Pestka et al. 1987; reviews, this Vol.). Furthermore, the role of cAMP in growth inhibition by IFNs has been demonstrated directly (Schneck et al. 1982; Okutani et al. 1991). In addition, the direct antiproliferative action of double-stranded RNA, involving also the enhancement of 2-5A, is mediated by cAMP as well (Hubbell et al. 1991). Moreover, since the inhibition of cell growth is generally accompanied by an increase in the cAMP level and in the activity of the cAMP-dependent protein kinase (Pastan et al. 1975; Friedman et al. 1976), the cAMP-induced enhancement of the 2-5A content evidently contributes to the general biochemical mechanisms connected with a depression of cell division.

The data illustrating the correlation between the activation of cAMP-dependent protein kinase and increase in 2-5A synthetase and fall in 2′-PDE activities were obtained for NIH 3T3 cells in the resting state (Itkes et al. 1985). This correlation corresponds to the data described above on the cAMP action in vitro and in cell culture. It is important that in this case the activity of cellular cAMP-dependent protein kinase was directly controlled; the observed effect was caused by its action, which agrees well with the mechanism discussed (Table 8).

It should be mentioned that in the resting NIH 3T3 cells the activation of cAMP-dependent protein kinase is controlled by the heat-stable protein inhibitor of the enzyme (Witehouse and Walsh 1982) rather than by the intracellular concentration of cAMP. This finding shows that investigations of the interrelation of the cAMP and 2-5A systems should involve the analysis of all the components of these systems including enzyme and inhibitor activities, levels of target proteins, etc.

Table 8. Dynamics of the cAMP-dependent protein kinase, 2-5A synthetase and 2′-PDE in NIH 3T3 cells in the resting state (calculated from data by Itkes et al. 1985; Itkes and Severin 1987)

Cells[a]	Protein kinase activity (%)	2-5A synthetase activity (cpm)	2′-PDE activity (cpm)	Content of 2-5A (mol/mg cell protein)
Control	100	500	510	0.3×10^{-14}
5 Days	550	1700	90	–
9 Days	600	4000	–	2.5×10^{-14}

[a]Time of incubation of the cell culture with the serum-depleted media is presented.

5 Summary

The data obtained are in good agreement with the hypothesis that cAMP is involved in the control of 2-5A metabolism, including the mediation of the regulation of 2-5A by IFNs; 2-5A, in turn, affects the intracellular cAMP level.

The general question originating from the data is that of a biochemical mechanism connecting the activation of the cAMP/2-5A system and the effect of depression of cell division. In my opinion, this universal effect is the result of the action of the known 2-5A-dependent mechanism, namely, RNase L (see review by Pestka et al. 1987), rather than any new 2-5A-stimulating enzyme. The RNase L activated by 2-5A decreases the total level of protein synthesis and accelerates the degradation of cellular RNA, resulting in the inhibition of cell growth. It should be mentioned that such activation of RNA turnover is generally characteristic for nondividing cells, especially for cells in the resting state (Epifanova et al. 1983). Thus, the regulatory system of cAMP/2-5A is involved evidently in the antiproliferative mechanism characteristic for the resting cells, controlling the variations in the levels of RNA turnover and protein synthesis.

Acknowledgements. We are grateful to *Molecular Genetics, Microbiology and Virology* and Edition "Medicine" (Moscow) for kind permission to reproduce figures.

References

Baeuerle PA, Baltimore D (1988) IxB: a specific inhibitor of the NFxB transcription factor. Science 242:540–546

Balandin IG, Krispin TI, Kalmakhelidze KA, Itkes AV, Solovjova MF, Kuznetzov VP (1988) Influence of interferons on cAMP system in cells. Mol Genet Microbiol Virol (Moscow) 10:40–44

Baldwin AS, Sharp P (1987) Binding of a nuclear factor to a regulatory sequence in the promoter of the mouse H–2K^b class I major histocompatibility gene. Mol Cell Biol 7:305–313

Becker PB, Ruppert S, Schutz G (1987) Genomic footprinting reveals cell type-specific DNA binding of ubiquitous factors. Cell 51:435–443

Bohnlein E, Lowenthal JW, Siekevitz M, Ballard DW, Franza BR, Greene WC (1988) The same inducible nuclear proteins regulate mitogen activation of both the interleukin-2 receptor alpha gene and type I HIV. Cell 53:827–836

Bokhonko AI, Turpaev KT, Itkes AV, Mamontova TV, Orlova TG, Kafiani CA, Severin ES (1988) Modulation of interferon production by L-929 cells treated with theophylline. Mol Genet Microbiol Virol (Moscow) 10:30–33

Bonnevienielsen V, Husum G, Kristiansen K (1991) Lymphocytic 2′,5′-oligoadenylate synthetase is insensitive to dsRNA and interferon stimulation in autoimmune BB rats. J Interferon Res 11:351–356

Chochung YS, Clair T, Tortora G, Yokozaki H (1991a) Role of site-specific cAMP analogs in the control and reversal of malignancy. Pharmacol Ther 50:1–33

Chochung YS, Clair T, Tortora G, Yokozaki H, Pepe S (1991b) Suppression of malignancy targeting the intracellular signal transducing proteins of cAMP. Life Sci 48:1123–1132

Cohen P (1982) The role of protein phosphorylation in neural and hormonal control of cellular activity. Nature 296:613–620

Epifanova OI, Terskikch VV, Polunovsky BA (1983) Resting cells. Nauka, Moscow

Etienne-Smekens M, Vandenbussche P, Content J, Dumont JE (1983) (2-5) Oligoadenylate in rat liver modulation after partial hepatectomy. Proc Natl Acad Sci USA 80:4609–4613

Friedman DL, Johnson RA, Zeiling CE (1976) The role of cyclic nucleotides in the cell cycle. Adv Cyclic Nucleotide Res 7:69–90

Friedman RM, Pastan I (1969) Interferon and cyclic-3′,5′-adenosine monophosphate: potentiation of antiviral activity. Biochem Biophys Res Commun 36:735–740

Fujita T, Miamoto M, Kimura Y, Hammer J, Taniguchi T (1989) Involvement of a *cis*-element that binds H2TF–1/NFxB like factor(s) in the virus-induced interferon-β gene expression. Nucleic Acids Res 17:3335–3346

Goodbourn S, Maniatis T (1988) Overlapping positive and negative regulatory domains of the human β-interferon gene. Proc Natl Acad Sci USA 85:1447–1451

Hubbele HR, Boyer JE, Roane P, Burch RM (1991) Cyclic AMP mediates the direct antiproliferative action of mismatched double-stranded RNA. Proc Natl Acad Sci USA 88:906–910

Imagava M, Chiu R, Karin M (1987) Transcription factor AP-2 mediates induction by two different signal transduction pathways: protein kinase C and cyclic AMP. Cell 51:251–260

Itkes AV, Kochetkova MN (1981) Activation of phosphodiesterase of cyclic adenosine monophosphate by oligoadenylate. Biochem Int 3:341–347

Itkes AV, Severin ES (1987) Regulation of the 2′,5′-oligoadenylate system by cyclic adenosine monophosphate-dependent phosphorylation. Adv Enzymol 59:213–240

Itkes AV, Krispin TI, Shloma DV, Balandin IG, Tunitskaya VL, Severin ES (1982) Cyclic AMP elevation induces 2′,5′-oligo (A) synthetase activity and antiviral and antiproliferative effects in human ovary carcinoma cells (CaOv). Biochem Int 5:389–398

Itkes AV, Kartasheva ON, Kafiani CA, Severin ES (1984a) Inhibition of 2′,-phosphodiesterase by cAMP-dependent protein kinase. Involvement of phosphorylation of protein inhibitor. FEBS Lett 176:65–68

Itkes AV, Turpaev KT, Kartasheva ON, Kafiani CA, Severin ES (1984b) Cyclic AMP-dependent regulation of activities of synthetase and phosphodiesterase of 2′,5′-oligoadenylate in NTH 3T3 cells. Mol Cell Biochem 58:165–171

Itkes AV, Kartasheva ON, Tunitskaya VL, Turpaev KT, Kafiani CA, Severin ES (1985) Activities of cAMP-dependent protein kinase and enzymes of 2′,5′-oligoadenylate metabolism in NIH 3T3 cells deepening into the resting state. Exp Cell Res 157:335–342

Jacobsen H, Krause D, Friedman RM, Silverman RH (1983) Induction of ppp $(Ap)_nA$-dependent RNase in murine JLC V9R cells during growth inhibition. Proc Natl Acad Sci USA 80:4954–4958

Kalmakhelidze KA, Krispin TI, Itkes AV, Solovjova MF, Balandin IG (1987) Influence of human interferons on the activity of phosphodiesterase of cAMP in the human ovary carcinoma cell culture (CaOv). Mol Genet Microbiol Virol (Moscow) 9:44–46

Kimchi A, Shure H, Revel M (1981) Anti-mitogenic function of interferon-induced (2′-5′) oligo (adenylate) and growth-related variations in enzymes that synthesize and degrade this oligonucleotide. Eur J Biochem 114:5–10

Knight M, Cayley PJ, Silverman RH, Wreshcher LH, Gilbert CS, Broun RE, Kerr IM (1980) Radiommune, radiobinding and HPLC analysis of 2-5A and related oligonucleotides from intact cells. Nature 288:189–192

Krishnan I, Baglioni C (1981) Elevated levels of (2′-5′) oligoadenylic acid polymerase activity in growth-arrested human lymphoblastoid Namalva cells. Mol Cell Biol 1:932–938

Krispin TI, Parfenova TI, Itkes AV, Balandin IG, Kuznetzov VP, Severin ES (1984) Antiviral activity of α, β and γ-interferons in the presence of theophylline. Involvement of 2′,5′-oligo (A) synthetase induction. Biochem Int 8:159–164

Maenhaut C, Roger PP, Reuse S, Dumont JE (1991) Activation of the cyclic AMP cascade as an oncogenic mechanism – the thyroid example. Biochimie 73:29–36

McKnight SL, Lane DL, Gluecksohn-Waelsch S (1989) Is the CCAAT/enhancer-binding protein central regulator of energy metabolism? Genes Dev 3:2021–2024

Metz R, Ziff E (1991) cAMP stimulates the C/EBP-related transcription factor rNFIL-6 to translocate to the nucleus and induce *c-fos* transcription. Genes Dev 5:1754–1766
Mitchell PJ, Wang C, Tjian R (1987) Positive and negative regulation of transcription in vitro: enhancer-binding protein AP–2 is inhibited by SV–40 T-antigen. Cell 50:847–861
Nabel G, Baltimore D (1987) An inducible transcription factor activates expression of human immunodeficiency virus in T-cells. Nature 326:711–713
Nagammine Y, Pearson D, Altus MA, Reich E (1984) cDNA and gene nucleotide sequence of porcine plasminogene activator. Nucleic Acids Res 12:9525–9541
Okutani T, Nishi N, Kagawa Y, Takasuga H, Takenaka I, Usui T, Wada F (1991) Role of cyclic AMP and polypeptide growth regulators in growth inhibition by interferon in PC–3 cells. Prostate 18:73–80
Pastan IH, Johnson GS, Anderson WB (1975) Role of cyclic nucleotides in growth control. Annu Rev Biochem 44:491–522
Pestka S, Langer JA, Zoon KC, Samuel CE (1987) Interferons and their actions. Annu Rev Biochem 56:727–777
Porter ACG, Chernajovsky Y, Dale TC, Gilbert CS, Stark GR, Kerr IM (1988) Interferon response element of the human gene 6–16. EMBO J 7:85–92
Roesler WJ, Vandenbark GR, Hanson RW (1988) Cyclic AMP and induction of gene transcription. J Biol Chem 283:9063–9066
Rutherford MN, Hunnigan GE, Williams BRG (1988) Interferon-induced binding of nuclear factors to promoter elements of the 2-5A synthetase gene. EMBO J 7:751–759
Schneck S, Rager-Zisman B, Rosen OM, Bloom BR (1982) Genetic analysis of the role of cAMP in mediating effects of interferon. Proc Natl Acad Sci USA 79:1879–1883
Shirakawa F, Mizel SB (1989) In vitro activation and nuclear translocation of NF-$\varkappa$B catalized by cyclic AMP-dependent protein kinase and protein kinase C. Mol Cell Biol 9:2424–2430
Silverman RH, Cayley PJ, Knight M, Gilbert CS, Kerr IM (1982) Control of ppp (A2′p)nA system in Hela cells. Effects of interferon and virus infection. Eur J Biochem 124:131–138
Stark G, Dower WJ, Schimke RT, Brown RE, Kerr IM (1979) 2–5A synthetase: assay, distribution and variation with growth or hormone status. Nature 278:471–473
Thompson WJ, Pratt ML, Strada SJ (1984) Biochemical properties of high-affinity cyclic AMP phosphodiesterase. Adv Cyclic Nucleotide Res 16:137–148
Tunitskaya VL, Itkes AV, Kochetkova MN, Severin ES (1983) Study on the process of activation of phosphodiesterase of cyclic adenosine monophosphate by 2′,5′-oligoadenylate. Biokchimia (Moscow) 48:1721–1725
Weber JM, Stewart RB (1975) Cyclic AMP potentiation of interferon antiviral activity and effect of interferon on cellular cyclic AMP levels. J Gen Virol 28:363–372
Wirth T, Baltimore D (1988) Nuclear factor NF-$\varkappa$B can interact functionally with its binding site to provide lymphoid-specific promoter function. EMBO J 7:3109–3113
Witehouse S, Walsh DA (1982) Purification of the physiological form of the inhibitor protein of the cAMP-dependent protein kinase. J Biol Chem 257:6028–6032

Regulation of HIV Replication in Monocytes by Interferon

J.A. Turpin, S.X. Fan, B.D. Hansen, M.L. Francis, L.M. Baca-Regen, H.E. Gendelman, and M.S. Meltzer[1]

1 Introduction

1.1 CD4$^+$ T-Cells, the HIV-Infected Cell in Blood

The hallmark of human immunodeficiency virus (HIV) infection is the progressive loss of CD4$^+$ T-cells over a prolonged interval. In the infected individual, two types of cells are infected by HIV: CD4$^+$ T-cells and tissue macrophages. Levels of HIV in blood and tissues are dependent upon and change with the stage of the infection. Acute infection, usually lasting weeks to months after initial exposure to the virus, is characterized by a substantial viremia in which HIV actively replicates within blood leukocytes and high titers of free virus are found in plasma (more than 10 000 infectious virions/ml blood) (Clark et al. 1991; Daar et al. 1991). The chronic subclinical phase of infection is notable for low levels of plasma viremia and of virus-infected cells (less than 100 infectious virions/ml blood) (Ho et al. 1989). The major reservoirs for HIV in blood are CD4$^+$ T-cells which are infected at a frequency of about 0.1 to 1% (Schnittman et al. 1989). HIV-infected CD4$^+$ T-cells have on average only one proviral DNA copy integrated into genomic DNA. Less than 0.1% of these infected cells is transcriptionally active at any given time (Harper et al. 1986; Simmonds et al. 1990). During subclinical infection, the frequency of blood cells that express HIV mRNA and presumably produce infectious virus is only 0.01% (Harper et al. 1986; Clarke et al. 1990; Daar et al. 1991). During subclinical disease when very few cells are producing virus in blood, it is likely that cells in tissue provide most of the actively replicating virus that maintains infection during the long latent interval of 8 to 12 years (Lifson et al. 1988).

1.2 Macrophages, the HIV-Infected Cell in Tissue

When body tissues are examined, active replication of HIV is found in the lymph nodes, lungs, and brain of infected individuals (Meltzer et al. 1990). Cells that

[1]HIV Immunopathogenesis Program, Department of Cellular Immunology, Walter Reed Army Institute of Research, Washington, DC 20307-5100, USA

Progress in Molecular and Subcellular Biology, Vol. 14
W.E.G. Müller/H.C. Schröder (Eds.)
© Springer-Verlag Berlin Heidelberg 1994

support virus replication in tissues are not the CD4$^+$ T-cells but rather tissue macrophages. Macrophages are a unique cell lineage in that they are all derived from blood monocytes but express distinctive functions and biochemical patterns induced by the tissue microenvironment (Nathan and Cohen 1985). Such functions comprise tissue-specific immune reactions and the maintenance of tissue integrity in the steady state and during inflammation. There are two tissue sites commonly associated with HIV-infected macrophages: brain macrophages (microglia, perivascular, and parenchymal macrophages) and the follicular dendritic cells (macrophages) of lymph node (Shaw et al. 1985; Koenig et al. 1986; Cameron et al. 1987; Vazeux et al. 1987). The frequency of HIV-infected brain and lymph node macrophages can exceed 10% in affected tissue. Unlike CD4$^+$ T-cells, which predominantly carry only the integrated latent virus, HIV infection of tissue macrophages is virtually all productive. Further, the high frequency of virus-infected brain macrophages directly correlates with tissue injury and the coincident clinical manifestations of AIDS-related dementia. Alveolar macrophages of lung are proposed by some as another reservoir for virus in HIV infection. Although infection of lung macrophages is well documented, the frequency of infected cells ranges from 0.01 to 10% in several studies (Plata et al. 1987; Clarke et al. 1990; Rose et al. 1991). Even though alveolar macrophage populations contain HIV DNA in both early and late-state disease, most infected cells are transcriptionally inactive. Macrophages of brain, lymph nodes, and perhaps other tissues (lung, skin, liver) act as a virus reservoir throughout disease and serve as a nidus for virus dissemination and spread of infection.

1.2.1 HIV Infection of Monocytes in Culture

There is little or no evidence for a significant fraction of infected monocytes in blood at any stage of HIV disease (Clarke et al. 1990; Olafsson et al. 1991; Oka et al. 1992). Most studies document a frequency of HIV-infected monocytes at no more than 0.001%. Nevertheless, blood monocytes are susceptible to HIV infection and permissive to active virus replication. These cells serve as a valid model for analysis of HIV interactions with tissue macrophages. Both susceptibility to infection and levels of virus replication in infected cells change with time in culture (Kalter et al. 1991). Freshly isolated blood monocytes are relatively resistant to HIV infection. In contrast, cells cultured 5 to 7 days before exposure to virus are readily infected and show high levels of virus replication. HIV-infected monocytes form multinucleated giant cells within 6 days of exposure to HIV-1_{ADA} at a multiplicity of infection of 0.01 to 0.1 infectious virus/cell. Viral particles in the HIV-infected monocyte are characteristically formed and localized within Golgi complex-derived intracellular vacuoles with few viral particles at the plasma membrane. In marked contrast, viral particles in HIV-infected T-cells are assembled and localized at the plasma membrane with little or no virions within the cell (Gendelman et al. 1988). HIV released from infected cells is transmitted to other cells primarily through cell-cell interactions and not through the fluid phase (Sato et al. 1992). As with T-cells, HIV infection of monocytes is completely dependent upon interactions between viral envelope gp120 and the CD4 receptor (Gomatos et al. 1990; Finbloom et al. 1991). HIV released from infected monocytes is bi-trophic in that it is capable of infecting both T-cells and monocytes (Gendelman et al. 1990a). This is in contrast to virus derived from T-cells which replicates poorly in monocytes but efficiently infects and replicates within other T-cells. The molecular basis of monocyte tropism in HIV isolates is independent of gp120-CD4 interactions and determined by changes in a few amino acids within the V3 loop of gp120 (Hwang et al. 1991). Still to be defined changes in virus entry controlled by the V3 loop determine whether or not any viral isolate will infect monocytes. T-cell tropic HIV will replicate in monocytes only if viral entry reactions are bypassed (e.g., transfection of T-cell tropic HIV DNA into monocytes). Clinical studies suggest that monocyte tropic HIV isolates are present at high frequency in early disease. T-cell tropic HIV is present late in disease and associated with poor prognosis (Popovic and Gartner 1987; Schuitemaker et al. 1991). Monocytes that express lower levels of CD4 may be the primary carriers of the DNA provirus (Schnizlein-Blick et al. 1992). This may be a direct result of HIV-induced down regulation of CD4 expression (Mann et al. 1990; Geleziunas et al. 1990) or a consequence of monocyte differentiation independent of infection (Valentin et al. 1991).

1.3 Changes in the Cytokine Network During HIV Infection

Several groups report abnormally high levels of cytokines in sera of AIDS patients (Enk et al. 1986; Lepe-Zuniga et al. 1987; Lahdevirta et al. 1988; Wright et al. 1988). Infection of monocytes with HIV is associated with marked cytopathic effects in 30 to 40% of cells in culture (formation of multinucleated giant cells, increased intracellular vacuolization, cell death, and lysis) (Gendelman et al. 1988; Orenstein et al. 1988). Similar changes occur in HIV-infected macrophages of brain and lymph nodes. Despite these dramatic changes, little or no change in cytokine gene expression or secretion is detected in these infected monocyte cultures (Meltzer et al. 1990; Molina et al. 1990a, b). Monocytes infected for 7 to 14 days with HIV-1_{ADA}, a monocyte tropic virus, do not express TNFα, IL-1α, IL-1β, IL-6, *c-fos* or *c-jun* mRNA. TGFβ-1 and IL-8 mRNA are expressed in control monocytes and HIV infection does not alter this constitutive expression. Further, the above cytokine gene expression in monocytes after induction by LPS is likewise not affected by HIV infection. Changes in cytokine gene expression is seen only when monocytes display extensive HIV-associated cytopathic effects. At these times, 30 to 50% of cells fuse to form multinucleated giant cells, and the capacity of LPS to induce cytokine gene expression is coincidentally decreased, as is β-actin gene expression. That there is no change in ribosomal RNA content in these cultures suggests that changes in cytokine transcription may be a reflection of increased cell death. Thus, HIV-infected monocytes remain relatively quiescent during virus infection:

secretion of biologically active autocrine cytokines by monocytes or the capacity to secrete such mediators after appropriate stimuli remain unaffected by HIV infection. Interestingly, even though HIV-infected monocytes secrete little or no cytokines per se, these infected cells are highly responsive to the transcriptional effects of exogenously added cytokines (Meltzer et al. 1990; Farrar et al. 1991; Matsuyama et al. 1991). The fact that HIV infection of monocytes by itself induces little or no cytokine response, yet these cells retain full capacity to respond to exogenous cytokines, suggests that the macrophage may be an "innocent bystander" in the infection process. That macrophages remain central to all immune responses and are full participants in the cytokine networks of tissues, provides ample opportunity for cytokine control of virus replication in these infected cells.

2 Interferons and HIV Infection

The interferons have a long history as both immunomodulatory cytokines and antiviral agents. Are interferons involved in the natural history of AIDS and HIV-associated disease? Can these antiviral cytokines be used therapeutically in infected individuals? During the acute seroconversion reaction, interferon activity is found in serum for 3 to 6 weeks after initial exposure to HIV. Coincident with the development of antibodies against HIV and the loss of p24 from serum, interferon activity returns to baseline and remains at this low level for years. Paradoxically, high levels of interferon-α (IFN-α) are commonly found again in serum of patients with late-stage HIV disease and indicate poor prognosis (Pomerantz and Hirsch 1987; von Sydow et al. 1991). In addition, certain surrogate markers for interferon activity, such as neopterin, β2-microglobulin, and 2′,5′-oligoadenylate synthetase activity directly correlate with serum p24 levels, HIV viremia, and poor prognosis. Early reports suggest that the IFN-α in serum of patients with AIDS appears to be acid-labile IFN-α (Destefano et al. 1982). Given that all purified and recombinant IFN-α species are stable after exposure to pH 2, the meaning (and some would say reproducibility) of the acid-labile IFN-α of AIDS is not clear. In the face of high levels of IFN-α in serum of patients with late-stage HIV disease, production of IFN-α (but not other cytokines) is inhibited in blood leukocytes from the same patients (Rossol et al. 1989). This inhibition is mediated by events at the transcriptional level, since neither IFN-α protein nor mRNA is detected in these cells. The nature of the stimulus for IFN-α production, the cells that produce these cytokines, the identities and relative amounts of IFN-α species produced, and the effectiveness of this cytokine response in control of virus infection are unknown at this time.

2.1 Identification of the Key Issues

Reports from numerous laboratories document potent antiviral effects of IFN-α, IFN-β, and IFN-γ in restriction of HIV replication in T-cell and myeloid cell lines and in blood leukocytes (Ho et al. 1985; Dolei et al. 1986; Hartshorn et al. 1987; Yamada et al. 1988; Michaelis and Levy 1989; Poli et al. 1989; Kornbluth et al. 1990; Brinchmann et al. 1991; Mace and Gazzolo 1991; Pitha 1991). All of these interferons may play some role in control of HIV replication in the infected individual. For many different virus infections, the major source of IFN-α in man is the blood monocyte and tissue macrophage (Roberts et al. 1979; Saksela et al. 1984). The IFN-α gene family consists of over 24 subtypes and contrasts with IFN-β and IFN-γ which are each products of a single gene (Pestka et al. 1987; Laurence 1990). With the large repertoire of IFN-α genes available to cells, it might not be unexpected that monocytes and lymphocytes produce different mixtures of the IFN-α subtypes (Goren et al. 1986). Indeed, Sendai or New Castle disease virus (NDV) induces different levels and subtypes of IFN-α in blood monocytes and lymphocyte subpopulations. A single IFN-α subtype, identified by molecular weight, accounts for more than 50% of the total IFN-α production of such treated monocytes. Further, the same population of monocytes exposed to different viruses produces different amounts and subtypes of IFN-α. Such mixtures of IFN-α induced by different viruses possess distinctly different capacities to effect antiviral activity against several viruses (Bell et al. 1983). Further, the capacity of cells to respond to IFN-α for induction of antiviral pathways is more dependent upon the infected cell type than on the subspecies of IFN-α produced (Samuel et al. 1982). Each of these factors (multiplicity of IFN-α genes, variable stimuli required for IFN-α production, the menu of IFN-α subtypes produced by a stimulated cell, and the effects of this menu on IFN-α-responsive cells) makes analysis of the role of IFN-α in HIV infection complex. To date, there are few answers and many questions.

2.1.1 What Induces IFN-α?

During initial infection with HIV, IFN-α appears in plasma coincident with and directly proportional to HIV viremia. Antibodies against HIV are detected in these patients by 4 to 6 weeks. With the development of this antibody response both IFN-α and HIV virus titers decrease to baseline and remain there for years. In late-state disease, IFN-α and HIV virus reappear in patient plasma and remain at high titer until death. The stimulus for IFN-α production and the nature of the producing cells at any stage of HIV infection are not yet clear. Culture fluids from freshly isolated blood leukocytes of HIV seronegative donors contain high levels of interferon activity 24 h after exposure to HIV virus. Interferon activity directly correlates with the number of monocytes in the cell preparation. Indeed, no activity is detected with these same cells after depletion of monocytes. Such interferon activity is active for protection of bovine MDBK cells against the cytopathic effects associated with vesicular stomatitis virus infection, retains full biological activity after exposure to pH 2, and is completely neutralized by antibodies specific for IFN-α. The capacity of HIV virus to induce IFN-α in monocytes (about 5000 to 10 000 IU/ml IFN-α) is cytokine-specific: levels of IFN-γ, IL-1β, IL-6, and IFN-α are at baseline and comparable to those of untreated control cells. HIV-infected T-cells and monocytes or T-cell and myeloid cell lines also induce IFN-α in monocytes (Gendelman et al. 1992). Indeed, such HIV-infected cells fixed in 4% paraformaldehyde retain full capacity to induce IFN-α. Similarly, the capacity of HIV virus to induce IFN-α is not directly related to virus replication. Both monocyte tropic and T-cell tropic HIV-1 strains induce IFN-α in monocytes even though T-cell tropic HIV-1 does not replicate in these cells. Further, culture fluids from monocytes exposed to virions from 8E5 cells (a constitutively infected T-cell line that produces *pol*-defective and non-infectious HIV-1 particles) or to HIV made noninfectious by heating at 56 °C for 60 min each contain high levels of IFN-α.

The capacity of HIV virus to induce IFN-α is completely inhibited by soluble CD4 or antibodies against CD4. Neither treatment affects the capacity of poly (I)·poly (C) or dengue virus to induce IFN-α in the same cultures. Antibodies against gp160 and gp120, but not gp41 or p24 also inhibit IFN-α production by monocytes exposed to HIV. The capacity of HIV-infected cells to induce IFN-α is also inhibited by antibodies against CD4 or gp120 and by soluble CD4 (Capobianchi et al. 1992; M. Francis, unpubl. observ.). Thus, the principal determinant for induction of IFN-α production for both HIV-infected cells and HIV virus is the envelope glycoprotein gp120. Antibodies against V3 loop determinants as well as the gp120-CD4 binding site inhibit IFN-α induction. Since virtually all of the neutralizing antibody activity in HIV-infected patients is directed against gp120 V3 loop determinants, the preceding observations suggest that potentially protective antiviral responses associated with IFN-α production in the HIV-infected patient are strongly inhibited by the development of antibodies against gp120. A sobering thought for vaccine development.

2.1.2 *What Is the Best Time for IFN-α Antiviral Activity?*

In assessing all the possible effects of IFN-α, and relating this to any potential clinical application, two treatment schedules are apparent. Can IFN-α induce an antiviral state in lymphocytes and monocytes when present at the time of infection? More importantly, can IFN-α affect HIV replication after the virus has integrated into the host genome and initiated a productive infection?

2.2 *IFN-α Antiviral Activity in T-Cells*

2.2.1 *Effects of IFN-α at the Time of Initial HIV Infection*

HIV-infected T-cells form syncytia within 72 h of infection and the infection progresses throughout the culture with increased levels of reverse transcriptase (RT) activity and p24 antigen until 10 to 15 days. After this time, both RT activity and p24 antigen levels drop coincident with cell death and lysis (Gendelman et al. 1990b, c). Addition of up to 10 000 IU/ml IFN-α to theseT-cell cultures at the time of infection and throughout the following 2 weeks has little or no effect on virus replication. The frequency of infected cells through 2 weeks of infection, levels of HIV DNA copies/culture, levels of HIV RNA/culture, and levels of p24 antigen in culture fluids are all unaffected by these high concentrations of IFN-α. In contrast, levels of RT activity in culture fluids are decreased to about 50 to 70% those of control cultures. Such decreased RT activity levels are apparent at 50 to 500 IU/ml IFN-α, but show no further decrease at 10 000 IU/ml IFN-α. Shirazi and Pitha (1992) show inhibition of HIV proviral DNA levels in certain T-cell lines treated with IFN-α. Such inhibition was directly related to the concentration of IFN-α and suggests some IFN-α-associated interference with an early stage of the virus life cycle before reverse transcription in these T-cell lines.

2.2.2 A Defect in HIV Assembly?

The preceding data document only modest effects of IFN-α on HIV replication in T-cells: no change in frequency of infected cells, HIV DNA or HIV RNA, and a two- to three-fold decrease in RT activity levels in culture fluids. More detailed analysis of the virions released from IFN-α-treated cells shows a 1000-fold decrease in infectivity compared to an equal number of viral particles from control-infected cells (Hansen et al. 1992). Further, quantitation of viral protein levels within these IFN-α-treated T-cells by radioimmunoprecipitation with pooled HIV-1 seropositive sera shows a direct correlation between IFN-α concentration and the amounts of HIV envelope and core proteins that accumulate within the cell. Processing of these proteins (relative amounts of gp160 and gp120 or p55 and p24 in IFN-α-treated and control cells) is unaffected by IFN-α treatment. These data document a 10-to 20-fold increase in viral protein levels within IFN-α-treated T-cells and a viral assembly defect similar to that previously described for murine retroviruses. Mouse cells infected with retrovirus type C (mouse mammary tumor virus, MMTV) and treated with interferon show several abnormalities in viral production: increased accumulation of fully assembled virus at the cell membrane, increased frequency of morphologically aberrant viral particles and particles without core RNA, and synthesis of abnormal proteins and glycoproteins (Chang and Friedman 1977; Chang et al. 1977; Bandyopadhyay et al. 1979; Maheshwari et al. 1980; Pitha et al. 1980; Yagi et al. 1980; Aboud and Hassan 1983). Immunofluorescence analysis of IFN-α-treated HIV-infected T-cells for gp120 reveals obvious changes in localization of this envelope glycoprotein. Assembly of HIV virions takes place at the inner side of the plasma membrane and results in an accumulation of viral envelope proteins in a characteristic halo pattern by immunofluorescence. Cells treated with IFN-α lose the characteristic halo pattern for gp120 and instead show diffuse expression of gp120 throughout the cell. It is likely that this assembly or transport defect for gp120 protein is related to the 1000-fold loss of virus infectivity. Indeed, examination of viral particles released from IFN-α treated HIV-infected T-cells shows a profound loss of gp120 to levels less than 10% of those of virus from control cells (Hansen et al. 1992). The 1000-fold loss of HIV infectivity is directly related to changes in gp120 content of the viral particle. Thus, like earlier studies on interferon treatment of MuLV- and MMTV-infected mouse cells, IFN-α treatment of HIV-infected T-cells induces a marked defect in virus assembly and loss of virion-associated envelope glycoproteins. Interactions between gp120 and CD4 on susceptible cells are prerequisite for infection. In HIV-infected T-cells, over 90% of gp160 is degraded through lysosomal-associated pathways before virus assembly (Willey et al. 1988). The exact mechanisms for the IFN-α-associated redistribution of gp120/160 within T-cells is not known. Related studies of intracellular transport pathways for vesicular stomatitis virus envelope glycoprotein in IFN-α-treated cells show marked accumulation of the glycoprotein in the *trans*-Golgi complex and suggest that mechanisms for such accumulation are related to defects in transport of viral proteins through the cell (Singh et al. 1988; Dedera et al. 1990).

2.3 IFN-α Antiviral Activity in Monocytes

The preceding account documents a very modest effect of IFN-α on the replication of HIV in T-cells: levels of HIV particles, proteins, mRNA, and DNA in cultures of IFN-α-treated and control HIV-infected T-cells are almost indistinguishable (less than twofold difference). These studies grossly underestimate the antiviral activity of IFN-α. Such antiviral activity is unrelated to virus replication but manifest by a 1000-fold decrease in infectivity of HIV particles. The antiviral effects of IFN-α on HIV-infected monocytes are very different from those on T-cells. For the purposes of discussion, we will divide these antiviral activities into two model systems: (1) addition of IFN-α to monocytes prior to or simultaneously with the HIV inoculum and (2) addition of IFN-α to monocytes with established and productive HIV infection.

2.3.1 Effects of IFN-α at the Time of Initial HIV Infection

Addition of 500 to 1000 IU/ml rIFN-α_{2b} simultaneously with the virus inoculum and continuously thereafter through 3 weeks completely inhibits infection of monocytes by the HIV virus (Gendelman et al. 1990b, c). In such treated monocytes, there is no evidence of HIV proviral DNA by polymerase chain reaction (PCR) amplification of the DNA in cell lysates using *LTR* and *gag* primers 24 h after infection. Thus, IFN-α induces an antiviral state that blocks an early stage of the viral life cycle prior to formation of provirus. Further, removal of IFN-α 3 or 9 days after infection does not rescue possible residual virions sequestered in these cells and DOES NOT permit subsequent productive infection. This suggests that viral particles or uncoated (naked) viral RNA are not preserved. Treatment of monocytes with IFN-α at the time of infection may induce a number of effective antiviral pathways that change: (1) expression of the CD4 virus receptor or any of numerous putative accessory receptors (FcR, mannosylated protein receptors, cell and extracellular matrix integrin receptors, class I and II MHC, and others) required for HIV binding and uptake; (2) membrane fluidity to prevent gp41-dependent fusion with the plasma membrane or other points in the entry pathway of the virus; (3) 2′,5′-oligoadenylate synthetase and RNase L pathways to effect rapid degradation of viral RNA and abort infection; or (4) signal transduction pathways involved in virus entry, uncoating, and reverse transcription.

2.3.2 Effects of IFN-α on Established Productive HIV Infection

2.3.2.1 The Window of Opportunity

Addition of IFN-α to HIV-infected monocytes effects dramatic downregulation of virus replication (Gendelman et al. 1990b, c). Levels of HIV viral particles, structural and regulatory proteins, and all virus-specific RNA (genomic, structural, and regulatory) in monocytes infected with HIV for 3 to 7 days and then treated with IFN-α decrease to baseline levels and cannot be detected 1 to 2 weeks after IFN-α treatment. Pointedly, the same cells still contain high levels of HIV DNA even 4 to 6 weeks after IFN-α treatment. Thus, HIV gene expression is completely inhibited by IFN-α treatment, but viral DNA is retained within the infected cell. Coincident with inhibition of HIV gene expression is a marked inhibition of infection-associated cytopathic effects: decreased frequency of multinucleated giant cells and reversion to or acquisition of a cell morphology comparable to that of uninfected control cells. A salient feature of this IFN-α-induced antiviral activity is what we call the "window of opportunity".

The replicative cycle of HIV in monocytes can be characterized by the sequential appearance of proviral DNA, viral-specific mRNA, and RT activity. Proviral DNA is first detected in cell lysates 12 to 24 h after infection (about 700 HIV DNA copies/1000 cells at a multiplicity of infection of 0.1 infectious $HIV\text{-}1_{ADA}$/cell) with progressive increase to a plateau at 6 to 9 days of infection (more than 20000 HIV DNA copies/1000 cells) (Rich et. al. 1992). HIV-specific mRNA is detected as early as 48 h by RNA-directed PCR and at 3 to 5 days after infection by Northern blot. Levels of HIV RNA reach a plateau by 9 to 12 days after infection. The first appearance of RT activity and infectious virus in culture fluids is coincident with the ability to detect viral RNA by Northern blot. RT activity levels plateau at 9 days after infection. At this time, the frequency of productively infected monocytes is 25 to 30% by in situ hybridization for HIV RNA. During the early phase of infection (1 to 7 days after addition of the HIV inoculum), 500 to 1000 IU/ml IFN-α is highly efficient at inducing a state of HIV gene expression that approaches "true microbiological latency". As infection progresses with time (7 to 14 days after infection), the efficacy of IFN-α decreases until it has virtually no effect on virus replication, regardless of how much is added.

Mechanisms for the refractoriness of HIV-infected monocytes to the antiviral effects of IFN-α with progressive infection are not yet clear. There are several possibilities. First, the virus might generally subvert the transcriptional machinery of the monocyte to inhibit initiation of any cellular transcriptional responses including IFN-α-induced gene expression. If this occurs, it is certainly not general since the ability of LPS to induce new transcriptional responses for cytokine gene expression is unaffected even at peak infection (discussed above). Further, TNFα, IL-1β, and IL-6 transcription and synthesis of biologically active proteins induced by poly(I)·poly(C) or NDV are also unaffected (Gendelman et al. 1990d). Second, and more likely, HIV might specifically inhibit induction of IFN-α-mediated pathways. Blood mononuclear leukocytes from patients with AIDS show significantly lower levels of RNase L activity and IFN-α receptor expression than an equal number cells from controls (Carter et al. 1987; Lau et al. 1988). Further, there is ample precedent with both DNA and RNA viruses for mechanisms that overcome (2′,5′)-oligoadenylate synthetase-mediated activation of RNase L and p68 kinase activation (Taylor and Grossberg 1990). Indeed, we previously showed that IFN-α gene expression is selectively inhibited in HIV-infected monocytes (Gendelman et al. 1990d). Freshly isolated monocytes produce more than 8000 IU/ml IFN-α in response to poly(I)·poly(C). With time in culture (3 to 4 weeks), these same cells gradually lose their capacity to produce IFN-α. This loss greatly accelerated by HIV infection so that 7 to 10 days after infection, these poly (I)·poly (C)-treated monocytes produce no IFN-α activity and express no IFN-α mRNA. Expression of other cytokine mRNAs (IL-1, IL-6, TNFα) and synthesis of biologically active cytokine activities are unaffected.

2.4 IFN-α-Induced Antiviral Pathways in HIV-Infected Monocytes

Attempts to correlate the known antiviral pathways induced by IFN-α with the potent effects of IFN-α on HIV replication in monocytes have not yet provided clear insight into the mechanism for antiviral activity. All effects of IFN-α on cells are mediated though the actions of inducible genes. IFN-α induces a number of proteins and enzymatic pathways that potentially contribute to antiviral activity (Peska et al. 1987; Staeheli 1990). Among the enzymes induced by IFN-α is 2′,5′-oligoadenylate synthetase which activates RNase L to degrade viral transcripts (Schröder et al. 1989). The p68 kinase is also induced by IFN-α and can be measured indirectly by phosphorylation of eIF-2α, a protein synthesis initiation factor that blocks translation when phosphorylated. Levels of 2′,5′-oligoadenylate synthetase mRNA and activity are exceedingly low in uninfected monocytes through 3 weeks of culture (5 nM/mg protein/h). These same activities are increased more than five-fold in HIV-infected monocytes 3 and 7 days after infection. Interestingly, these changes appear independent of IFN-α in that there is no interferon activity in these culture fluids and no IFN-α or IFN-β mRNA in cell lysates. Further, evidence for other IFN-α effects such as expression of the IFN-α-inducible human gene products 6–16 and 9–27 is absent. Levels of 2′,5′-oligoadenylate synthetase activity in both uninfected and

HIV-infected monocytes increase with exogenously added IFN-α to comparable levels (30 to 40 nM/mg protein/h). The significance of 2′,5′-oligoadenylate synthetase activity in the HIV-infected monocyte is not clear. Examination of these same cells for RNase L and p68 kinase activity shows values at baseline with and without IFN-α treatment. Both activities were present at high levels in IFN-α-treated HeLa cells as control. Interestingly, HIV-infected monocytes with high levels of 2′,5′-oligoadenylate synthetase activity (but no evidence for IFN-α production) are protected against vesicular stomatitis virus cytopathic effects to the same extent as uninfected cells treated with IFN-α. Thus, HIV infection per se apparently induces a potent state of antiviral activity in infected monocytes in the apparent absence of IFN-α (Baca et al. 1992).

Analysis of HIV mRNA in IFN-treated monocytes in the presence of actinomycin D shows a ca. two-fold increase in mRNA degradation with time. Increased rates of degradation are evident for genomic, structural, and regulatory mRNA species. However, analysis of such cellular mRNA as β-actin shows about the same rate of degradation after IFN-α treatment. Thus, it is unlikely that increased mRNA degradation alone can explain the near absence of HIV RNA 3 to 5 days after IFN-α treatment. IFN-α might induce a direct or indirect transcriptional inhibition of HIV gene expression. Indeed, direct analysis of HIV gene transcription in isolated nuclei from IFN-α-treated HIV-infected monocytes shows a two- to three-fold decrease in synthesis of nuclear HIV transcripts. These data support the hypothesis that a major action of IFN-α for antiviral activity against HIV is transcriptional inhibition of HIV gene expression.

2.5 IFN-α-Induced Latency in HIV-Infected Monocytes

HIV-infected monocytes treated with IFN-α maintain high levels of proviral DNA but show little or no evidence of viral gene transcription. Such microbiological latency is maintained in the continuous presence of IFN-α for 6 to 8 weeks after infection. At 3 weeks after infection, HIV proviral DNA is easily detected by PCR at levels unchanged from those present at the time of IFN-α treatment. Is this apparent latent state permanent or dependent upon continuous IFN-α treatment? Removal of IFN-α 1 or 2 weeks after induction of latency (e.g., little or no HIV RNA in cell lysates) initiates productive virus replication with the reappearance of RT activity in culture fluids and HIV-associated cytopathic effects. HIV replication to levels that approximate those of the initial productive infection takes about 3 to 6 days. Thus, even though an antiviral state is induced by IFN-α treatment that near completely inhibits HIV gene expression, this state requires continuous restimulation by exogeneous IFN-α to maintain efficacy.

2.6 Transcriptional Mechanisms for IFN-α-Induced Antiviral Activity

In addition to the three structural genes common to all retroviruses (*env*, *pol*, and *gag*), the HIV-1 genome contains at least six other genes that regulate the

HIV life cycle. Such regulation is complex and only partially defined. For example, the Tat protein increases expression from the HIV long terminal repeat (LTR) to increase expression of all viral genes. Rev differentially enhances expression of the unspliced or single-spliced messages that encode structural genes and genomic RNA. The mRNA for the regulatory proteins Tat, Rev, and Nef are doubly spliced. Most of the regulatory events that occur during HIV infection occur at the LTR. LTR-directed viral gene expression occurs as a final common pathway controlled by interactions at multiple enhancer and inhibitor motifs within the LTR.

That IFN-α induces potent inhibition of HIV gene expression in monocytes suggests that a major site for antiviral activity is directed at the control of the HIV LTR. IFN-α may induce cellular factors that directly or indirectly interfere with enhancer or inhibitor elements within the LTR for control of virus transcription. Thus, IFN-α may induce novel nuclear binding factors that directly compete or amplify viral regulatory factors at the LTR or IFN-α may alter the relative quantities or effects of such virus regulatory factors as Tat, Nef, or Rev to indirectly act at the LTR.

2.6.1 At the LTR – Are NF-κB and Sp1 the Culprits?

Transcription of HIV genes is effected by a tightly controlled balance between positive and negative cell- and virus-derived factors that interact at the LTR (Steffy and Wong-Staal 1991). Two nuclear binding factors in particular are positive regulators of HIV transcription: NF-κB and Sp1 (Jones et al. 1986; Griffin et al. 1989; Ross et al. 1991). The NF-κB complex consists of multiple proteins with homology to the *rel* oncogene family that can associate in homodimeric, heterodimeric, or heterotetrameric forms (Ghosh et al. 1990; Kieran et al. 1990; Ruben et al. 1991). Proteins unrelated to the NF-κB complex can also interact with NF-κB motifs on the LTR (Baldwin et al. 1990; Ron et al. 1991). These other proteins interact and form complexes that are capable of both positive and negative regulation of LTR-directed gene transcription (Muchardt et al. 1992). Such regulation is described in a T-cell line infected with a *tat*-defective HIV provirus clone where IFN-α alters the binding pattern of NF-κB to the LTR (Popik and Pitha 1991). Thus, one possible mechanism for IFN-α control of HIV transcription may revolve around alterations in the balance of the components available for NF-κB complexes.

2.6.2 At the LTR and Beyond – Tat and Rev

HIV carries within its genome information for synthesis of a number of proteins important in the control of virus replication and progression though its life cycle (Vaishnav and Wong-Staal 1991; Zack et al. 1990; Cullen 1991, for comprehensive reviews). These proteins bind to specific regions on viral DNA and mediate effects at transcriptional and posttranscriptional levels. One region, the *trans*-acting response element (TAR), is present in the 5′ leader of all HIV transcripts and contains a large repeat that forms a stem-loop structure associated with initiation and control of viral transcription. This region is the normal target of Tat, a viral protein that upregulates viral transcription in cooperation with NF-κB and Sp1 (Berkhout et al. 1990). The TAR loop mimics double-stranded RNA and can bind and activate 2′,5′-oligoadenylate synthetase and p68 kinase activities (Edery et al. 1989; SenGupta and Silverman 1989). Such activation could potentially inhibit HIV replication through increased degradation of viral mRNA. But these enzyme activities are induced in T-cells and macrophages by productive HIV infection per se in the absence of IFN-α. Are these effective antiviral pathways? Do the modestly increased levels of these enzymes induced by exogenous IFN-α effect the antiviral response? If such pathways are important in the antiviral activity of IFN-α, then they must specifically affect the mRNA for one or more regulatory proteins and not all viral proteins.

Inhibition of p68 kinase activity is also described during HIV infection of T-cells (Roy et al. 1990). In these experiments, expression of *tat* in HIV-infected T-cells and *tat*-expressing HeLa cells correlates with inhibition of p68 kinase activity in both IFN-α-treated and control cells. Tat protein is able to displace and inhibit activation of 2′,5′-oligodenylate synthetase activity from the TAR region and provide a possible mechanism for IFN-α nonresponsiveness (Schröder et al. 1990). Thus, levels at Tat within cells may be important in determining whether or not the p68 kinase and 2′,5′-oligoadenylate synthetase (and subsequently RNase L) are activated.

Another HIV-derived regulatory protein, Rev, is intimately involved in the regulation of Tat (Felber et al. 1990). Rev regulates the amount of Tat, so that increases in Rev correlate with suppression of Tat and suppression of viral transcription. In addition, Rev plays a central role in control of HIV mRNA expression, the type of infection (latent or productive), and the level of virus replication (Malim et al. 1988; Felber et al. 1989). Rev functions by modulating expression of the two main categories of HIV mRNA: multispliced species consisting of Tat, Nef and Rev, and unspliced mRNA consisting of structural genes and full length genomic RNA. Levels of Rev proteins are closely correlated with the progression of infection in both T-cells and macrophages, and also correlate with production of viral particles (Robert-Guroff et al. 1990; Klotman et al. 1991). Such correlations center around Rev's function of stabilizing unspliced mRNA and allowing its transport out of the nucleus away from the splicedsomes. Both Rev and Tat interact with cellular proteins and may require them for their function and production (Marciniak et al. 1990; Nelbock et al. 1990; Trono and Baltimore 1990; Vaishnav et al. 1991). IFN-α treatment of productively infected monocytes downregulates unspliced *env* and *gag* transcripts and spliced mRNA expression (Meylan et al. 1991). The ability of IFN-α to regulate Rev protein levels and subsequently Tat suggests that these proteins or their associated cellular factors could be prime candidates for mechanistic studies.

2.6.3 *A Model for IFN-α Action?*

The previously described window of opportunity may then represent the interactions of IFN-α-mediated antiviral pathways and those pathways that control viral replication. The ability of IFN-α to suppress HIV replication may reflect the ability of 2′,5′-oligoadenylate synthetase and p68 kinase to bind to the TAR stem-loop structure and prevent viral transcription. Removal of IFN-α from treated cells and subsequent reversal of a latent state could then be accounted for by the loss of this blocking activity. This could be further aided by modulation of NF-κB and Sp1 patterns and actions of IFN-α on cellular proteins required for Rev and Tat function. Such interactions create a situation in which viral transcription is blocked physically, but the required proteins for the function of the LTR or Tat and Rev are either missing or altered in action or proportions. The "closing" of the window of opportunity may reflect the proportionate increases of Rev and Tat associated with a productive infection. The increase in Rev which signals increases in structural mRNAs and Tat proteins could stoichiometrically overcome early transcriptional blocking effects of *P68* kinase and 2′,5′-oligoadenylate synthetase. In this case even though an antiviral state is achieved by activation of 2′,5′-oligoadenylate synthetase and *P68* kinase by other pathways in the cell, physical interference at the Tat binding structure prevents these factors from becoming fully effective.

3 Conclusion and Future Directions

IFN-α regulates HIV infection in patients and in culture systems. IFN-β and IFN-γ have not yet been studied as intensely as IFN-α. In general, both IFN-β and IFN-γ inhibit HIV replication, but under certain conditions both can also upregulate HIV expression (Koyahagi et al. 1988). The inability of T-cells from HIV-infected individuals to produce IFN-γ and the high levels of IFN-γ in serum correlate with progression of disease (Cauda et al. 1987/1988). The antiviral effects of IFN-γ on HIV-infected monocytes are about ten-fold more potent than IFN-α for inhibition of HIV DNA levels, mRNA synthesis, p24 expression, and RT activity (S. X. Fan, unpubl.). Undoubtedly, interferons play a role in HIV infection. The key questions center around determining the mechanism(s) of interferon action and how it interacts through its inducible pathways with the HIV life cycle. This information will be invaluable for designing alternate approaches to combat HIV infection.

References

Aboud M, Hassan Y (1983) Accumulation and breakdown of RNA-deficient intracellular viron particles in interferon-treated NIH 3T3 cells chronically producing Moloney murine leukemia virus. J Virol 45:489–495

Baca LM, Genis P, Zhou A, Silverman RH, Larner A, Meltzer MS, Gendelman HE (1992) Regulation and antiviral activities of interferon α inducible genes in human immunodeficiency virus infected monocytes. J Exp Med (submitted)

Baldwin AS, LeClai KP, Harinder S, Sharp PA (1990) A large protein containing zinc finger domains binds to related sequence elements in the enhancers of the class I major histocompatability complex and kappa immunoglobulin genes. Mol Cell Biol 10:1406–1414

Bandyopadhyay AK, Chang EH, Levi CC, Friedman RM (1979) Structural abnormalities in murine leukemia viruses produced by interferon treated cells. Biochem Biophys Res Commun 87:983–988

Bell DM, Roberts NJ Jr, Hall CB (1983) Different antiviral spectra of human macrophage interferon activities. Nature 305:319–321

Berkhout B, Gatignol A, Rabson AB, Jeang K-T (1990) TAR-independent activation of the HIV-1 LTR: evidence that TAT requires specific regions of the promoter. Cell 62:757–767

Brinchmann JE, Gandernack G, Vartdal F (1991) In vitro replication of HIV-1 in naturally infected $CD4^+$ T cells is inhibited by rIFN-α_2 and by a soluble factor secreted by activated $CD8^+$ T cells but not by rIFN-β, rIFN-γ or TNF-α. J AIDS 4:480–488

Cameron PU, Dawkins RL, Armstrong JA, Bonifacio E (1987) Western blot profiles, lymph node ultra-structure and viral expression in HIV-infected patients: a correlative study. Clin Exp Immunol 68:465–478

Capobianchi MR, Ankel H, Ameglio F, Paganelli R, Pizzoli PM, Dianzani F (1992) Recombinant glycoprotein 120 of human immunodeficiency virus is a potent interferon inducer. AIDS Res Hum Retroviruses 8:575–579

Carter WA, Brodsky I, Pellegrino MG, Henriques HF et al. (1987) Clinical, immunological, and virological effects of ampligen, a mismatched double-stranded RNA, in patients with AIDS or AIDS-related complex. Lancet i:1286–1292

Cauda R, Tyring SK, Tamburrini E, Ventura G, Tambarello M, Ortona L (1987/1988) Diminished interferon gamma production may be the earliest indicator of infection with the human immunodeficiency virus. Viral Immunol 1:247–258

Chang EH, Friedman RM (1977) A large glycoprotein of Moloney murine leukemia virus derived from interferon treated cells. Biochem Biophys Res Commun 77:392–397

Chang EH, Mayers MW, Wong KY, Friedman RM (1977) The inhibitory effect of interferon on a temperature–sensitive mutant of moloney murine leukemia virus. Virology 77:625–636

Clark SJ, Saag MS, Don Decker W et al. (1991) High titers of cytopathic virus in plasma of patients with symptomatic primary HIV infection. New Eng J Med 324:954–960

Clarke JR, Krishnan V, Bennett J, Mitchell D, Jeffries DJ (1990) Detection of HIV in human lung macrophages using the polymerase chain reaction. AIDS 4:1133–1136

Cullen BR (1991) Regulation of gene expression in the human immunodeficiency virus type 1. Adv Virus Res 40:1–17

Daar ES, Moudgil T, Meyer RD, Ho DD (1991) Transient high levels of viremia in patients with primary human immunodeficiency virus type 1 infection. New Engl J Med 324:961–964

Dedera DN, Vander Hayden N, Ratner L (1990) Attenuation of HIV-1 infectivity by an inhibitor of oligosaccharide processing. AIDS Res Hum Retroviruses 6:785–794

Destefano A, Firedman R, Friedman-Kien A, Goedert JJ et al. (1982) Acid-labile human leukocyte interferon in homosexual men with Kaposi's sarcoma and lymphadenopathy. J Infect Dis 146:451–459

Dolei A, Fattorossi A, D'Amelio R, Aiuti F, Dainzani F (1986) Direct and cell-mediated effects of interferon-α and γ on cells chronically infected with HTLV-III. J Interferon Res 6:543–549

Edery I, Petryshyn R, Sonenberg N (1989) Activation of double-stranded RNA-dependent kinase (dsl) by the TAR region of HIV-2 mRNA: a novel translational control mechanism. Cell 56:303–312

Enk C, Gerstoft J, Molles S, Remvig L (1986) Interleukin 1 activity in the acquired immunodeficiency syndrome. Scand J Immunol 23:491–497

Farrar WL, Korner M, Clouse KA (1991) Cytokine regulation of human immunodeficiency virus expression. Cytokine 3:531–542

Felber BK, Hadzopoulou-Cladaras M, Cladaras C, Copeland T, Pavlakis GN (1989) Rev protein of human immunodeficiency virus type 1 affects stability and transport of the viral mRNA. Proc Natl Acad Sci USA 86:1495–1499

Felber BK, Drysdale CM, Pavlakis GN (1990) Feedback regulation of human immunodeficiency virus type 1 expression by the rev protein. J Virol 64:3734–3741

Finbloom DS, Hoover DL, Meltzer MS (1991) Binding of recombinant HIV coat gp 120 to human monocytes. J Immunol 146:1316–1321

Geleziunas R, Bour S, Boulerice F, Hiscott J, Wainberg MA (1990) Diminution of CD4 surface protein but not CD4 messenger RNA levels in monocytic cells infected by HIV-1. AIDS 5:29–33

Gendelman HE, Orenstein JM, Martin MA, Ferrua C et al. (1988) Efficient isolation and propagation of human immunodeficiency virus on recombinant colony-stimulating factor 1-treated monocytes. J Exp Med 167:1428–1441

Gendelman HE, Baca L, Husayni H et al. (1990a) Macrophage–human immunodeficiency virus interaction: viral isolation and target cell tropism. AIDS 4:221–228

Gendelman HE, Baca L, Turpin JA, Kalter DC, Hansen BD, Orenstein JM, Diffenbach CW, Friedman RM, Meltzer MS (1990b) Regulation of HIV replication in infected monocytes by IFN-α: mechanisms for viral restriction. J Immunol 145:2669–2676

Gendelman HE, Baca L, Turpin JA, Kalter C, Hansen BD, Orenstein JM, Friedman RM, Meltzer MS (1990c) Restriction of HIV replication in infected T cells and monocytes by interferon-α. AIDS Res Hum Retroviruses 6:1045–1049

Gendelman HE, Friedman RM, Joe S, Baca LM, Turpin JA, Dveksler G, Meltzer MS, Dieffenbach C (1990d) A selective defect of interferon α in human immunodeficiency virus-infected monocytes. J Exp Med 172:1433–1442

Gendelman HE, Baca LM, Kubrak CA, Genis P, Burrous S, Friedman RM, Munch D, Meltzer MS (1992) Induction of interferon α in peripheral blood mononuclear cells by human immunodeficiency virus (HIV)-infected monocytes: restricted antiviral activity of the HIV-induced interferons. J Immunol 148:422–429

Ghosh S, Gifford AM, Riviere LR, Tempst P, Nolan GP, Baltimore D (1990) Cloning of the p50 DNA binding subunit of NF-κ B: homology to rel and dorsal. Cell 62:1019–1029

Gomatos PJ, Stamatos NM, Gendelman HE et al. (1990) Lack of effect for recombinant soluble CD4 on infection of human monocytes by HIV. J Immunol 144:4183–4188

Goren T, Fischer DG, Rubinstein M (1986) Human monocytes and lymphocytes produce different mixtures of α-interferon subtypes. J Interferon Res 6:323–329

Griffin GE, Leung K, Folks TM, Kunkel S, Nabel GJ (1989) Activation of HIV gene expression during monocyte differentiation by induction of NF-κ B. Nature 339:70–73

Hansen BD, Nara PL, Maheshwas RK, Sidhu G, Bernbaum JG, Hoekzema D, Meltzer MS, Gendelman HE (1992) Loss of infectivity by progeny virus from interferon α-treated human immunodeficiency virus type 1-infected T cells is associated with defective assembly of envelope gp120. J Virol 66:7543–7548

Harper ME, Marselle LM, Gallo RC, Wong-Staal F (1986) Detection of lymphocytes expressing human T-lymphotropic virus type III in lymph nodes and peripheral blood from infected individuals by in situ hybridization. Proc Natl Acad Sci USA 83:772–776

Hartshorn KL, Neumeyer D, Vogt MW, Schooley RT, Hirsch MS (1987) Activity of interferons alpha, beta and gamma against human immunodeficiency virus replication in vitro. AIDS Res Hum Retroviruses 3:125–133

Ho DD, Rota TR, Kaplan JC, Hartshorn KL, Andrews CA, Schooley TR, Hirsch MS (1985) Recombinant human interferon Alfa-a suppresses HTLV-III replication in vitro. Lancet 1: 602–604

Ho DD, Moudgil T, Alam M (1989) Quantitation of human immunodeficiency virus type 1 in the blood of infected persons. N Engl J Med 321:1621–1625

Hwang SS, Boyle TJ, Lyerly HK, Cullen BR (1991) Identification of envelope V3 loop as the primary determinant of cell tropism in HIV. Science 253:71–74

Jones KA, Kadonaga JT, Luciw PA, Tijan R (1986) Activation of the AIDS retrovirusus promoter by the cellular transcription factor, Sp1. Science 232:755–759

Kalter DC, Nakamura M, Turpin JA et al. (1991) Enhanced HIV replication in MCSF-treated monocytes. J Immunol 146:298–306

Kieran M, Blank V, Logeat F, Vandekerckhove J, Lottspeich F, Le Bail O, Urban MB, Kourilsky P, Baeuerle PA, Israel A (1990) The DNA binding subunit of NF-κ B is identical to factor KBF-1 and homologous to the rel oncogene product. Cell 62:1007–1018

Klotman ME, Kim S, Buchbinder A, DeRossi A, Baltimore D, Wong-Staal F (1991) Kinetics of expression of multiply spliced RNA in early human immunodeficiency virus type 1 infection of lymphocytes and monocytes. Proc Natl Acad Sci USA 88:5011–5015

Koenig S, Gendelman HE, Orenstein JM et al. (1986) Detection of AIDS virus in macrophages in brain tissue from AIDS patients with encephalopathy. Science 233:1089–1093

Kornbluth RS, Oh PS, Munis JR, Cleveland PH, Richman DD (1990) The role of interferons in the control of HIV replication in macrophages. Clin Immunol Immunopathol 54:200–219

Koyahagi Y, O'Brien WA, Zhao JQ, Golde DW, Gasson JC, Chen ISY (1988) Cytokines alter production of HIV-1 from primary mononuclear phagocytes. Science 241:1673–1675

Lahdevirta J, Maury CPJ, Teppo AM, Repo H (1988) Elevated levels of circulating cachectin/tumor necrosis factor in patients with acquired immunodeficiency syndrome. Am J Med 85:289–291

Lau AS, Read SE, Williams BRG (1988) Downregulation of interferon α but not γ receptor in vivo in the acquired immunodeficiency syndrome. J Clin Invest 82:1415–1421

Laurence J (1990) Immunology of HIV infection, I: biology of the interferons. AIDS Res Hum Retroviruses 6:1149–1156

Lepe-Zuniga JL, Mansell PWA, Hersh EM (1987) Idiopathic production of interleukin-1 in the acquired immunodeficiency syndrome. J Clin Microbiol 25:1695–1700

Lifson AR, Rutherford GW, Jaffe HW (1988) The natural history of human immunodeficiency virus infection. J Infect Dis 158:1360–1367

Lynn WS, Tweedale A, Cloyd MW (1988) Human immunodeficiency virus (HIV-1) cytotoxicity: perturbation of the cell membrane and depression of phospholipid synthesis. Virology 163:43–51

Mace K, Gazzolo L (1991) Interferon-regulated viral replication in chronically HIV 1-infected promonocytic U937 cells. Res Virol 142:213–220

Maheshwari RK, Manerjee DK, Waechter CJ, Olden K, Friedman RM (1980) Nature 287:454–458

Malim MH, Hauber J, Fenrick R, Cullen BR (1988) Immunodeficiency virus rev *trans*-activator modulates the expression of the viral regulatory genes. Nature 335:181–183

Mann DL, Gartner S, LeSane F, Blattnere WA, Popovic M (1990) Cell–surface antigens and function of monocytes and a monocyte-like cell line before and after infection with HIV. Clin Immunol Immunopathol 54:174–183

Marciniak RA, Garcia-Blanco MA, Sharp PA (1990) Identification and charcterization of a HeLa nuclear protein that specifically binds to the *trans*-activating-response (tar) element of human immunodeficiency virus. Proc Natl Acad Sci USA 87:3624–3628

Matsuyama T, Kobayashi N, Yamamoto N (1991) Cytokines and HIV infection: is AIDS a tumor necrosis factor disease? AIDS 5:1405–1417

Meltzer MS, Skillman DR, Hoover DL, Hanson BD, Turpin JA, Kalter DC, Gendelman HE (1990) Macrophages and the immunodeficiency virus. Immuno Today 11:217–223

Meylan PR, Guatelli JC, Munis JR, Kornbluth RS, Richman DD (1991) Interferons inhibit the replication of HIV-1 in macrophages by reducing both viral DNA and regulatory gene mRNAs. Int Conf AIDS 7:60 (Abstr TU.A.20)

Michaelis B, Levy JA (1989) HIV replication can be blocked by recombinant human interferon beta. AIDS 3:27–31

Molina J-M, Scadden DT, Amirault C, Woon A, Vannier E, Dinarello CA, Groopman JE (1990a) Human immunodeficiency virus does not induce interleukin-1, interleukin-6 or tumor necrosis factor in mononuclear cells. J Virol 64:2901–2906

Molina J-M, Schindler R, Ferriana R, Sakaguchi M, Vannier E, Dinarello CA, Groopman JE (1990b) Production of cytokines by peripheral blood monocytes/macrophages infected with human immunodeficiency virus type 1 (HIV-1). J Infect Dis 161:888–893

Muchardt C, Seeler J-S, Nirula A, Shurland D-L, Gaynor RB (1992) Regulation of human immunodeficiency virus enhancer function by PRDII-BF-1 and c-rel gene products. J Virol 66:244–250

Nathan CF, Cohen ZA (1985) Cellular components of inflammation: monocytes and macrophages. In: Kelly WN, Harris ED, Ruddy S, Sledge CB (eds) Text book of rheumatology, vol 1. Saunders, New York, p 144

Nelbock P, Dillon PJ, Perkins A, Rosen CA (1990) A cDNA for a protein that interacts with the human immunodeficiency virus tat transactivator. Science 248:1650–1653

Oka S, Urayama K, Hirabayashi Y, Kimura S, Mitamura K, Shimada K (1992) Human immunodeficiency virus DNA copies as a virologic marker in a clinical trial with β-interferon. J AIDS 5:707–711

Olafsson K, Smith MS, Marshburn P, Carter SG, Haskill S (1991) Variation of HIV infectibility of macrophages as a function of donor, stage of differentiation, and site of origin. J AIDS 4:154–164

Orenstein JM, Meltzer MS, Phillips T, Gendelman HE (1988) Cytoplasmic assembly and accumulation of human immunodeficiency virus types 1 and 2 in recombinant human colony-stimulating factor-1-treated human monocytes: an ultrastructural study. J Virol 62:2578–2586

Pestka S, Langer JA, Zoon KC, Samuel CE (1987) Interferons and their actions. Annu Rev Biochem 56:727–777

Pitha PM (1991) Multiple effects of interferon on HIV-1 replication. J Interferon Res 11:313–318

Pitha PM, Fernie B, Maldarelli F, Hattman T, Wivel NA (1980) Effect of interferon on mouse leukemia virus (MuLV) V. Abnormal proteins in virons of Rauscher MuLV produced in the presence of interferon. J Gen Virol 46:97–110

Plata F, Autran B, Martins LP et al. (1987) AIDS virus-specific cytotoxic T lymphocytes in lung disorders. Nature 328:248–251

Poli G, Orenstein JM, Kinter A, Folks TM, Fauci AS (1989) Interferon-α but not AZT suppresses HIV expression in chronically infected cell lines. Science 244:575–577

Pomerantz RJ, Hirsch MS (1987) Interferon and human immunodeficiency virus infection. Interferon 9:114–127

Popik W, Pitha PM (1991) The inhibition by interferon of Herpes simplex virus type 1-activated transcription of fat defective virus. Proc Natl Acad Sci USA 88:9573–9577

Popovic M, Gartner S (1987) Isolation of HIV from monocytes but not T lymphocytes. Lancet ii:916

Rich EA, Chen ISY, Zack JA, Leonard ML, O'Brien WA (1992) Increased susceptibility of differentiated mononuclear phagocytes to productive infection with human immunodeficiency virus-1 (HIV). J Clin Invest 89:176–183

Robert–Guroff M, Popovic M, Gartner S, Markham P, Gallo RC, Reitz MS (1990) Structure and expression of tat-, rev-, and nef-specific transcripts of human immunodeficiency virus type 1 infected lymphocytes and monocytes. J Virol 64:3391–3398

Roberts NJ Jr, Douglas RG Jr, Simons RM, Diamond EE (1979) Virus induced interferon production by human macrophages. J Immunol 123:365–369

Ron D, Braiser AR, Haebner JF (1991) Angiotensin gene-inducible enhancer-binding protein I, a member of a new family of large nuclear proteins that recognize NF-κ B sites through a zinc finger motif. Mol Cell Biol 11:2887–2895

Rose RM, Krivine V, Pinkston P, Gillis JM, Huang A, Hammer SM (1991) Frequent identification of HIV DNA in broncoalveolar lavage cells obtained from individuals with the acquired immunodeficiency syndrome. Am Rev Respir Dis 143:850–854

Ross EK, Buckler-White AJ, Rabson AB, Englund G, Martin MA (1991) Contribution of NF-κ B and Sp1 binding motifs to the replicative capacity of human immunodeficiency virus type 1: distinct patterns of viral growth are determined by T-cell types. J Virol 65:4350–4358

Rossol S, Voth R, Laubenstein HP, Müller WEG, Schröder HC, Meyer zum Büschenfeld KH, Hess G (1989) Interferon production in patients infected with HIV-1. J Infect Dis 159:815–821
Roy S, Katze MG, Parkin NT, Edery I, Hovanessian AG, Sonenberg N (1990) Control of the interferon-induced 68-kilodalton protein kinase by the HIV-1 tat gene product. Science 247:1216–1219
Ruben S, Dillon PF, Schreck R, Henkel T, Chen CH, Maher PA, Baeuerle PA, Rosen CA (1991) Isolation of a rel-related human cDNA that potentially encodes a 65 kD subunit of NF-κ B. Science 251:1490–1493
Saksela E, Virtanen I, Hovi T, Secher DS, Cantell K (1984) Monocyte is the main producer of human leukocyte α-interferons following Sendai virus induction. Prog Med Virol 30:78–86
Samuel CE, Khutson GS, Masters PS (1982) Mechanism of interferon action: ability of cloned human type-alpha interferons to induce protein phosphorylation and inhibit virus replication is specified by the host cell rather than the interferon subspecies. J Interferon Res 2:97–108
Sato H, Orenstein J, Dimitrov D, Martim M (1992) Cell to cell spread of HIV occurs within minutes and may not involve the participation of virus particles. Virology 186:712–724
Schnizlein-Blick CT, Sherman MR, Boggs DL, Leemhuis TB, Fife KH (1992) Incidence of HIV infection in monocyte subpopulations characterized by CD4 and HLA-DR surface density. AIDS 6:151–156
Schnittman SM, Sallipopoulos MC, Lane HC (1989) The reservoir for HIV in human peripheral blood is a T cell that maintains expression of CD4. Science 245:305–308
Schröder HC, Wenger R, Kuchino Y, Müller WEG (1989) Modulation of nuclear matrix-associated 2,5-oligoadenylate metabolism and ribonuclease L activity in H9 cells by human immunodeficiency virus. J Biol Chem 264:5669–5673
Schröder HC, Ugarkovic D, Wenger R, Reuter P, Okamoto T, Müller WEG (1990) Binding of tat protein to TAR region of human immunodeficiency virus type-1 blocks TAR-mediated activation of (2′-5′)oligoadenylate synthetase. AIDS Res Hum Retroviruses 6:659–672
Schuitemaker H, Kootstra NA, de Goede REY, De Wolf F, Miedema F, Tersmette M (1991) monocytrophic human immunodeficiency virys type 1 (HIV) variants detectable in all stages of HIV infection lack T cell line tropism and syncytium-inducing ability in primary T cell culture. J Virol 65:356–363
SenGupta N, Silverman RH (1989) Activation of interferon-regulated, dsRNA-dependent enzymes by human immunodeficiency virus-1 leader RNA. Nucleic Acids Res 17:969–976
Shaw GM, Harper ME, Hahn BH et al. (1985) HTLV-III infection in brains of children and adults with encephalopathy. Science 227:177–182
Shirazi Y, Pitha PM (1992) Alpha interferon inhibits early stages of human immunodeficiency virus type 1 replication cycle. J Virol 66:1321–1328
Simmonds P, Balfe P, Peutherer JF, Ludlam CA, Bishop JO, Leigh Brown AJ (1990) Human immunodeficiency virus-infected individuals contain provirus in small numbers of peripheral mononuclear cells at low copy number. J Virol 64:864–872
Singh UK, Maheshwari RK, Damewood GP IV, Stephensen CB, Oliver C, Freidman RM (1988) Interferon alters intracellular transport of vesicular stomatitis virus glycoprotein. J Biol Reg Homeo Agents 2:53–62
Staeheli P (1990) Interferon-induced proteins and the antiviral state. Adv Virus Res 38:147–200
Steffy K, Wong-Staal F (1991) Genetic regulation of human immunodeficiency virus. Microbiol Rev 55:193–205
Taylor JL, Grossberg SE (1990) Recent progress in interferon research: molecular mechanisms of regulation, action and virus circumvention. Virus Res 15:1–26
Trono D, Baltimore D (1990) A human cell factor is essential for HIV-1 rev action. EMBO J 9:4155–4160
Vaishnav YN, Wong-Staal F (1991) The biochemistry of AIDS. Annu Rev Biochem 60:577–630
Vaishnav YN, Vaishnav M, Wong-Staal F (1991) Identification and characterization of a nuclear factor that specifically binds to the rev response element (RRE) of human immunodeficiency virus type 1 (HIV-1). New Bio 3:142–150

Valentin A, von Gegerfelt A, Matsuda S et al. (1991) In vitro maturation of mononuclear phagocytes and susceptibility to HIV-1 infection. J AIDS 4:751–759

Vazeux R, Brousse N, Jarry A et al. (1987) AIDS subacute encephalitis. Identification of HIV-infected cells. Am J Pathol 126:403–410

von Sydow M, Sonnerborg A, Gaines H, Stronnegard O (1991) Interferon-alpha and tumor necrosis factor-alpha in serum of patients in various stages of HIV infection. AIDS Res Hum Retroviruses 4:375–380

Willey RL, Bonifacino JS, Potts BJ, Martin MA, Klausner RD (1988) Biosynthesis, cleavage, and degradation of the human immunodeficiency virus type 1 envelope glycoprotein gp160. Proc Natl Acad Sci USA 85:9580–9584

Wright SC, Jewett A, Mitsuyasu R, Bonavida B (1988) Spontaneous cytotoxicity and tumor necrosis factor production by peripheral blood monocytes from AID patients. J Immunol 141:99–104

Yagi MJ, King NW Jr, Bekesi G (1980) Alterations of mouse mammary tumor virus glycoprotein with interferon treatment. J Virol 34:225–233

Yamada O, Hattori N, Kurimura T, Kita M, Tsunataro K (1988) Inhibition of growth of HIV by human natural interferon in vitro. AIDS Res Hum Retroviruses 4:287–294

Zack JA, Arrigo SJ, Chen ISY (1990) Control of expression and cell tropism of human immunodeficiency virus type 1. Adv Virus Res 38:125–146

Transmembrane Signaling by IFN-α

L.M. Pfeffer, S.N. Constantinescu, and C. Wang[1]

1 Introduction

1.1 Background

In order to understand the mechanism whereby human interferon alpha (IFN-α) transduces its signal through the plasma membrane, we first must describe the multiple biological effects that are induced by the cytokine. IFN-α was the first cytokine described, purified, and cloned and has remarkable effects on cells. IFN is part of the host defense and is produced in response to microbial infection, tumor formation, and antigen stimulation (Pfeffer 1987). IFN-α was first recognized by its ability to protect cells against a wide variety of RNA and DNA viruses. However, IFN-α also has antimicrobial, antiproliferative, antitumor, and immunomodulatory actions. IFNs do not produce these effects directly, but rather elicit their manifold biological effects by first binding to specific receptors on the surface of target cells. This interaction transduces a signal to the nucleus that results in the induction of some genes (the exact number is not known) and the downregulation of others (Aguet 1980; Rubinstein and Orchansky 1986; Pfeffer 1987; Langer and Pestka 1988). Gene induction by IFN is transient, involving both a switch-on and a switch-off mechanism, and is then followed by a period of desensitization. In addition to gene induction, IFNs regulate the expression of many key proteins at posttranscriptional levels (i.e., receptors for insulin and EGF, 2′,5′-oligoadenylate synthetase) (Pfeffer 1987; Pfeffer et al. 1987a; Eisenkraft et al. 1991). The complex regulation of cellular protein content by IFN-α ultimately results in the biological effects of IFN-α. IFNs are highly effective molecules with the occupancy of only a few receptors per cell triggering a biological response in target cells. The alterations in some 50-100 cellular proteins are responsible for the IFN-induced changes in cell behavior.

The four major antigenic types of IFNs, IFN-α, -β, -ω, and -γ, are defined by the cellular source of their production. While type I IFNs (IFN-α, -β, and -ω) compete with each other for cellular binding and thus apparently share a common complex cell surface receptor, the receptor(s) for type II IFN (IFN-γ) seems

[1] Department of Pathology, University of Tennessee Health Science Center, 800 Madison Avenue (576 BMH), Memphis, TN 38163, USA

Progress in Molecular and Subcellular Biology, Vol. 14
W.E.G. Müller/H.C. Schröder (Eds.)
© Springer-Verlag Berlin Heidelberg 1994

to be a distinct entity (Branca et al. 1982). The human genes for binding components for type I and II IFNs are located on chromosomes 21 and 6, respectively (Langer and Pestka 1988). However, additional signal transduction and/or binding subunits appear to be necessary for biological activity of both type I and II IFNs (Jung et al. 1987; Langer and Pestka 1988; Pelligrini et al. 1992). Although there is a great deal of overlap in the biological effects induced by all IFNs, some are selectively induced by each type of IFN (Revel and Chebath 1986; Pestka et al. 1987).

IFN-α binding to cell surface receptors is an important event, but binding alone is not sufficient for induction of antiviral and antiproliferative activity. For example, several IFN-resistant cell lines display significant levels of high affinity IFN receptors that are structurally identical to the receptors on IFN-sensitive cells, while they do not transduce a biological signal (Kessler et al. 1988; Vanden Broecke and Pfeffer 1988). Thus, recent investigation has focused on the immediate biological consequences of IFN-receptor interactions. The initial interaction with its receptors leads to formation of an "activated" high affinity IFN-receptor complex, which is insoluble in nonionic detergents (Mogensen and Bandu 1983; Zoon et al. 1986; Pfeffer et al. 1987b). As with many other polypeptide ligands, after cell surface binding, IFN-α is internalized by receptor-mediated endocytosis, where it is degraded intracellularly, and released from the cell (Zoon et al. 1983). Study of the immediate biological events associated with transmembrane signaling has been aided by the identification of the IFN-stimulated gene (ISG) family, whose transcription is rapidly activated upon IFN addition. The genes induced by type-I IFNs are distinct from, yet overlap, genes activated by type-II IFNs (Friedman et al. 1984; Larner et al. 1986), as well as those induced by other polypeptide ligands (i.e., growth factors). This review will focus on the recent evidence concerning the transmembrane signaling pathways activated by IFN-α. In the past few years it has become apparent that the interaction of IFN-α with its receptor activates several transmembrane signaling mechanisms. For example, IFN-α activates phospholipid turnover resulting in the production of choline metabolites and arachidonic acid. IFN-α also stimulates two distinct types of protein kinase, PTK and PKC.

1.2 Transcriptional Activation by IFN-α and the Role of DNA-Binding Factors

The cellular response to IFN is dependent in part on the de novo synthesis of IFN-induced proteins involved in the antiviral and antiproliferative actions. Among these proteins are a class of genes, the ISG family, whose transcription is rapid and yet transient even in the continued presence of IFN-α. A major step toward the dissection of the signaling pathway of IFN-α was the cloning of a number of ISGs from human cells. Their activation is one of the earliest physiologically relevant responses induced by IFN-α in target cells. For example, transcription of one ISG, ISG15, in many cells is detectable within minutes and reaches maximal levels within 1–2 h of IFN addition, and, in general, does not require protein synthesis (Larner et al. 1986). However, in some cell types the induction of ISGs can be blocked by inhibitors of protein synthesis (Improta et al. 1992).

Deletion and point mutation analysis have identified a homologous and highly conserved *cis*-acting DNA sequence of 14–18 bp in the 5′ promoter-enhancer region of ISGs, the IFN-stimulus response elements (ISRE), which is both necessary and sufficient for ISG transcriptional activation by IFN-α (Friedman and Stark 1985; Reich et al. 1987). Regulation of enhancer elements is, in general, mediated through the interaction of sequence-specific nuclear proteins. Gel retardation analysis of nuclear proteins has shown that IFN-α treatment induces the appearance of three distinct DNA-binding protein complexes that specifically recognize this promoter element (Kessler et al. 1988; Levy et al. 1989). There are three such factors, termed the IFN-stimulated gene factors (ISGFs): (1) ISGF1 is constitutively present and detectable in cells not treated with IFN-α; (2) ISGF2 is induced slowly and requires protein synthesis for its appearance; and (3) ISGF3 is activated very rapidly without a protein synthesis requirement.

There are several lines of evidence to suggest that ISGF3 is apparently a transcription factor for the ISGs: (1) it appears in the cytoplasm within minutes of IFN treatment before its appearance in the nucleus (Dale et al. 1989); (2) its appearance does not require new protein synthesis; (3) purified ISGF3 can stimulate ISRE-dependent in vitro transcription (Fu et al. 1990); (4) its rise and fall correlate with the cycle of ISG transcription; and (5) its absence in a variety of IFN-resistant cells correlates with the lack of ISG transcription in these cells (Kessler et al. 1988). ISGF3 is composed of two subunits, ISGF3α and ISGF3γ. The γ-component is a single 48-kDa polypeptide, which binds DNA with low affinity but high specificity for the ISRE (Levy et al. 1989; Fu et al. 1990). ISGF3γ is constitutively present in cells at various levels, but its amount can be increased by IFN-γ treatment in at least some cells. The α-component consists of three polypeptides and exists as an inactive cytosolic complex, requiring IFN-α binding for activation. ISGF3α combines with ISGF3γ in the cytoplasm to modulate the DNA binding affinity of ISGF3γ, and then translocates to the nucleus where the activated ISGF3 complex binds to the ISRE, initiating transcription (36,38). ISGF3γ, but not α, is inactivated by N-ethylmaleimide (NEM) treatment, which provides an important basis for reconstitution assays for ISGF3 subunits (Levy et al. 1989). Extracts of IFN-α-treated cells exposed to NEM contain ISGF3α, while extracts of IFN-γ-treated cells contain only the γ-component. Combinations of these extracts reconstitute ISGF3 activity, as determined by binding to the ISRE in gel retardation analysis.

It must be kept in mind that all IFN-α-responsive genes may not be regulated through a common signaling pathway. Recent evidence suggests that at least two IFN-α-dependent signaling pathways operate for the induction of IFN-α-responsive genes. Expression of the adenovirus oncoprotein E1A blocks the IFN-α-induced transcriptional activation of several classical ISGs (ISG15,

ISG54, P68 kinase, 2-5 A synthetase) in various cell lines (Kessler et al. 1988; Gutch and Reich 1991). Although E1A expression in HeLa cells does indeed block the transcription of these ISGs, E1A expression does not affect the transcriptional activation of the ISGF2 (IRF-1) in response to IFN-α (Haque and Williams 1992). Thus, these results suggest that IFN-inducible genes are activated through multiple signaling pathways.

1.3 Multisubunit Structure of the IFN-α Receptors

In order to understand the transmembrane signaling mechanism through which IFN-α works, we must first understand the basic structure of the IFN-α receptor. Initially, analysis of results with purified IFN-α subtypes suggested that IFN-α subtypes may differ in antiviral, antiproliferative, and immunomodulatory activities by several thousand-fold (Pfeffer 1987). Thus, it was apparent that there were either multiple IFN-α receptors or that the IFN-α receptor had a complex structure which permits varied interactions. Efforts to solubilize and isolate the IFN-α receptor have been unrewarding. Binding assays on the detergent-solubilized cell extracts have yielded only low-affinity binding sites, which suggests that other components (subunits) are necessary to obtain high-affinity binding (L.M. Pfeffer, unpubl. data). A 95–100 kDa IFN-α receptor binding subunit from lymphoblastoid cells has been solubilized and characterized (Faltynek et al. 1983).

Affinity cross-linking of ^{125}I-IFN-α to cell surface receptors has demonstrated that there is a complex series of bands between 110 and 160 kDa, as well as slower migrating species which probably reflect associations of several IFN-α receptor subunits (Vanden Broecke and Pfeffer 1988). Extraction of IFN-α receptors after ligand binding with detergent yields a complex that migrates with an M_r of 600 kDa after gel filtration (Eid and Mogenson 1990). ^{125}I-labeling of the isolated IFN-α receptor identified a 95-kDa binding protein.

A major step toward understanding the structure of the IFN-α receptor has been the cloning of a cDNA for a human IFN-α receptor subunit (Uze et al. 1990). Human DNA was transfected into mouse cells and then cells were selected for the ability of IFN-α8 to protect the cells against VSV infection. A 5-kb human DNA was isolated from secondary transfectants and used to detect an mRNA present in human cells. This human DNA was then used to clone a cDNA from the human Daudi (B-cell lymphoma) library. When transfected into mouse cells, this cDNA transfers the ability of human IFN-α8 to protect mouse cells against VSV infection. Other IFN-α subtypes or IFN-β were unable to protect the transfected mouse cells. It was also suggested that transfected cells bound IFN-α8, but not other IFN-α subtypes. However, IFN-α8 binding was only obtained when assays were performed at 37 °C, while cells bind IFN-α8 efficiently at 4 °C. The cDNA (2784 bp) contains a single open reading frame which encodes a ~ 63 500 dalton transmembrane protein with 15 potential N-linked glycosylation sites. The short intracellular domain (100 amino acids) does not possess sequences similar to those of other proteins involved in signal transduction, such as intrinsic protein kinase activity. Sequence analysis of this protein has suggested that it is a member of the cytokine receptor superfamily, with a duplicated ligand binding domain (Bazan 1990). Transfection of this cDNA into IFN-α-resistant mouse and human cell lines restores cellular sensitivity to the antiviral action of IFN-α and induces high basal levels of ISG expression (Pfeffer et al. 1992). These effects occur without a detectable increase in IFN-α binding, suggesting that this subunit may function in the signal transduction pathway of IFN-α (Colamonici et al. 1993; Pfeffer et al. 1992).

Strong evidence for a multisubunit IFN-α receptor structure has been provided by the development of monoclonal antibodies (mAb) against the IFN-α2 receptor (Colamonici et al. 1990, 1992). The mAb detect a 110-kDa protein (α-subunit) in surface-iodinated cells and in Western blots, and also immunoprecipitate 130- and 230-kDa bands after cross-linking cells with ^{125}I-IFNα2. The 230-kDa band apparently reflects the association of the 110-kDa protein with other receptor subunit(s). Further evidence that the α-subunit is a component of the IFN-α receptor has been provided by the finding that IFN-α pretreatment downregulates the level of the α-subunit, as determined by immunoprecipitation of surface-iodinated material and by fluorescence-activated cell sorting with mAb (Colamonici et al. 1992). That other subunits of the IFN-α receptor exist is inferred by the result that these mAb do not recognize all the IFNα-receptor complexes observed after affinity cross-linking of IFN-α2 to cell surface receptors.

In addition, cross-linking studies with IFN-α2 have illustrated that there are two predominant patterns of IFN-α receptors (Colamonici et al. 1992). The predominant form is observed on many cell types, including cells highly sensitive to the antiproliferative and antiviral action of IFN-α, and is composed of at least two subunits, with a molecular weight of 110 kDa (α-subunit) and 90 kDa. In contrast, the variant receptor form is found in some cells resistant to the antiproliferative and antiviral effects of IFN-α; it consists of the α-subunit and a 55-kDa subunit.

2 The Roles of PKC and PTK in Transmembrane Signaling by IFN-α

2.1 *Signal Transduction by Polypeptide Ligands*

The binding of polypeptide ligands to cell surface receptors activates distinct, but overlapping pathways of signal transduction. Several lines of evidence have indicated that IFN-induced changes in protein phosphorylation may play an important role in IFN action. IFN-α has to amplify the signal generated by the occupancy of only a few cell surface receptors. Ligand-mediated activation of protein kinases results in a cascade of phosphorylations (or dephosphorylations) of cytoplasmic and membrane proteins and is associated with the transcriptional activation of specific genes. Evidence for a role of protein kinases in transcriptional regulation has been provided by studies in which activators or inhibitors of the enzymes have been found to mimic or block ligand-induced effects on gene transcription.

A variety of cell membrane receptors contain intracellular protein tyrosine kinase catalytic domains, which are activated by ligand binding. However, many cell surface receptors do not contain intrinsic enzyme activity, and their mechanism for receptor signaling is poorly understood (Bazan 1990). Recently, occupancy of the T-cell antigen receptor and some cytokine receptors has been found to be associated with the activation cytoplasmic PTKs, including $p59^{fyn}$, $p56^{lck}$, $p52^{lyn}$, $p62^{c-yes}$. Although many of the specific nonreceptor PTKs have not yet been identified, tyrosine phosphorylation of a number of cellular proteins is an important early step in the signal transduction pathway activated by many members of the cytokine superfamily.

2.2 The Role of DAG and PKC in IFN-α Signaling

2.2.1 Rapid Changes in Lipid Hydrolysis and DAG in IFN-α Signaling

Early results have shown that IFN-α rapidly modifies the membrane fluidity of target cells (Pfeffer et al. 1981) and also stimulates phosphatidylcholine and cholesterol synthesis (Pfeffer et al. 1985). Recent evidence has established that IFN-α rapidly induces transient increases in the cellular DAG content (Yap et al. 1986a, b; Pfeffer et al. 1990, 1991). The magnitude of these increases correlates with the IFN-induced inhibition of cell proliferation and of virus replication (Yap et al. 1986a, b). Although IFN-α increases intracellular DAG levels, the lipid source for the increase in DAG has been a matter of controversy. According to the classical scheme, inositol phosphate hydrolysis results in the production of both inositol trisphosphate, which leads to increased endogenous Ca^{2+} concentration, and DAG, the endogenous PKC activator. In Daudi lymphoblastoid cells, IFN-α was reported to induce a transient increase in inositol phosphates (Yap et al. 1986a, b). However, IFN-α treatment of human cells results in no measurable changes in the intracellular concentration of Ca^{2+} and in no inositol phospholipid turnover (Pfeffer et al. 1990, 1991). Recently, other cellular lipids (e.g., phosphatidylcholine, PC) have been shown to serve as alternative sources of DAG (Exton 1990). In HeLa cells, IFN-α activates PC turnover, resulting in DAG production (Pfeffer et al. 1990). Increased PC hydrolysis is detectable within minutes of IFN treatment of HeLa cells, as well as the appearance of PC breakdown products, phosphorylcholine, and phosphatidic acid.

In Daudi cells, the involvement of IFN-induced inositol phospholipid turnover has also been ruled out as a source for the increase in cellular DAG mass, but lipid source for DAG is unclear (Pfeffer et al. 1991). A phospholipase D (PLD) pathway seems to be involved in Daudi cells, since the IFN-α-induced DAG increase may be inhibited by propanolol, a phosphatidic acid phosphhydrolase inhibitor (L.M. Pfeffer and S.N. Constantinescu, unpubl. data). Interestingly, in Daudi and MOLT-16 cells the IFN-α-activated PLD pathway may be a consequence of PTK activation, since both genistein and herbimycin (potent PTK inhibitors) abolish the IFN-α-induced DAG increase. In addition, since IFN-α has been reported to induce DAG production in nuclei (Cataldi et al. 1990), it will be important to correlate this with the IFN-α-induced translocation of PKC-ε to perinuclear areas (L.M. Pfeffer and S.N. Constaninescu, unpubl. data). Finally, transfection of the cloned IFN-α receptor subunit into the IFN-resistant human K-562 cell line restores IFN-α responsiveness (Colamonici et al. 1992b) and also results in higher constitutive levels of DAG as compared to empty vector-transfected cells (L.M. Pfeffer, S.N. Constantinescu, and O.R. Colamonici, unpubl. data). The overall conclusion from these studies is that IFN-α activates cellular phospholiphase(s), resulting in the production of the lipid-derived second messenger DAG, in the absence of inositol phospholipid turnover and an increase in endogenous Ca^{2+}.

IFN-α treatment of mouse Balb/c 3T3 cells results in a rapid and transient stimulation of phospholipase A_2 (PLA_2), as determined by arachidonic acid production in IFN-α-treated cells (Hannigan and Williams 1991). Inhibition of PLA_2 activity by p-bromophenylacyl bromide blocks ISRE-specific factor binding to the ISRE of the 2,5-oligoadenylate synthetase gene, as well as ISRE-dependent transcription. Furthermore, perturbations of arachidonic acid metabolism with inhibitors of cyclooxygenase and lipoxygenase activity modulate factor binding to the ISRE. Whether the actions of these drugs are at the level of activation and/or translocation of ISRE-specific factors is at present unknown.

2.2.2 IFN-α and Activation of PKC

Several laboratories have concentrated on a role of protein kinase C (PKC) in the signal transduction pathway of IFN-α, since IFN-α increases DAG concentration in many cell types and since DAG is the physiological activator of PKC (Fig. 1). Protein kinase C (PKC) is a family of related serine/threonine kinases that occupy important regulatory roles in a variety of cellular responses. PKC has been shown to play a crucial role in the signal transduction pathway elicited by a variety of growth factors, hormones, and neurotransmitters (Nishizuka 1988). Molecular cloning analyses have revealed that there are at least eight different PKC isozymes (Nishizuka 1988) that can be divided into two major groups. The conventional PKC groups consists of four isotypes, PKC-α, -βI, -βII, and -γ, which are characterized by the presence of the Ca^{2+}-binding C2 domain and by the requirement for Ca^{2+}, phospholipids, and DAG for activation (Sekiguchi et al. 1988). The more recently discovered nonconventional δ, ε, ζ, and η isotypes of PKC seems to be independent of Ca^{2+} for activation and lack the putative Ca^{2+}-binding C2 domain (Schaap et al. 1989). Within the PKC family, the various isotypes exhibit tissue-specific distributions, differing requirements of activators and cofactors, and differing substrate specificities (Ono et al.

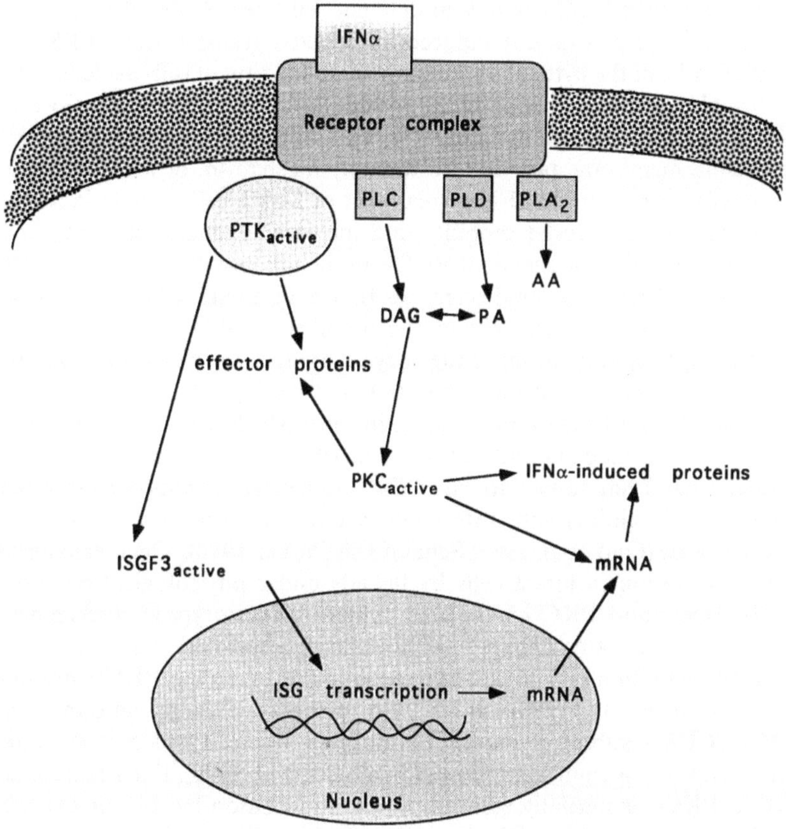

Fig. 1. Current representation of transmembrane signaling by IFN-α. Occupancy of the IFN-α receptor is followed by a series of signaling events; the chronology of these events is not known. These signaling events include: (1) activation of an unknown phospholipase, leading to the DAG production without increased intracellular Ca^{2+} concentrations; (2) the DAG production thereby results in the selective activation of calcium-independent isotypes of PK; and (3) activation of PTK, resulting in the phosphorylation of the 113-, 91-, and 84-kDa components of ISFG3α and activation of the ISG-specific transcription factor, ISGF3. The activated ISGF3 translocates to the nucleus, where it binds to the ISRE and activates the transcription of ISGs, resulting in the production of ISG mRNA

1988; Schaap and Parker 1990). Thus, it is likely that individual isotypes of PKC have distinct physiological roles in transducing cellular events.

Recent studies have indicated a role of PKC activation in the signal transduction pathway of IFN-α. Pretreatment of cells with the selective PKC inhibitor, H-7 and staurosporine, or chronic exposure to phorbol esters (which can downregulate some PKC isotypes) can inhibit the action of IFN-α in human fibroblasts, as well as in HeLa cells (Akai and Larner 1989; Osada et al. 1990; Pfeffer et al. 1990, 1991; Reich and Pfeffer 1990). Direct evidence for a role of PKC activation in IFN-α action has been provided by the discovery that IFN-α activates selective PKC isotypes in different cell lines (Pfeffer et al. 1990, 1991). In HeLa cells, IFN-α rapidly induced the selective translocation of PKC-β, but not PKC-α from the cytosol to the particulate fraction of cells as determined by immunoblot analysis (Pfeffer et al. 1990). The IFN-induced translocation of PKC-β was observed within 5 min of IFN-α addition. However, translocation to the plasma membrane may not be the only mechanism of PKC activation in intact cells. For example, IFN-α treatment of Daudi cells results in a selective time-dependent decreased mobility and increased immunoreactivity in SDS-PAGE of the Ca^{2+}-independent PKC-ε in the absence of translocation (Pfeffer et al. 1991). PKC-α is unaffected by IFN-α in these cells. The changes in immunoreactivity and electrophoretic mobility of PKC-ε apparently reflect activation (phosphorylation) of the enzyme. Indeed, in ^{32}P-labeled Daudi cells, IFN-α rapidly induces the phosphorylation of PKC-ε (S.N. Contantinescu, D.J. MacEwan, L.M. Pfeffer, unpubl. data). In contrast, IFN-α treatment of an IFN-resistant Daudi subclone did not result in PKC-ε activation.

PKC-ε has been shown to have Ca^{2+}-independent phorbol ester binding activities and exhibits substrate specificity distinct from other characterized PKC isotypes (Ono et al. 1988; Schaap and Parker 1990). The mechanism(s) of PKC-ε activation in intact cells by ligands under physiological conditions is poorly understood. PKC-ε is the predominant PKC isotype in chicken neurons, where insulin activates PKC-ε by a mechanism associated with a change in the electrophoretic mobility of the enzyme, but not involving PKC translocation (Heidenreich et al. 1990). It is also present in large amounts in the CD4 + /CD8 + subset of murine thymocytes, where it becomes activated by plant lectins or calcium ionophores (Strulovici et al. 1990). Furthermore, unlike PKC-α, PKC-ε is partially resistant to downregulation by chronic exposure to phorbol esters in many cells, including Daudi cells (Heidenreich et al. 1990; Strulovici et al. 1990; Pfeffer et al. 1991). These findings suggest that PKC-ε may play a unique role in signal transduction. Indeed, Daudi cells have both PKC-α and -ε, but PKC-ε appears to be linked to the biological effects of IFN-α (Pfeffer et al. 1991). This correlates well with the finding that IFN-α treatment of Daudi cells results in DAG production in the absence of an increase in intracellular Ca^{2+}.

2.2.3 Involvement of PKC in the Posttranscriptional Effects of IFN-α

Several reports have suggested that PKC may play an important role in posttranscriptional regulation of gene expression modulated by IFN-α (Akai and Larner 1989; Faltynek et al. 1989; Reich and Pfeffer 1990). Pretreatment of Daudi cells with phorbol esters (downregulating some of the PKC isotypes), or treatment with the protein kinase inhibitor H-7, inhibit the induction of 2,5-oligoadenylate synthetase (OAS) gene expression and enzymatic activity, but do not inhibit OAS transcription (Faltynek et al. 1989). Furthermore, although H-7 was also found to inhibit antiviral activity of IFN-α and induction of ISG mRNA expression by IFN-α in Daudi and HeLa cells, H-7 did not inhibit the transcriptional activation of ISG15 and ISG54 by IFN-α (Reich and Pfeffer 1990; Pfeffer et al. 1991). In addition, long-term pretreatment with phorbol esters not only blocks the accumulation of mRNA for IFN-responsive genes in several cell types, but also blocks the ability of IFN-α to protect cells against virus infection (Akai and Larner 1989; Pfeffer et al. 1990). For example, prolonged phorbol ester treatment of HeLa cells downregulates PKC-β (the IFN-activated PKC), blocks steady-state accumulation of ISG mRNAs, and blocks induction of antiviral activity of IFN-α (Pfeffer et al. 1990; Reich and Pfeffer 1990). In Daudi cells, however, phorbol esters induce PKC-ε activation in the absence of its rapid downregulation, although PKC-α becomes rapidly downregulated (Pfeffer et al. 1991). It is interesting that although IFN-α treatment of IFN-resistant Daudi cells activates ISG15 and ISG54 gene expression, IFN-α does not activate PKC-ε in resistant cells or result in high induced levels of ISG mRNA (Kessler et al. 1988; L.M. Pfeffer, C. Wang, S.N. Constantinescu, unpubl. data). Thus, it is tempting to speculate that PKC-ε might be involved in the posttranscriptional processes necessary to maintain high levels of ISG mRNAs.

Although activation of PKC in HeLa and Daudi cells by phorbol esters or cell-permeable DAGs does not induce the expression of IFN-responsive genes, they are able to mimic several aspects of IFN action, i.e., induce antiviral activity and prime cells for IFN production (Sehgal et al. 1987; Constantinescu et al. 1990). For example, phorbol esters can potentiate the induction of HLA class I antigen expression induced by IFN-α in human T-cells (Ersalimsky et al. 1989). Thus, taken together, the results obtained with protein kinase inhibitors and activators strongly suggest that PKC activation is an important element in transmembrane signaling by IFN-α.

One overall concern is the reliance on the use of inhibitors and activators in such studies since the specificity of their action in cultured cells is not absolute. However, treatment with a specific PKC inhibitor, a peptide pseudosubstrate of PKC, markedly reduces the antiviral protection afforded by IFN-α in Daudi cells and HeLa cells (L.M. Pfeffer, unpubl. data), providing further support for a role of PKC as an important element in IFN-α action. However, PKC activation is not sufficient, nor is it the only pathway involved in IFN-α signal transduction. Together, these studies suggest a complex and multifaceted role of PKC in the enhanced expression of ISGs, through posttranscriptional regulation, and in the induction of antiviral activity by IFN-α. In addition, these findings highlight the cell type differences in the cellular response to IFN-α.

2.3 *The Role of Tyrosine Phosphorylation and PTK in IFN-α Signaling*

2.3.1 *Complementation with the TYK2 PTK*

In order to dissect the early steps in IFN-α signaling, mutants in the IFN-responsive system have been isolated through the use of an IFN-inducible drug-selectable marker. The 2fTGH human cell line was constructed, in which the *E. Coli* guanine phosphoribosyl transferase (*gpt*) gene is controlled tightly by the upstream region of an ISG (the 6–16 gene). The *gpt* gene was chosen since there are established protocols to select either for or against its expression in mammalian cells. After mutagenesis of 2fTGH cells and selection in 6-thioguanine and IFN-α for failure to induce *gpt* expression, unresponsive mutants comprising four complementation groups in the IFN-responsive pathway have been isolated.

One such IFN-α-unresponsive mutant cell line 11,1 is defective in IFN-α, but not in IFN-γ inducible gene expression (Pelligrini et al. 1992). This defect apparently resides in IFN-α signaling as these mutant cells bind IFN-α, but are unable to produce active ISGF3α. Using DNA transfer, this defect in 11,1 cells was complemented by a cosmid, encoding a single message which is greatly reduced in mutant 11,1 cells (Velazquez et al. 1992). Cloning of cDNAs from a Daudi library has identified this message to be virtually identical to the nonreceptor PTK, *tyk2*, of previously unknown function. This finding strongly suggests a role of this PTK in early signaling events of IFN-α. Receptors that lack intrinsic PTK activity have been shown to become physically associated with receptors upon ligand binding. Interestingly, in vitro kinase assays of the JAK-1 protein, which is highly related to the *tyk2* kinase, have indicated that serine/threonine as well as tyrosine residues are recognized by this kinase as acceptor sites on proteins.

2.3.2 Tyrosine Phosphorylation of ISGF3α and ISG Transcriptional Activation

The involvement of protein phosphorylation in the signaling pathway that leads to ISG activation was first suggested by the finding that staurosporine completely blocks the transcriptional activation of ISGs in HeLa and Daudi cells (Reich and Pfeffer 1990; Pfeffer et al. 1991). To assess the effect of staurosporine on the IFN-induced appearance of ISRE-specific DNA-binding proteins, gel retardation assays on cellular extracts were performed. Staurosporine was found to abolish the appearance of ISGF3α not only in nuclear, but also in cytoplasmic extracts of IFN-α-treated HeLa cells (Reich and Pfeffer 1990). These results suggest that staurosporine blocks an early step in transmembrane signaling by IFN-α. Specifically, one which is necessary for the activation of latent ISGF3α by IFN-α, rather than merely its IFN-induced translocation to the nucleus (see Fig. 1). However, at that time it was not known what kinase activated a component of ISFG3. We presumed that the kinase was PKC, however, we were aware that another kinase may phosphorylate ISGF3.

Further evidence for an important role of PTK activity has been provided by the finding that components of the ISGF3 transcription factor are rapidly phosphorylated on tyrosine residues in response to IFN-α treatment (Fig. 1). We had previously suggested that one likely substrate of an IFN-activated kinase is the latent transcriptional factor, ISGF3, which is activated and translocated to the nucleus in response to IFN-α. Transcription factors that exist in an inactive form prior to ligand-induced signaling are found in many instances in the cytoplasm of mammalian cells. Protein purification and subsequent cloning have revealed that ISGF3α consists of three proteins (113, 91, and 84 kDa) that reside in a latent form in the cytoplasm until they become activated by IFN-α and thereby translocate to the nucleus. Western blot analysis and immunoprecipitation with antiphosphotyrosine antibodies have demonstrated that the three ISGF3α proteins are phosphorylated on tyrosine residues rapidly after IFN-α addition to cells (Fu 1992; Schindler et al. 1992). Phosphoamino acid analysis of ^{32}P-labeled ISGF3α proteins has confirmed that they are indeed phosphorylated on tyrosine residues upon IFN-α treatment. Tyrosine phosphorylation of these proteins is inhibited by the PKC and PTK inhibitor staurosporine, and by the PTK inhibitor genistein. Furthermore, coimmunoprecipitation experiments have revealed that IFN-α treatment induces the three ISGF3α proteins to form a complex (Schindler et al. 1992). In addition, phosphatase treatment of these proteins inhibits the formation of the ISGF3 complex in vitro (Fu 1992). Additionally, in vitro kinase assays show tyrosine and serine/threonine kinase activity in ISFG3α immunoprecipitates.

Sequence analysis of the ISGF3α proteins show high conservation of residues within *src* homology (SH) domains, domains which are commonly present in a number of PTK-regulated proteins (Fu 1992). SH domains are believed to play a role in the docking of regulatory proteins to the cytoskeleton. This is of particular importance considering the early studies that show coupling of IFN-α receptors to the cytoskeleton in IFN-α-sensitive, but not in IFN-α-resistant Daudi cell lines (Pfeffer et al. 1987b).

2.3.3 Rapid Tyrosine Phosphorylation in Response to IFN-α

In contrast to reports of rapid changes in PKC activity and the phosphorylation of ISGF3α components, relatively little is known about the phosphorylation of cellular proteins as a consequence of these rapid changes induced by IFN-α. IFN selectively activates the phosphorylation of several proteins (including elF-2α), while inhibiting the phosphorylation of others (Roberts et al. 1976; Soslau et al. 1984). Using two-dimensional gel electrophoretic analysis of ^{32}P-labeled cellular proteins, the phosphorylation of a few proteins was observed within 5 min of IFN-α addition to HeLa and Daudi cells (L.M. Pfeffer and M.A. Marino, unpubl. observ.).

Another likely substrate for an IFN-induced kinase is the IFN-α receptor itself. Most, if not all, cytokine receptors appear to be phosphorylated at tyrosine or serine/threonine residues. Although several studies have suggested that receptor phosphorylation alone is necessary for their downregulation, receptor phosphorylation has been associated with other biologically significant events. Immunoblotting and immunoprecipitation of ^{32}P-labeled proteins with an mAb against the α-subunit of the IFN-α receptor have shown that this component is rapidly phosphorylated on tyrosine residues in response to IFN-α addition to highly IFN-α-sensitive cells, an effect blocked by genistein (potent PTK inhibitor) (Platanias and Colamonici 1992). In contrast, no IFN-α-induced tyrosine phosphorylation was observed in the highly IFN-resistant U-937 cell line. These results provide strong evidence that the phosphorylation of the 110-kDa α-subunit plays a role in the transmembrane signaling pathway through the IFN-α receptor. However, since this subunit has not been purified or cloned, it remains unclear if it has intrinsic PTK activity, or if its phosphorylation results from the action of an IFN-α-activated PTK.

As already discussed above, another substrate of an IFN-activated kinase is the latent transcriptional factor, ISGF3, which is activated and translocated to the nucleus in response to IFN-α. The NaF-dependent mechanism of translocation of cytoplasmic ISGF3 to the nucleus is completely unexplained (Levy et al. 1989), but early studies showing that IFN-α induces the cytoskeletal coupling of IFN-α-receptor complexes strongly suggest a role for the cytoskeleton in early signaling events (Pfeffer et al. 1987b). Regulation of transcription involves the phosphorylation-dependent modulation of transcription factor activity by: (1) release and/or translocation of sequestered factors in the cytoplasm; (2) alteration in DNA binding activity of nuclear transcription factors; and (3) modulation of the interaction of factor transactivation domains with transcriptional machinery (Hunter and Karin 1992). Thus, it is not surprising that transcriptional activation of some ISGs by IFN-α is dependent on the IFN-induced phosphorylation of ISGF3 or other factors via PTK or PKC activation. In fact, before the identification of the ISGF3α polypeptides as tyrosine phosphorylated proteins, DNA-protein interactions at ISRE were recognised to involve the phosphorylation of nuclear protein(s) (Roy and Lebleu 1990).

2.4 Specificity of Signaling for Different Ligands

How is it possible that PKC and PTK activation can serve as a signaling pathway for such a variety of different ligands, each generating highly specific cellular responses? Where does the specificity reside? For IFN-α, signaling specificity resides in: (1) specific phosphorylation of latent transcription factors and thereby the activation of unique promoter-enhancer sequences within ISGs (Fu 1992; Schindler et al. 1992); and (2) activation of selective PKC isotypes (Pfeffer et al. 1990, 1991), which are involved in both transcriptional and posttranscriptional events (Fig. 1). Type-I IFN selectively activates PKC-β in HeLa cells, while epidermal growth factor, which stimulates HeLa cell proliferation, has no effect on the subcellular distribution of PKC-β (Pfeffer et al. 1990). This observation suggests that specificity may reside in the particular lipid species, which is hydrolyzed in response to IFN-α treatment. Because the actual DAG species that results from lipid hydrolyzed by phospholipases will clearly vary in both acyl chain length and saturation, it will be important to determine the specific preferences of the isotypes of PKC for the actual forms of DAG produced in response to various stimuli. It will also be important to determine how activation of specific PKC isotypes relate to the cellular response. Interestingly, in vitro studies indicate that PKC-β and -ε, which become activated during IFN-α treatment of intact cells, are not sensitive to changes in calcium concentrations under physiological conditions (Sekiguchi et al. 1987). Thus, one general feature of signaling by IFN-α in Daudi and HeLa cells is the generation of DAG in the absence of inositol phospholipid turnover and cytosolic Ca^{2+} elevation, and the subsequent activation of calcium-independent isotypes of PKC (Pfeffer et al. 1990, 1991). However, the involvement of other protein kinases in determining specificity cannot be ruled out, since PKC is known to affect the activity of other protein kinases. Interestingly, the mechanisms by which IFN-α receptor interaction results in PTK activation are unknown.

2.5 Analogies of Transmembrane Signaling Through the IFN-α Receptor with That of Other Receptors

Many ligand receptor transmembrane signaling mechanisms involve coupling to several independent signaling pathways. For example, platelet-activating factor receptor-coupled signal transduction pathways have been shown to be a composite of activation of phospholipid turnover (via PLC, PLD, PLA_2), and protein kinase (PKC and PTK) action (Skula 1992).

Another highly analogous system of signaling events initiated by ligand-receptor binding is the T-cell receptor (TCR). The TCR, like the IFN-α receptor, is a multisubunit transmembrane complex (Klausner and Samelson 1991). TCR ligand stimulation activates PKC through DAG production involving the activation of the inositol phospholipid specific PLC-γ1. Interestingly, the TCR also transduces signals through the activation of one or more PTKs, with multiple, rapid (within 10 s), protein tyrosine phosphorylation events triggered by ligand stimulation (Klausner and Samelson 1991). The activation of PKC and PTK is linked, as activation of PLC-γ1 involves tyrosine phosphorylation. Furthermore, recent evidence suggests that the TCR is functionally coupled to multiple nonreceptor PTKs in T-cells. Although these PTKs may be indirectly activated through TCR occupancy, the most frequently cited are the two *src*-related PTKs, $p56^{lck}$ and $p59^{fyn}$. Thus, although transmembrane signaling through the IFN-α receptor appears to involve the *tyk2* PTK, the involvement of other PTKs in IFN-α signaling cannot be ruled out.

3 Conclusions

IFN-α binds to discrete cell surface receptors on target cells, and thereby alters gene expression. Although our current understanding of the IFN-α transmembrane signaling mechanism is far from complete, recent studies suggest that it involves the activation of at least two distinct classes of protein kinases, protein kinase C (PKC) and protein tyrosine kinases (PTK). Inhibitors of PKC and PTK block IFN-α-induced transcription of a distinct set of genes, the IFN-stimulated genes (ISGs). PKC activation by IFN-α is selective for calcium-independent isotypes, and is associated with the IFN-α-induced production of diacylglycerol (DAG) in the absence of inositol phospholipid turnover and increases in intracellular calcium concentrations. IFN-α induces rapid changes in protein tyrosine phosphorylation. The proteins phosphorylated on tyrosine include components of ISGF3, the latent transcription factor for ISGs. At present, the induction of the multiple biological effects by IFN-α seems to reflect: (1) PTK-mediated direct transcriptional effects and (2) transcriptional and posttranscriptional regulation of the specific IFN-α-induced effector proteins, which is mediated by IFN-α-activated PKC isotypes.

References

Aguet M (1980) High affinity binding of ^{125}I-labeled mouse interferon to a specific cell surface receptor. Nature 284:459–461

Akai H, Larner AC (1989) Phorbol ester-mediated down-regulation of an interferon-inducible gene. J Biol Chem 264:3252–3255

Bazan JF (1990) Structural design and molecular evolution of a cytokine receptor superfamily. Proc Natl Acad Sci USA 87:6934–6938

Branca AA, Faltynek CR, D'Alessandro SB, Baglioni C (1982) Interaction of interferon with cellular receptors: internalization and degradation of cell-bound interferon. J Biol Chem 257:13291–13296

Cataldi A, Miscia S, Lisio R, Rana R, Cocco L (1990) Transient shift of diacylglycerol and inositol lipids induced by interferon in Daudi cells: evidence for a different patterns between nuclei and intact cells. FEBS Lett 269:465–468

Colamonici OR, D'Alessandro F, Diaz MO, Gregory SA, Neckers LM, Nordan R (1990) Characterization of three monoclonal antibodies that recognize the interferon α 2 receptor. Proc Natl Acad Sci USA 87:7230–7234

Colamonici OR, Pfeffer LM, D'Alessandro F, Platanias L, Rosolen A, Nordan R, Cruciani RA, Diaz MO (1992) Multichain structure of the IFN-α receptor on hematopoietic cells. J Immunol 148:2126–2132

Colamonici OR, Porterfield B, Domanski P, Constantinescu S, Pfeffer LM (1993) Complementation of the IFNα response in resistant cells by expression of the cloned subunit of the IFN-α receptor: A central role of this subunit in IFNα signaling. (submitted)

Constantinescu SN, Cernescu C, Balta F, Maniu H, Popescu LM (1990) The priming effect of human interferon-α is mediated by protein kinase C. J Interferon Res 10:589–597

Dale TC, Imam AM, Kerr IM, Stark GR (1989) Rapid activation by IFN α of a latent DNA-binding protein present in the cytoplasm of untreated cells. Proc Natl. Acad Sci USA 86:1203–1207

Eid P, Mogenson K (1990) Detergent extraction of the human interferon-alpha/beta receptor: a soluble form capable of binding interferon. Biochem Biophys Acta1034:114–121

Eisenkraft BL, Nanus DM, Albino AP, Pfeffer LM (1991) α-Interferon down-regulates epidermal growth factor receptors on renal carcinoma cells: relation to cellular responsiveness to the antiproliferative action of α-interferon. Cancer Res 51:5881–5887

Ersalimsky JD, Kefford RF, Gilmore DJ, Milstein C (1989) Phorbol esters potentiate the induction of class I HLA expression by induction of class I HLA expression by interferon α. Proc Natl Acad Sci USA 86:1973–1976

Exton JH (1990) Signalling through phosphatidylcholine breakdown. J Biol Chem 265:1–4

Faltynek CR, Branca AA, McCandless S, Baglioni C (1983) Characterization of an interferon receptor on human lymphoblastoid cells. Proc Natl Acad Sci USA 80:3269–3273

Faltynek CR, Princler GL, Gusella GI, Varesio L, Radzioch D (1989) A functional protein kinase C is required for induction of 2-5A synthetase by recombinant interferon-α A in Daudi cells. J Biol Chem 264:14305–14311

Friedman RL, Stark GR (1985) α-Interferon-induced transcription of HLA and metallothionein genes containing homologous upstream sequences. Nature 314:637–639
Friedman RL, Manly SP, McMahon M, Kerr IM, Stark GR (1984) Transcriptional and post-transcriptional regulation in interferon-induced gene expression in human cells. Cell 38:745–755
Fu X-Y (1992) A transcription factor with SH2 and SH3 domains is directly activated by an interferon α-induced cytoplasmic protein tyrosine kinase(s). Cell 70:323–335
Fu X-Y, Kessler DS, Veals SA, Levy DE, Darnell JE (1990) ISGF-3, the transcriptional activator induced by interferon α, consists of multiple interacting polypetide chains. Proc Natl Acad Sci USA 87:8555–8559
Gutch MJ, Reich NC (1991) Repression of the interferon signal transduction pathway by the adenovirus E1A oncogene. Proc Natl Acad Sci USA 88:7914–7918
Hannigan GE, Williams BRG (1991) Signal transduction by interferon-α through arachidonic acid metabolism. Science 251:204–207
Haque SJ, Williams BR (1992) Evidence for two distinct interferon-α-signalling pathways. J Interferon Res 12:S105
Heidenreich KA, Toledo SP, Brunton LL, Watson MJ, Daniel-Issakani S, Strulovici B (1990) Insulin stimulates the activity of a novel protein kinase C, PKC-ε, in cultured fetal chick neurons. J Biol Chem 265:15076–15082
Hunter T, Karin M (1992) Regulation of transcription by phosphorylation. Cell 70:375–387
Improta T, Pine R, Pfeffer LM (1992) Interferon-γ potentiates the antiviral activity and the expression of interferon-stimulated genes induced by interferon-α in U-937 cells. J Interferon Res 12:87–94
Jung V, Rashidbaigi A, Jones C, Tischfield JA, Shows TB, Pestka S (1987) Human chromosome 6 and 21 are required for sensitivity to human interferon gamma. Proc Natl Acad Sci USA 84:4151–4155
Kessler DS, Pine R, Pfeffer LM, Levy DE, Darnell JE (1988) Cells resistant to interferon are defective in activation of a promoter-binding factor. EMBO J 7:3779–3783
Klausner RD, Samelson LE (1991) T cell antigen receptor activation pathways: the tyrosine kinase connection. Cell 64:875–878
Langer J, Pestka S (1988) Interferon receptors. Immunology 9:875–878
Larner AC, Chaudhuri A, Darnell JE (1986) Transcriptional induction by interferon: new protein(s) determine the length and extent of induction. J Biol Chem 261:453–459
Levy DE, Kessler DS, Pine R, Darnell JE (1989) Cytoplasmic activation of ISGF3, the positive regulator of interferon-stimulated transcription, reconstituted in vitro. Genes Dev 3:1362–1371
Mogensen EE, Bandu MT (1983) Kinetic evidence for an activation step following the binding of human interferon α2 to the membrane receptors of Daudi cells. Eur J Biochem 134:355–364
Nishizuka Y (1988) Studies and prospectives of the protein kinase C family and its implications for cellular regulation. Nature 334:661–665
Ono Y, Fujii T, Ogita K, Kikkawa U, Igarishi K, Nishizuka Y (1988) The structure, expression, and properties of additional members of the protein kinase C family. J Biol Chem 263:6927–6932
Osada S, Mizuno K, Saido TC, Akita Y, Suzuki K, Kuroki T, Ohno S (1990) A phorbol ester receptor/protein kinase, nPKC-eta, a new member of the protein kinase C family predominantly expressed in lung and skin. J Biol Chem 265:22434–22440
Pelligrini S, John J, Shearer M, Kerr IM, Stark GR (1992) Use of a selectable marker regulated by alpha interferon to obtain mutants in the signaling pathway. Mol Cell Biol 9:4605–4612
Pestka S, Langer JA, Zoon KC, Samuel CE (1987) Interferons and their actions. Annu Rev Biochem 56:727–777
Pfeffer LM (1987) Mechanisms of interferon action. CRC Press, Boca Raton
Pfeffer LM, Landsberger FR, Tamm I (1981) Beta-interferon-induced time-dependent changes in the plasma membrane lipid bilayer of cultured cells. J Interferon Res 1:613–620
Pfeffer LM, Kwok BCP, Landsberger FR, Tamm I (1985) Interferon stimulates cholesterol and phosphatidylcholine synthesis but inhibits cholesterol ester synthesis in HeLa-S3 cells. Proc Natl Acad Sci USA 82:2417–2421

Pfeffer LM, Donner DB, Tamm I (1987a) Interferon-α down-regulates insulin receptors in lymphoblastoid (Daudi) cells. J Biol Chem 262:3665–3670

Pfeffer LM, Stebbing N, Donner DB (1987b) Cytoskeletal association of human α-interferon-receptor complexes in interferon-sensitive and resistant lymphoblastoid cells. Proc Natl Acad Sci USA 84:3249–3253

Pfeffer LM, Strulovici B, Saltiel AR (1990) Interferon-α selectively activates the β isoform of protein kinase C through phosphatidylcholine hydrolysis. Proc Natl Acad Sci USA 87:6537–6541

Pfeffer LM, Eisenkraft BL, Reich NC, Improta T, Baxter G, Daniel-Issakani S, Strulovici B (1991) Transmembrane signaling by interferon α involves diacyglycerol production and activation of the ε isoform of protein kinase C in Daudi cells. Proc Natl Acad Sci USA 88:7988–7992

Pfeffer LM, Constaninescu SN, Wang C, Colamonici OR (1992) Evidence for a role of the cloned IFNα receptor in the signal transduction of IFNα. J Interferon Res 12:S56

Platanias LC, Colamonici OR (1992) Interferon α induces rapid tyrosine phosphorylation of the α subunit of its receptor. J Biol Chem 267:24053–24057

Reich NC, Pfeffer LM (1990) Evidence for involvement of protein kinase C in the cellular response to interferon α. Proc Natl Acad Sci USA 87:8761–8765

Reich NC, Evans B, Levy DE, Fahey D, Knight E, Darnell JE (1987) Interferon-transcription of a gene encoding a 15-KD protein depends on an upstream enhancer element. Proc Natl Acad Sci USA 84:6394–6398

Revel M, Chebath J (1986) Interferon activated genes. Trends Biochem Sci 11:166–170

Roberts WJ, Hovanessian AR, Brown RE, Clemens MJ, Kerr IM (1976) Interferon-mediated protein kinase and low-molecular-weight inhibitor of protein synthesis. Nature 264:477–480

Roy C, Lebleu B (1990) DNA-protein interactions at the interferon-responsive promoter; evidences for an involvement of phosphorylation. Nucleic Acids Res 18:2125–2131

Rubinstein M, Orchansky P (1986) The interferon receptors. CRC Crit Rev Biochem 21:249–275

Schaap D, Parker PJ (1990) Expression, purification, and characterization of protein kinase C-ε. J Biol Chem 265:7301–7307

Schaap D, Parker PJ, Bristol A, Kriz R, Knopf J (1989) Unique substrate specificity and regulatory properties of PKC-ε: a rationale for diversity. FEBS Lett 243:351–357

Schindler C, Shuai K, Prezioso VR, Darnell JE Jr (1992) Interferon-dependent tyrosine phosphorylation of a latent cytoplasmic transcription factor. Science 257:809–813

Sehgal PB, Walter Z, Tamm I (1987) Rapid enhancement of β2-interferon/B-cell differentiation factor BSF-2 gene expression in human fibroblasts by diacylglycerols and the calcium ionophore A23187. Proc Natl Acad Sci USA 84:3663–3667

Sekiguchi K, Tsukuda M, Ase K, Kikkawa U, Nishizuka Y (1987) Three distinct forms of rat brain protein kinase C: differential response to unsaturated fatty acids. Biochem Biophys Res Commun 145:797–802

Sekiguchi K, Tsukuda M, Ase K, Kikkawa U, Nishizuka Y (1988) Model of activation and kinetic properties of three distinct forms of protein kinase C from rat brain. J Biochem (Tokyo) 103:759–765

Skula SD (1992) Platelet-activating factor receptor and signal transduction mechanisms. FASEB J 6:2296–2301

Soslau G, Bogucki AR, Gillespie D, Hubbell HR (1984) Phosphoproteins altered by antiproliferative doses of human interferon-β in a bladder carcinoma cell line. Biochem Biophys Res Commun 119:941–948

Strulovici B, Daniel-Issakani S, Baxter G, Knopf J, Sultzman L, Cherwinski H, Nestor JJ, Webb DR, Ransom J (1990) Distinct mechanisms of regulation of protein kinase Cε by hormones and phorbol esters. J Biol Chem 266:168–173

Uze G, Lutfalla G, Gresser I (1990) Genetic transfer of a functional human interferon α receptor into mouse cells: cloning and expression of its cDNA. Cell 60:225–234

Vanden Broecke D, Pfeffer LM (1988) Characterization of interferon-α binding sites on human cell lines. J Interferon Res 8:803–811

Velazquez L, Fellous M, Stark GR, Pelligrini S (1992) A protein tyrosine kinase in the interferon α/β signaling pathway. Cell 70:313–322
Yap WH, Teo TS, McCoy E, Tan YH (1986a) Rapid and transient rise in diacylglycerol concentration in Daudi cells exposed to interferon. Proc Natl Acad Sci USA 83:7765–7769
Yap WH, Teo TS, Tan YH (1986b) An early event in the interferon-induced transmembrane signalling process. Science 234:355–358
Zoon KC, Arnheiter H, Zur Nedden D, Fitzgerald DJP, Willingham MC (1983) Human interferon alpha enters cells by receptor-mediated endocytosis. Virology 130:195–203
Zoon KC, Arnheiter H, Fitzgerald DJP (1986) Procedures for measuring receptor-mediated binding and internalization of human interferon. Methods Enzymol 119:332–339

Photolabeling of the Enzymes of the 2-5A Synthetase/RNase L/p68 Kinase Antiviral Systems with Azido Probes

R.J. SUHADOLNIK[1]

1 Introduction

This review describes approaches to the photoaffinity labeling of 2′,5′-oligoadenylate (2-5A) synthetase, RNase L, and p68 kinase employing azido probes with photolabile groups on carbon-2 or carbon-8 of adenine or inosine nucleotides. The covalent cross-linking of 2- or 8-azidoATP to 2-5A synthetase, 2- and 8-azido analogs of 2-5A to RNase L, and azido dsRNAs to 2-5A synthetase and p68 kinase is described. In addition, the newly discovered role of the 2-5A molecule as an inhibitor of HIV-1 reverse transcriptase (RT) is discussed.

Photoaffinity labeling with azidopurine analogs has been used successfully to define structure/function relationships related to nucleotide binding sites of proteins. Of the many photoaffinity probes that have been used to identify catalytic and allosteric binding domains of enzymes, those with azido groups covalently linked to purine or pyrimidine rings are some of the most efficient and selective probes (Potter and Haley 1983). The elegant contributions of Haley and coworkers have resulted in a highly refined technology whereby the photoaffinity labeled peptides can be isolated by metal chelate chromatography (Salvucci et al. 1992).

The 2- and 8-azido purine nucleosides and nucleotides and 5-azido pyrimidine nucleosides and nucleotides have been widely used with enzymes requiring nucleosides and nucleotides at catalytic or allosteric domains. The azido group is an appropriate reagent because it is rapidly converted to the singlet and triplet nitrene following low intensity UV irradiation for a few seconds at 0 °C. This nitrene-free radical covalently links to amino acid residues in the peptide binding domain. Although aromatic azides have been used in photoaffinity labeling, covalent cross-linking occurs via the azepine nucleophile and not the nitrene-free radical (Fig. 1; Potter and Haley 1983; Albini et al. 1991). Prolonged exposure to UV light with azido photoprobes leads to non-specific covalent cross-linking, particularly if reactive tyrosine residues are present. Therefore, critical to studies on the identification of nucleotide binding

[1] Department of Biochemistry, Temple University School of Medicine, Philadelphia, PA 19140, USA

Progress in Molecular and Subcellular Biology, Vol. 14
W.E.G. Müller/H.C. Schröder (Eds.)
© Springer-Verlag Berlin Heidelberg 1994

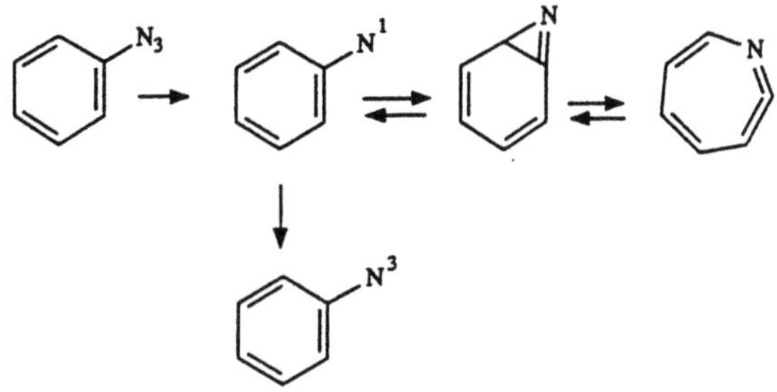

Fig. 1. Photodecomposition of aryl azide to singlet and triplet nitrene and formation of azepine. (Albini et al. 1991)

domains of enzymes is the synthesis and characterization of photoprobes with strategically placed azido groups. These azido photoprobes include: 2- and 8-azido purine nucleotides, the corresponding oligonucleotides, 5-azidopyrimidine nucleotides, and 2- and 8-azido double-stranded RNAs.

The 2-5A synthetase/RNase L/p68 kinase system is widely accepted as part of the antiviral mechanism of interferon (for reviews, see Pestka 1986; Samuel 1991; Kerr and Stark 1992; Sen and Lengyel 1992). The 2-5A synthetase/RNase L pathway is also important in the regulation of cell growth (Stark et al. 1979; Etienne-Smekens et al. 1983; Jacobsen et al. 1983; Krause et al. 1985; Wells and Mallucci 1985; Rysiecki et al. 1989). 2-5A synthetase and p68 kinase are intracellular enzymes activated by dsRNA; RNase L is activated by 2-5A. 2-5A synthetase catalyzes the synthesis of 2′,5′-oligoadenylates from ATP in a non-processive (dissipative) mechanism (Justesen et al. 1980). Four isoforms of 2-5A synthetase have been identified following treatment of human cells with interferon (p40, p46, p69, and p100) (Hovanessian 1991; Marié and Hovanessian 1992; Witt et al. 1993). There appears to be a single gene for the human 40- and 46-kDa 2-5A synthetases that has been mapped to chromosome 11. The 69- and 100-kDa synthetases are encoded by different genes. 2-5A synthetase (110 kDa) has also been isolated from rabbit reticulocytes (Wu and Eslami 1983; Ferbus et al. 1984). Most recently, a 100-kDa 2-5A synthetase has been isolated from human trophoblast cells (Zhang et al. 1993).

2-5A binds to and activates an endoribonuclease, RNase L, which degrades single-stranded viral or cellular RNAs. This degradation inhibits protein synthesis and interferes with viral replication (Brown et al. 1976; Clemens and Williams 1978; Williams et al. 1978, 1979a; Hovanesssian et al. 1979; Hovanessian and Wood 1980; Katze and Agy 1990). The activation requirements of 2-5A-dependent RNases from different sources vary. For example, RNase L from peripheral blood mononuclear cells and kidney cells is activated by 2-5A trimer and tetramer (Suhadolnik et al. 1983a; Krause and Silverman 1993), whereas 2-5A-dependent RNase from rabbit reticulocytes is activated by 2-5A tetramer, but not 2-5A trimer (Krause and Silverman 1993; Williams et al. 1979b).

Another enzyme induced by interferon is the p68 kinase (also referred to as PKR). The p68 kinase was reported simultaneously from the laboratories of Kerr, Revel, and Lengyel (Lebleu et al. 1976; Zilberstein et al. 1976; Hovanessian and Kerr 1979). Like 2-5A synthetase, the interferon-induced p68 kinase is dependent on dsRNA for activation. Once activated, the p68 kinase undergoes autophosphorylation. The autophosphorylated kinase catalyzes the phosphorylation of the α-subunit of initiation factor eIF-2 at serine 51 (Samuel 1979; Pathak et al. 1988). Phosphorylation of eIF-2α inhibits protein synthesis at the initiation step of translation. In addition to poly (I)-poly(C), poly(I)-poly $(C_{12}U)$, and HIV-TAR RNA also activate 2-5A synthetase and p68 kinase (Edery et al. 1989; SenGupta and Silverman 1989; Katze and Agy 1990; Li et al. 1990; Roy et al. 1990; Schröder et al. 1990; Hovanessian 1991; Gunnery et al. 1992). In addition to the regulation of mRNA and protein synthesis, the p68 kinase has been reported to have a tumor suppressor function (Koromilas et al. 1992; Meurs et al. 1993). The molecular cloning of murine and human dsRNA-dependent 2-5A synthetases and p68 kinase has been reported (Benech et al. 1985; Ghosh et al. 1991; Hovanessian 1991; Katze et al. 1991; Patel and Sen 1992; Thomis and Samuel 1992; Thomis et al. 1992).

2 Photoaffinity Labeling of the ATP Binding Domain of 2-5A Synthetase by 2- and 8-AzidoATP

2-5A synthetase has a broad substrate specificity and stereoselectivity (Doetsch et al. 1981; Ferbus et al. 1981; Hughes et al. 1983; Lee and Suhadolnik 1985; Kariko et al. 1987a, b; Suhadolnik et al. 1987). However, the ATP binding domain and the 2′-adenylation domain of 2-5A synthetase have not been specifically identified. The photoaffinity probes, $[\gamma\text{-}^{32}P]$ 2-azidoATP and $[\alpha\text{-}^{32}P]$ 8-azidoATP, have been used to investigate the binding of ATP to highly purified 2-5A synthetase from rabbit reticulocyte and human recombinant 40-kDa 2-5A synthetase (Suhadolnik et al. 1988b; Mordechai et al. 1992).

One of the requirements for a useful photoaffinity probe for identifying the nucleotide sequence at the binding domain of an enzyme is that the photoprobe be a competitive inhibitor of the natural substrate. 2- and 8-azidoATP are competitive inhibitors of the conversion of ATP to 2-5A (Suhadolnik et al. 1988b). The extent of photolabeling of 2-5A synthetase by $[\alpha\text{-}^{32}P]$8-azidoATP was maximal at 10 s of UV irradiation (Fig. 2). Photolabeling of 2-5A synthetase was saturated at 1.5 and 2.0 mM, respectively, with $[\gamma\text{-}^{32}P]$ 2-azidoATP and $[\alpha\text{-}^{32}P]$8-azidoATP (Suhadolnik et al. 1988b). With 2-azidoATP and 8-azidoATP, 50% photolabeling occurs at 0.25 and 0.5 mM, respectively (Fig. 3). Computer analysis of the curvilinear Scatchard plots using the methodology of Julin and Lehman (1987) suggests two classes of binding sites on 2-5A synthetase. The dissociation constants for 2-azidoATP were 5 and 380 μM, respectively, and for 8-azidoATP they were 9 and 100 μM, respectively. Competition assays with 8-azidoATP and various nucleotides were reported; the order of competition for photoinsertion of [α-^{32}P]8-azidoATP into the 2-5A synthetase was ATP > 2′dATP = 3′dATP > CTP > ITP > AMP > NAD$^+$ > UTP > UMP > CMP. The 50% competition with NAD$^+$ was expected in view of the report of acceptor and donor sites of 2-5A synthetase (Ferbus et al. 1981). Further characterization of the 2′-adenylation domain of 2-5A synthetase is underway with 8-azidoNAD$^+$.

It is noteworthy that 2′dATP, which is not a substrate for 2-5A synthetase, inhibits photoincorporation of 8-azidoATP as efficiently as does ATP (Suhadolnik et al. 1988b). Based on the inverse relationship between cell proliferation and activity of 2-5A synthetase (Jacobsen et al. 1983; Wells and Mallucci 1985), 2′dATP may be a regulator of 2-5A synthetase under physiological conditions. The competition of 3′dATP (cordycepin 5′-triphosphate) for photoinsertion of 8-azidoATP was not surprising because 3′dATP, an analog of ATP, is a substrate for 2-5A synthetase and is converted to 2′,5′-cordycepin trimer and tetramer 5′-triphosphates (Doetsch et al. 1981; Suhadolnik et al. 1983b; Eppstein et al. 1985; Nyilas et al. 1986).

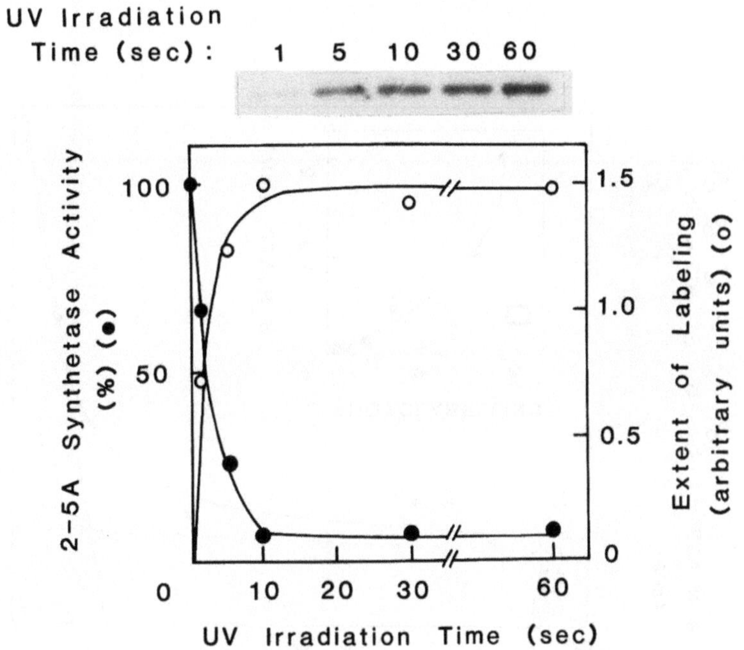

Fig. 2. Photoincorporation of 8-azidoATP into 2-5A synthetase and subsequent inactivation of the enzyme. 2-5A synthetase was UV-irradiated in the presence of 8-azidoATP and assayed for enzyme activity (●) or photolabeling (○). Activity of the enzyme without UV irradiation is referred to as 100%. (Sudadolnik et al. 1988b)

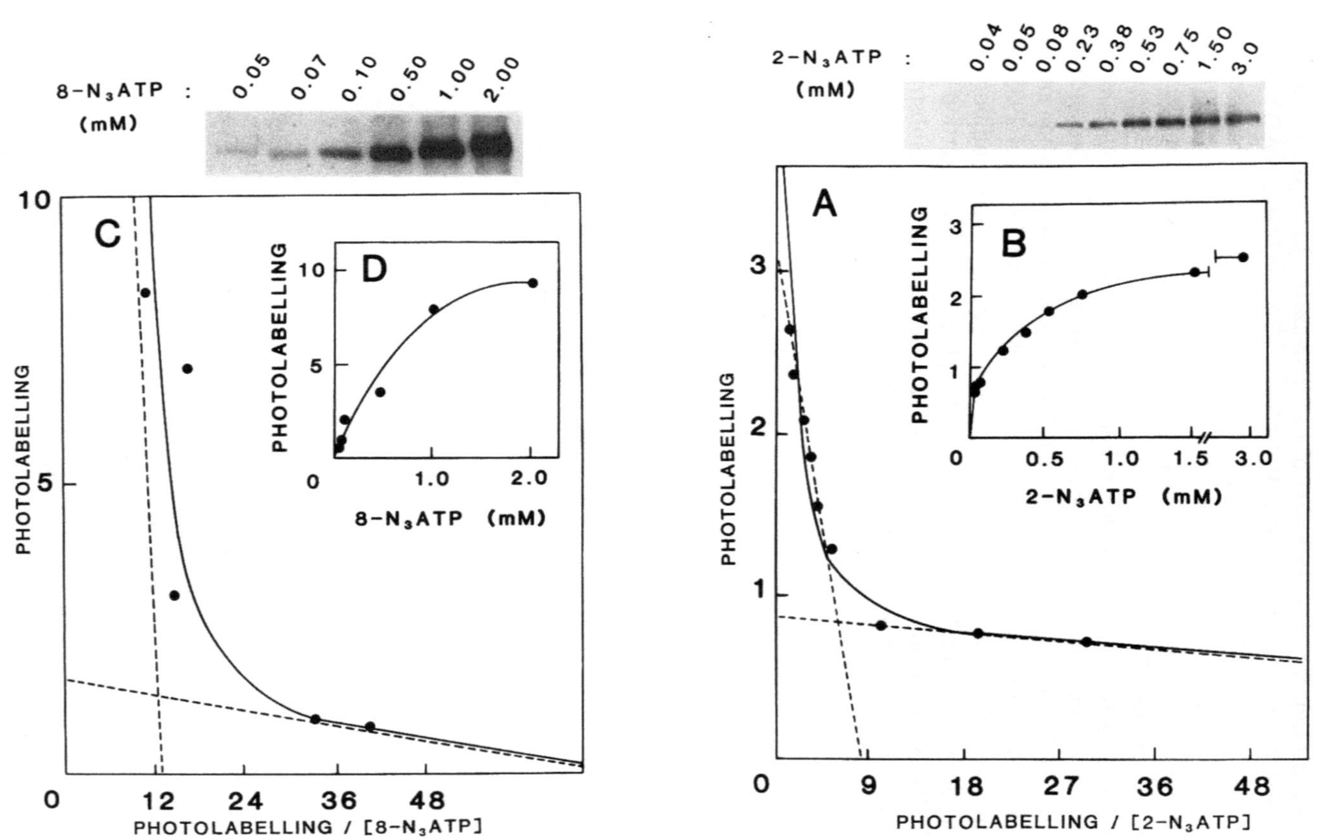

Fig. 3. Scatchard plots from saturation labeling experiments with [γ-^{32}P]2-azidoATP (**A**) and [α-^{32}P]8-azidoATP (**C**). Saturation curves of 2-5A synthetase by [γ-^{32}P]2-azidoATP (**B** and autoradiogram *insert*) and [α-^{32}P]8-azidoATP (**D** and autoradiogram *insert*). Photolabeling is expressed in arbitrary units. (Suhadolnik et al. 1988b)

In the absence of dsRNA, 2-5A synthetase does not convert ATP to 2-5A. [α-^{32}P]8-azidoATP is covalently photoinserted into 2-5A synthetase in the absence of dsRNA, indicating that dsRNA is not required for binding of ATP to the enzyme (Suhadolnik et al. 1988b). However, addition of dsRNA increased the amount of [α-^{32}P]8-azidoATP photoinserted into 2-5A synthetase, thereby increasing the binding affinity of the enzyme for ATP. It is also possible that only one of the two nucleotide binding sites (donor or acceptor) is available for binding 8-azidoATP in the absence of dsRNA. Binding of dsRNA may result in a conformational change in the enzyme with both sites becoming available for binding and thus photoinsertion of [α-^{32}P]8-azidoATP is increased.

The amino acid regions of the cloned murine and human 40-kDa 2-5A Synthetases required for enzyme activity have been identified by progressive mutation deletion (Ghosh et al. 1991). Residues 320-344 of the murine 2-5A synthetase are essential for enzyme activity. Photolabeling technology with [α-^{32}P]8-azidoATP followed by proteolysis and amino acid microsequencing has revealed that the photoinsertion occurs in the peptide fragment composed of amino acids 309-322 (Mordechai et al. 1992; Suhadolnik et al. unpubl.). The amino acid sequence 309-322 of the murine and human 2-5A synthetases shows 100% homology (GGGDPKGWRQLQE) (Ghosh et al. 1991).

3 Photoaffinity Labeling of RNase L and 2-5A Binding Proteins by 2- and 8-Azido 2′,5′-Adenylate Photoprobes

As part of continuing investigations related to the study of the binding and activation processes of RNase L, the 2- and 8-azido photoprobes of 2-5A have been synthesized enzymatically and/or chemically (Suhadolnik et al. 1988a). The enzymatic synthesis of the 2′,5′-2- and 8-azido adenylate trimer 5′-triphosphates from 2- and 8-azidoATP is shown in Fig. 4. The arrows indicate the position of the photoreactive azido groups. These two photoprobes were designed as photoaffinity probes of 2-5A to identify the amino acids in the binding domain of RNase L and to elucidate differences in levels of RNase L and/or other 2-5A binding proteins in normal and interferon-treated cells. The binding affinity of 2′,5′-8-azido-p_3A_3, determined in competition assays with 2′,5′-p_3A_4 [^{32}P]pCp, was 2×10^{-9} M (IC_{50}), similar to authentic 2′,5′-p_3A_3. Ribosomal RNA cleavage assays showed that the 2- and 8-azido photoprobes were able to bind to and activate RNase L to cleave 28S and 18S rRNA to specific cleavage products similar to authentic 2′,5′-p_3A_3 (Fig. 5).

The 2- and 8-azido 2-5A photoprobes have been used to identify nucleotide binding sites of RNase L and other 2-5A binding proteins in extracts of interferon-treated L929 cells (Suhadolnik et al. 1988a). The 2-azido 2′,5′-p_3A_3 photoprobe labeled only one protein (185 kDa), whereas the 8-azido 2′,5′-p_3A_3 photoprobe labeled six proteins (46, 63, 80, 89, 109, and 158 kDa) (Fig. 6, lanes 2-5). Only one protein (80 kDa) is covalently cross-linked with 2′,5′-p_3A_4 [^{32}P]pCp (Fig. 6, lane 6), as reported by Floyd-Smith et al. (1982). The

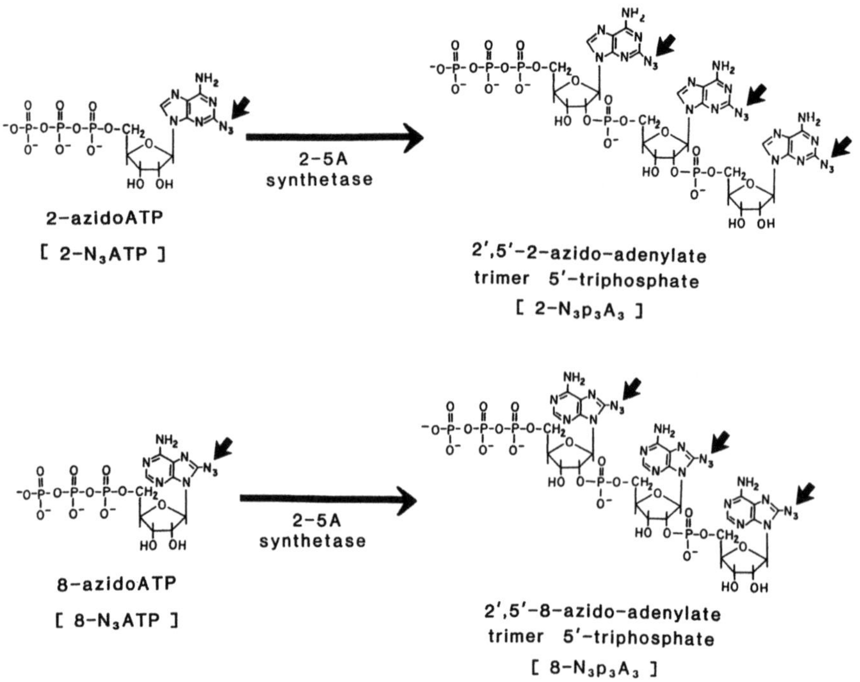

Fig. 4. Structures of photoaffinity probes of 2-5A enzymatically synthesized from 2-azidoATP and 8-azidoATP. (Suhadolnik et al. 1988a)

differences in photolabeling patterns observed with the 2- and 8- azido photoprobes of 2-5A might be explained by the strategic placement of the azido groups on the 2-5A molecule. Whereas 2-azidoATP exists in the anti-conformation, which is the naturally occurring form of nucleotides, the 8-azidoATP exists in the syn-conformation (Woody et al. 1988). Furthermore, the distance between C-2 and C-8 in the purine ring is about 6 Å (Garin et al. 1986). Therefore, the 2- and 8-azido groups in the 2-5A photoprobes may be in juxtaposition to different amino acid residues in the three-dimensional binding domain and thus may form covalent bonds with different amino acids following UV irradiation. The advantage of mapping the nucleotides in the 2-5A binding domains is that the 2- and 8-azidoadenylate trimer 5′-triphosphate photoprobes contain three azido groups capable of reacting with more than one amino acid in the 2-5A binding domain.

Further delineation of the structural requirements of the 2-5A molecule for binding to Rnase L was approached by the chemical synthesis of adenylyl-(2′,5′)-adenylyl-(2′,5′)-8-azidoadenosine with an azido group on only one adenylate residue (Charubala et al. 1989). Enzymatic phosphorylation of this monosubstituted 8-azido 2-5A photoprobe was accomplished to yield [^{32}P]p5′ A2′ p5′ A2′ p5′ A8-azidoA (specific activity 3000 Ci/mmol). The binding

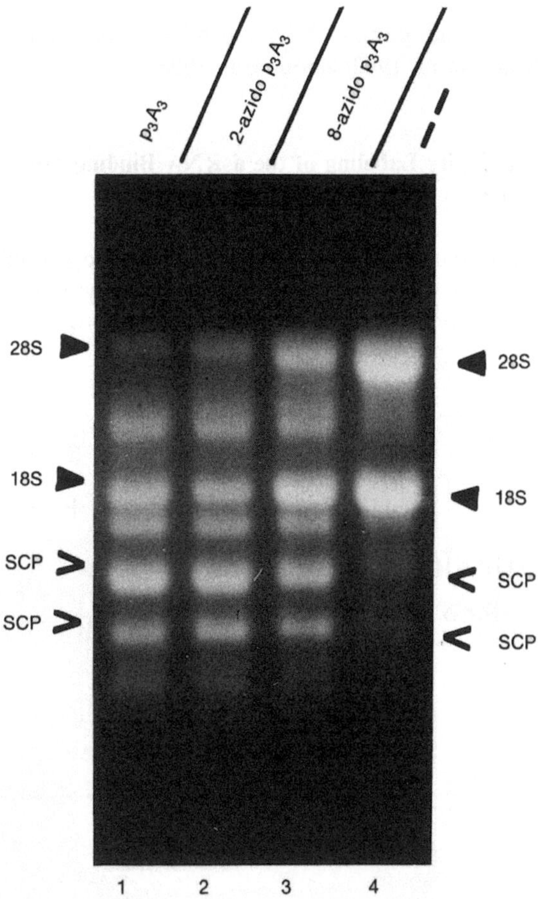

Fig. 5. Ribosomal RNA cleavage assays with 2- and 8-azido photoprobes of 2-5A. L929 cell extracts were incubated in the absence (*lane* 4) or presence of 2′,5′-p_3A_3 (*lane* 1), 2′,5′ -2-azido-p_3A_3 (*lane* 2) or 2′,5′-8-azido-p_3A_3 at 1×10^{-8} M final concentration in the absence of UV light. The positions of 28S and 28S rRNA and specific cleavage products (*SCP*) are indicated. (Suhadolnik et al. 1988a)

properties of this 8-azido photoprobe were the same as those of authentic 2-5A. Similarly, activation of RNase L by this photoprobe was the same as authentic 2-5A, as determined both by hydrolysis of poly (U) [^{32}P]pCp with partially purified RNase L and by hydrolysis of 28S and 18S rRNA in RNA cleavage assays. The photolabeling of L929 cell extracts with the [^{32}P]2′,5′-p-A-A-8-azidoA photoprobe resulted in the specific photolabeling of one protein (80 kDa). This compares with the labeling of six 2-5A binding proteins with the 2′,5′-A-A-8-azidoA 5′-triphosphate (Suhadolnik et al. 1988a). Studies are underway using the high specific activity [^{32}P]2′,5′-p-A-A-8-azidoA, proteolytic digestion and microsequencing to identify the amino acid(s) to which the 2-5A molecule is convalently linked when bound to RNase L. The molecular cloning of murine and human 2-5A-dependent RNases has been recently reported (Salehzada et al. 1992; Zhou et al. 1993).

4 Photoaffinity Labeling of the dsRNA Binding Domain of 2-5A Synthetase by Azido dsRNAs

The technique of photoaffinity labeling has been applied to the analysis of the allosteric site of 2-5A synthetase by the enzymatic synthesis of poly ([^{32}P]I, 8-azidoI)-poly (C) and poly ([^{32}P]I, 8-azidoI)-poly ($C_{12}U$) (Li et al. 1990). The

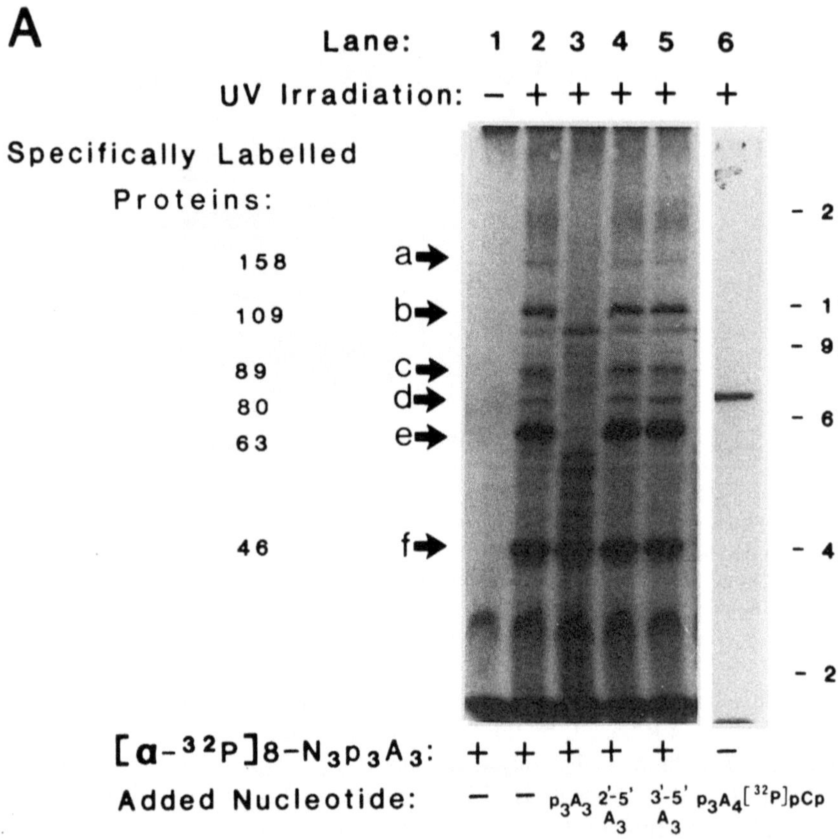

Fig. 6. Photoaffinity labelling of cytoplasmic L929 cell extracts by 8-azido-p_3A_3. **A** Interferon-treated L929 cell extracts were incubated with [α-^{32}P]2′,5′-8-azido-p_3A_3 at 1×10^{-6} M in the absence (*lanes* 1 and 2) or presence of 2′,5′-p_3A_3 at 1×10^{-4} M (*lane* 3), 2′,5′-A_3 at 5×10^{-4} M (*lane* 4) or 3′,5′-A_3 at 5×10^{-4} M (*lane* 5). Photolabeling of the cell extract was also done with 2′,5′-p_3A_4 [^{32}P]pCp (*lane* 6). **B** Densitometric trances of *lanes 2-5* of the autoradiogram in **A**. (Suhadolnik et al. 1988b)

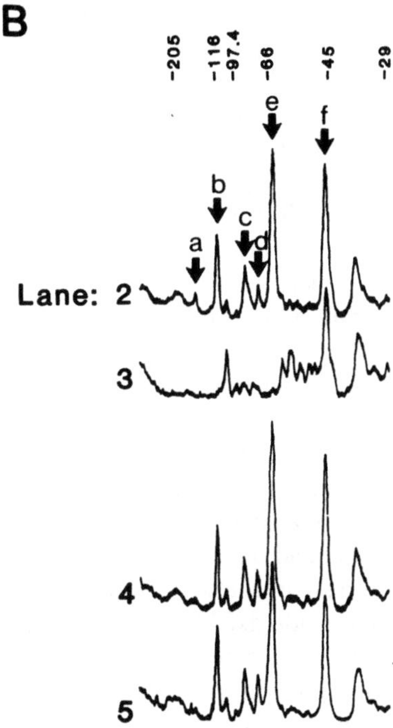

Fig. 6. (*Cont.*)

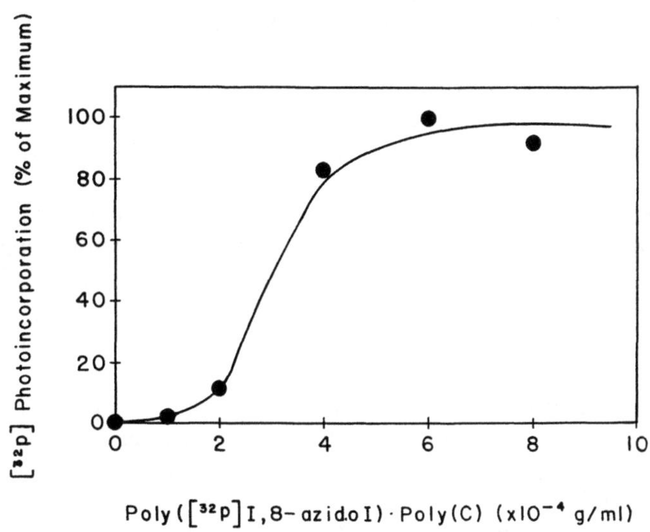

Fig. 7. Saturation of poly ([^{32}P]I, 8-azidoI)-poly (C) photoincorporation. 2-5A synthetase photolabeling mixtures were UV irradiated in the presence of 0, 1, 2, 4, 6 and 8×10^{-4} g/ml poly ([^{32}P]I, 8-azidoI)-poly (C). (Li et al. 1990)

covalent cross-linking of poly ([^{32}P]I, 8-azidoI)-poly(C) and poly ([^{32}P]I, 8azidoI)-poly (C_{12}U) to highly purified 2-5A synthetase from rabbit reticulocyte lysates and human recombinant 40-kDa 2-5A synthetase has been reported (Li et al. 1990; Suhadolnik, unpubl. results). Both of these azido dsRNAs mimic the parent molecules, poly(I)-poly(C) and poly(I)-poly(C_{12}U). In the absence of UV light, the azido dsRNA photoprobes bind to and activate 2-5A synthetase to convert ATP to 2-5A. Saturation of the allosteric binding sites of 2-5A synthetase by poly ([^{32}P]I, 8-azidoI)-poly (C) is observed at 6×10^{-4} g/ml (Fig. 7). The stoichiometry of protein modification by poly (I, 8-azidoI)-poly (C) at saturating concentrations is 0.6 pmol of dsRNA/pmol of enzyme. Studies to identify the amino acids in the dsRNA binding domain of the recombinant 40-kDa 2-5A synthetase as determined by the photoinsertion, proteolytic digestion and microsequencing with azido dsRNA are underway. These results will be compared with the report that the dsRNA binding region of recombinant murine 2-5A synthetase resides between residues 104 and 158, as determined by progressive deletion mutations (Ghosh et al. 1991). The identification of the amino acid residues in the dsRNA binding domain was accomplished by progressive deletion mutations.

The azido dsRNAs are also able to activate the dsRNA-dependent p68 kinase; however, at concentrations of 1×10^{-4} to 1×10^{-5} g/ml, the azido dsRNAs inhibit p68 kinase (Li et al. 1990). Activation and inhibition of p68 kinase by low and high concentrations of dsRNA and HIV-1 TAR RNA has also been reported (Hovanessian and Kerr 1979; Gunnery et al. 1992).

5 Photoaffinity Labeling of HIV-1 Reverse Transcriptase

Antiretroviral activity of the 2-5A molecule was first observed when 2-5A and structurally and stereochemically modified 2-5A derivatives inhibited HIV-1 reverse transcriptase (Montefiori et al. 1989). Subsequently, treatment of HIV-1-infected H9 cells with 2′,5′-cordycepin trimer core or 5′-monophosphate resulted in almost total inhibition of virus production with no cytostatic activity (Müller et al. 1991). 2′,5′-Cordycepin trimer 5′-monophosphate was shown to specifically inhibit binding of HIV-1 reverse transcriptase (RT) to the natural primer, [^{32}P]tRNA$^{lys.3}$. Under equilibrium binding conditions, 2′,5′-oligoadenylate 5′-triphosphates and derivatives are noncompetitive inhibitors of primer/HIV-1 reverse transcriptase (RT) complex formation as determined by photochemical cross-linking with recombinant p66/p66 homodimer or p66/p51 heterodimer forms of HIV-1 RT (Fig. 8 Sobal et al. 1993). Phosphorothioate substitution in the 2′,5′-internucleotide linkages yielded the most effective inhibitors (K_is of 7–13 μM). Studies are currently underway to identify the 2-5A binding domain of the p66/p66 homodimer and p66/p51 heterodimer forms of HIV-1 RT. Photolabeling studies are currently underway to identify the 2-5A binding domain of the p66/p66 homodimer and p66/p51 heterodimer forms of HIV-1 RT.

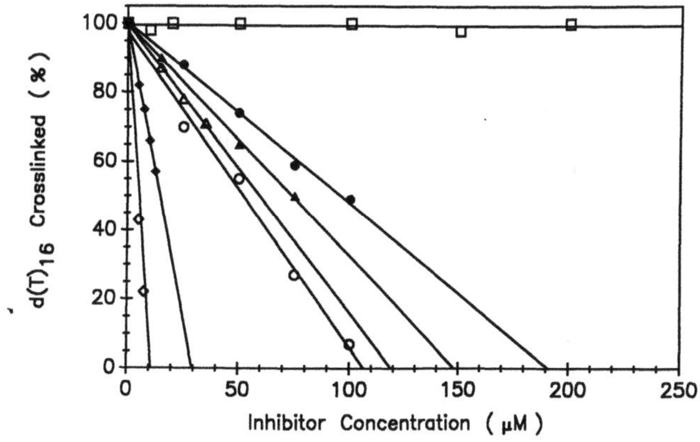

Fig. 8. Competition for primer binding to HIV-1 RT by 2-5A and 2-5A derivatives. Inhibition on $d(T)_{16}$-RT complex formation was determined by photochemical cross-linking assays with the P66/P51 heterodimer form of HIV-1 RT, under equilibrium binding conditions. Reaction mixtures contained HIV-1 RT and $[^{32}P]5'$ p-$d(T)_{16}$ with 2′,5′-p_3A_3 (●), 2′,5′-p_3A_4 (○), 2′,5′-p_3 $(3'dA)_2$ (▲), 2′,5′ -p_3 $(3'dA)_3$ (△), 2′,5′ -$p_3A_3\alpha S$ (◆), 2′,5′ -$p_3A_4\alpha S$ (◇), 2′,5′ -A_3, 2′,5′-pA_3, 3′,5′-A_3, 3′,5′-pA_3, 3′,5′ -p_3A_3 or ATP (□). (Sobal et al. 1993)

The covalently linkage of $tRNA^{lys.3}$ and $p(dT)_{16}$ to HIV-1 RT has been accomplished by UV irradiation (Müller et al. 1991; Sobol et al. 1991; Mitina et al. 1992). V8 protease hydrolysis and microsequencing have revealed that the polynucleotide binding site of HIV-1 RT is in close proximity to residues in the peptide comprising amino acids 195–300 (Sobol et al. 1991). More recently, Kumar et al. (1993) has completed detailed domain mapping studies of proteolytic cleavage of nucleic acid binding sites with HIV-1 RT. Nucleic acid binding capacity was found to occur in at least two domains in the N-terminal half of p66.

In summary, the studies above describe the application of photoaffinity labeling to the study of the nucleotide binding domains of 2-5A synthetase, RNase L, p68 kinase, and HIV-1 RT. Identification of the donor, acceptor, and dsRNA binding sites of these enzymes will provide valuable information concerning the formation of a productive complex between the enzyme, substrate, and allosteric activator. Finally, the antiretroviral activity of 2-5A and 2-5A derivatives adds a new dimension to our understanding of the physiological roles of the 2-5A molecule in mammalian cells. The studies described indicate that 2-5A and metabolically stable 2-5A derivatives are an interesting class of inhibitors of the HIV-1 life cycle. The 2′,5′-phosphorothioates in particular may be potentially useful in therapy of HIV-1 infection in that they can activate the natural cellular antiviral defense system and simultaneously inhibit HIV-1 replication.

References

Albini A, Bettinetti G, Minioli G (1991) Chemistry of nitrenes generated by the photocleavage of both azides and a five-membered heterocycle. J Am Chem Soc 113:6928–6934

Benech P, Mory Y, Revel M, Chebath J (1985) Structure of two forms of the interferon-induced(2′-5′) oligo A synthetase of human cells based on cDNAs and gene sequences. EMBO J 4: 2249–2256

Brown GE, Lebleu B, Kawakita M, Shaila S, Sen GC, Lengyel P (1976) Increased endonuclease activity in an extract from mouse Ehrlich ascites tumor cells which had been treated with a partially purified interferon preparation: dependence on dsRNA. Biochem Biophys Res Commun 69:114–122

Charubala R, Pfleiderer W, Sobol RW, Li SW, Suhadolnik RJ (1989) Chemical synthesis of adenylyl-(2′ → 5′)-adenylyl (2′ → 5′)-8-azidoadenosine, and activation and photoaffinity labelling of RNase L by [^{32}P]p5′A2′p5′A2′p5′N_3^{8}A. Helv Chim Acta 72:1354–1361

Clemens MJ, Williams BRG (1978) Inhibition of protein synthesis by pppA2′p5′A2′p5′A: a novel oligonucleotide synthesized by interferon-treated L cell extracts. Cell 13:565–572

Doetsch P, Wu JM, Sawada Y, Suhadolnik RJ (1981) Synthesis and characterization of (2′-5′) ppp3′dA(p3′dA)$_n$, an analogue of (2′–5′) pppA (pA)$_n$. Nature 291:355–358

Edery I, Petryshyn R, Sonenberg N (1989) Activation of double-stranded RNA-dependent kinase (dsI) by the TAR region of HIV-1 mRNA: a novel translation control mechanism. Cell 56: 303–312

Eppstein DA, Van Der Pas MA, Schryver BB, Sawai H, Lesiak K, Imai J, Torrence PF (1985) Cordycepin analogs of ppp5′A2′p5′A2′p5′A (2-5A) inhibit protein synthesis through activation of the 2-5A-dependent endonuclease, J Biol Chem 260:3666–3671

Etienne-Smekens M, Vandenbussche P, Content J, Dumont JE (1983) (2′-5′) Oligoadenylate in rat liver: modulation after partial hepatectomy. Proc Natl Acad Sci USA 80:4609–4613

Ferbus D, Justesen J, Besançon F, Thang MN (1981) The 2′5′ oligoadenylate synthetase has a multifunctional 2′5′ nucleotidyl-transferase activity. Biochem Biophys Res Commun 100: 847–856

Ferbus D, Justesen J, Bertrand H, Thang MN (1984) 2′5′ Oligoadenylate synthetase in the maturation of rabbit reticulocytes. Mol Cell Biochem 62:51–55

Floyd-Smith G, Yoshi O, Lengyel P (1982) Interferon action: covalent linkage of (2′-5′) pppApApA (^{32}P) pCp to (2′-5′) (A)$_n$-dependent ribonucleases in cell extracts by ultraviolet irradiation. J Biol Chem 257:8584–8587

Garin J, Boulay F, Issartel JP, Lunardy J, Vignais PV (1986) Identification of amino acid residues phosolabeled with 2-azido [α-^{32}P] adenosine diphosphate in the β subunit of beef heart mitochondrial F_1-ATPase. Biochemistry 25:4431–4437

Ghosh SK, Kusari J, Bandyopadhyay SK, Samanta H, Kumar R, Sen GC (1991) Cloning, sequencing, and expression of two murine 2′-5′-oligoadenylate synthetases. Structure-function relationships. J Biol Chem 266:15293–15299

Gunnery S, Green SR, Mathews MB (1992) Tat-responsive region RNA of human immunodeficiency virus type 1 stimulates protein synthesis in vivo and in vitro: relationship between structure and function. Proc Natl Acad Sci USA 89:11557–11561

Hovanessian AG (1991) Interferon-induced and double-staranded RNA-activated enzymes: a specific protein kinase and 2′,5′-oligoadenylate synthetases. J Interferon Res 11:199–205

Hovanessian AG, Kerr IM (1979) The (2′5′) oligoadenylate pppA2′p5′A2′p5′A synthetase and protein kinase(s) from interferon-treated cells. Eur J Biochem 93: 515–526

Hovanessian AG, Wood JN (1980) Anticellular and antiviral effects of ppA(2′p5′A)$_n$, Virology 101: 81–90

Hovanessian AG, Wood JN, Meurs E, Montagnier L (1979) Increased nuclease activity in cells treated with pppA2′p5′A2′p5′A. Proc Natl acd Sci USA 76: 3261–3265

Hughes BG, Srivastava PC, Muse DD, Robins RK (1983) 2′,5′-Oligoadenylates and related 2′,5′-oligonucleotide analogues. 1. Substrate specificity of the interferon-induced murine 2′,5′-oligoadenylate synthetase and enzymatic synthesis of oligomers. Biochemistry 22:2116–2126

Jacobsen H, Krause D, Friedman RM, Silverman RH (1983) Induction of ppp (A2′ p)$_n$-dependent RNase in murine JLS-V9R cells during growth inhibition. Proc Natl Acad Sci USA 80: 4954–4858

Julin DA, Lehman IR (1987) Photoaffinity labeling of the recBCD enzyme of *Escherichia coli* with 8-azidoadenosine 5′-triphosphate. J Biol Chem 262: 9044–9051

Justesen J, Ferbus D, Thang MN (1980) Elongation mechanism and substrate specificity of 2′5′ oligoadenylate synthetase. Ann N Y Acad Sci 350:510–521

Kariko K, Sobol RW, Suhadolnik L, Li SW, Reichenbach NL, Suhadolnik RJ, Charubala R, Pfleiderer W (1987a) Phosphorothioate analogues of 2′, 5′ -oligoadenylate. Enzymatically synthesized 2′,5′-phosphorothioate dimer and trimer: unequivocal structural assignment and activation of 2′,5′-oligoadenylate-dependent endoribonuclease. Biochemistry 26:7127–7135

Kariko K, Li SW, Sobol RW, Suhadolnik RJ, Charubala R, Pfleiderer R (1987b) Phosphorothioate analogues of 2′,5′-oligoadenylate. Activation of 2′,5′-oligoadenylate-dependent endoribonuclease by 2′, ′ -phosphorothioate cores and 5′-monophosphates. Biochemistry 26:7136–7142

Katze MG, Agy MB (1990) Regulation of viral and cellular RNA turnover in cells infected by eukaryotic viruses including HIV-1. Enzyme 44:332–346

Katze MG, Wambach M, Wong M-L, Garfinkel M, Meura E, Chong K, Williams BRG, Hovanessian AG, Barber GN (1991) Functional expression and RNA binding analysis of interferon-induced, dsRNA activated 68,000 M_r protein kinase in a cell-free system. Mol Cell Biol 11:5497–5505

Kerr IM, Stark GR (1992) The antiviral effects of the interferons and their inhibition. J Interferon Res 12:237–240

Koromilas AE, Roy S, Barber GN, Katze MG, Sonenberg N (1992) Malignant transformation by a mutant of the IFN-inducible dsRNA-dependent protein kinase. Science 257:1685–1689

Krause D, Silverman RH (1993) Tissue-related and species-specific differences in the 2-5A oligomer size requirement for activation of 2-5A-dependent RNase. J Interferon Res 13:13–16

Krause D, Silverman RH, Jacobsen H, Leisy SA, Dieffenbach CW, Friedman RM (1985) Regulation of ppp (A2′p)$_n$)A-dependent RNase levels during interferon treatment and cell differentiation. Eur J Biochem 146: 611–618

Kumar A, Kim H-R, Sobol RW, Becerra SP, Lee B-J, Hatfield DL, Suhadolnik RJ, Wilson SH (1993) Mapping of nucleic acid binding in proteolytic bomdins of HIV-1 reverse transcriptase. Biochemistry (in press)

Lebleu B, Sen GC, Shaila S, Carer B, Lengyel P (1976) Interferon, dsRNA and protein phosphorylation. Proc Natl Acad Sci USA 73:335–341

Lee C, Suhadolnik RJ (1985) 2′,5′-Oligoadenylates chiral at phosphorus: enzymatic synthesis, properties, and biological activities of 2′,5′-phosphorothioate trimer and tetramer analogues synthesized form (Sp)-ATPαS. Biochemistry 24:551–555

Li SW, Moscow JJ, Suhadolnik RJ (1990) 8-Azido double-stranded RNA photoaffinity probes. Enzymatic synthesis, characterization, and biological properties of poly (I, 8-azidoI)-poly (C) and poly (I, 8-azidoI)-poly (c_{12}U) with 2′,5′-oligoadenylate synthetase and protein kinase. J Biol Chem 265:5470–5474

Marié I, Hovanessian AG (1992) The 69-kDa 2-5A synthetase is composed of two homologous and adjacent functional domains. J Biol Chem 267:9933–9939

Meurs EF, Galabru J, Barber GN, Katze MG, Hovanessian AG (1993) Tumor suppressor function of the interferon-induced double-stranded RNA-activated protein kinase. Proc Natl Acad Sci USA 90:232–236

Mitina RL, Doonin SV, Dobrikov MI, Tabatadze DR, Levina AS, Lavrik OI (1992) Human immunodeficiency virus type 1 reverse transcriptase. Affinity labeling of the primer binding site. FEBS Lett 312:249–251

Montefiori DC, Sobol RW, Li SW, Reichenbach NL, Suhadolnik RJ, Charubala R, Pfleiderer W, Modliszewski A, Robinson WE, Mitcell WM (1989) Phophorothioate and cordycepin analogues of 2′,5′-oligoadenylate: inhibition of human immunodeficiency virus type 1 reverse transcriptase and infection in vitro. Proc Natl Acad Sci USA 86:7191–7194

Mordechai E, Chebath J, Suhadolnik RJ (1992) Characterization of human recombinant 40 kDA 2′, 5′-oligoadenylate synthetase activation by fructose 1, 6-bisphosphate. J Interferon Res 12 (Suppl 1): S199 (Abstr 7.22)

Müller WEG, Weiler BE, Charubala R, Pfleiderer W, Leserman L, Sobol RW, Suhadolnik RJ, Schröder HC (1991) Cordycepin analogues of 2′,5′-oligoadenylate inhibit human immunodeficiency virus infection via inhibition of reverse transcriptase. Biochemistry 30:2027–2033

Nyilas A, Vrang L, Drake A, Oberg B, chattopadhyaya J (1986) The cordycepin analogue of 2, 5A and its threo isomer. Chemical synthesis, conformation and biological activity, Acta Chem Scand B00: 678–688

Patel RC, Sen GC (1992) Identification of the double-stranded RNA-binding domain of the human interferon-inducible protein kinase. J Biol Chem 267:7671–7679

Pathak VK, Schindler D, Hershey JWB (1988) Generation of a mutant form of protein synthesis initation factor eIF-2 lacking the site of phosphorylation of eIF-2 kinases. Mol Cell Biol 8:993–995

Pestka S (ed) (1986) Interferons. Part C. Methods Enzymol 119

Potter RL, Haley BE (1983) Photoaffinity labeling of nucleotide binding sites with 8-azidopurine analogs: techniques and applications. Methods Enzymol 91:613–633

Roy S, Katze MG, Parkin NT, Edery I, Hovanessian AG, Sonenberg N (1990) Control of the interferon-induced 68-kilodalton protein kinase by the HIV-1 tat gene product. Science 247:1216–1219

Roy S, Agy M, Hovanessian AG, Sonenbergn, Katze MG (1992) The integrity of the stem structure of human immunodeficiency virus type 1 tat-responsive sequence RNA is required for interaction with the interferon-induced 68,000-M_r protein kinase. J Virol 65:632–640

Rysiecki G, Gewert DR, Williams BRG (1989) Constitutive expression of a 2′,5′-oligoadenylate synthetase cDNA results in increased antiviral activity and growth suppression. J Interferon Res 9:649–657

Salehzada T, Silhol M, Steff AM, Lebleu B, Bisbal C (1992) Multimeric structure of 2′-5′ oligoadenylate dependent RNase L. J Interferon Res 12 (Suppl 1): S84 (Abstr W8-2)

Salvucci ME, Chavan AJ, Haley BE (1992) Identification of peptides from the adenine binding domains of ATP and AMP in adenylate kinase: isolation of photoaffinity-labeled peptides by metal chelate chromatography. Biochemistry 31:4479–4487

Samuel CE (1979) Phosphorylation of protein synthesis initiation factor eIF-2 in interferon-treated human cells by a ribosome-associated kinase possessing site-specificity similar to hemin-regulated rabbit reticulocyte kinase. Proc Natl Acad Sci USA 76:600–604

Samuel CE (1991) Antiviral actions of Interferon. Interferon-regulated cellular proteins and their surprisingly selective antiviral activities. Virology 183:1–11

Schröder HC, Ugarkovic D, Wenger R, Okamoto T, Müller WEG (1990) Binding of tat protein to TAR region of human immunodeficiency virus type 1 blocks TAR-mediated activation of (2′-5′) oligoadenylate synthetase. AIDS Res Hum Retroviruses 6:659–672

Schröder HC, Suhadolnik RJ, Pfleiderer W, Charubala R, Müller WEG (1992) (2′-5′) Oligoadenylate and intracellular immunity against retrovirus infection. Int J Biochem 24:55–63

Sen GC, Lengyel P (1992) The interferon system. A bird's eye view of its biochemistry. J Biol Chem 267:5017–5020

SenGupta DN, Silverman RH (1989) Activation of interferon-regulated, dsRNA-dependent enzymes by human immunodeficiency virus-1 leader RNA. Nucleic Acids Res 17:969–978

Sobol RW, Suhadolnik RJ, Kumar A, Lee BJ, Hatfield DL, Wilson SH (1991) Localization of a polynucleotide binding region in the HIV-1 reverse transcriptase: implications for primer binding. Biochemistry 30:10623–10631

Sobol RW, Fisher WL, Reichenbach NL, Kumar A, Beard WA, Wilson SH, Charubala R, Pfleiderer W, Suhadolnik RJ (1993) HIV-1 reverse transcriptase: inhibition by 2′,5′-oligoadenylates. Biochemistry (in press)

Stark G, Dower WJ, Schimke RT, Brown RE, Kerr IM (1979) 2-5A synthetase: assay, distribution and variation with growth or hormone status. Nature 278:471–473

Suhadolnik RJ, Flick MB, Mosca JD, Sawada Y, Doetsch PW, Vonderheid ED (1983a) 2′,5-Oligoadenylate synthetase from cutaneous T-cell lymphoma: biosynthesis, identification, quantitation, molecular size of the 2′,5-oligoadenylates, and inhibition of protein synthesis. Biochemistry 22:4153–4158

Suhadolnik RJ, Devash Y, Reichenbach NL, Flick MB, Wu JM (1983b) Enzymatic synthesis of the 2′,5′-A_4 tetramer analog, 2′,5′ -ppp3′dA(p3′dA)$_3$, by rabbit reticulocyte lysates: binding and activation of the 2′,5′-A_n dependent nuclease, hydrolysis of mRNA, and inhibition of protein synthesis. Biochem Biophys Res Commun 111:205–212

Suhadolnik RJ, Lee C, Kariko K, Li SW (1987) Phosphorothioate analogues of 2′,5′-oligoadenylate. Enzymatic synthesis, properties, and biological activities of 2′,5′-phosphorothioates from adenosine 5′-*O*-(2-thiotriphosphate) and adenosine 5′-*O*-(3-thiotriphosphate). Biochemistry 26: 7143–7149

Suhadolnik RJ, Kariko K, Sobol RW, Li SW, Reichenbach NL, Haley BE (1988a) 2- and 8-Azido photoaffinity probes. 1. Enzymatic synthesis, characterization, and biological properties of 2- and 8- azido photoprobes of 2-5A and photolabeling of 2-5A binding proteins. Biochemistry 27:8840–8846

Suhadolnik RJ, Li SW, Sobol RW, Haley BE (1988b) 2- and 8- Azido photoaffinity probes. 2. Studies on the binding process of 2-5A synthetase by photosensitive ATP analogues. Biochemistry 2: 8846–8851

Thomis DC, Samuel CE (1992) Mechanism of interferon action: autoregulation of RNA-dependent P1/eIF-2α protein kinase (PKR) expression in transfected mammalian cells. Proc Natl Acad Sci USA 89:10837–10841

Thomis DC, Doohan JP, Samuel CE (1992) Mechanism of interferon action: cDNA structure, expression and regulation of the interferon-induced, RNA-dependent P1/EIF-2α protein kinase from human cells. Virology 188:33–46

Wells V, Mallucci L (1985) Expression of the 2-5A system during the cell cycle. Exp Cell Res 159: 27–36

Williams BRG, Kerr Im, Gilbert CS, White CN, Ball LA (1978) Synthesis and breakdown of pppA2′p5′A2′p5′A and transient inhibition of protein synthesis in extracts from interferon-treated and control cells. Eur J Biochem 92:455–462

Williams BRG, Golgher RR, Brown RE, Gilbert CS, Kerr IM (1979a) Natural occurrence of 2–5A in interferon-treated EMC virus-infected L cells. Nature 282: 582–586

Williams BRG, Golgher RR, Kerr IM (1979b) Activation of a nuclease by pppA2′p5′A2′p5′A in intact cells. FEBS Lett 105: 47–52

Witt PL, Marié I, Robert N, Irizarry A, Borden EC, Hovanessian AG (1993) Isoforms p69 and p100 of 2′,5-oligoadenylate synthetase induced differentially by interferons in vivo and in vitro. J Interferon Res 13:17–23

Woody AYM, Evans RK, Woody RW (1988) Characterization of a photoaffinity analog of UTP, 5-azido-UTP, for analysis of the substrate binding site on *E. coli* RNA polymerase. Biochem Biophys Res Commum 150: 917–924

Wu JM, Eslami B (1983) Synthesis and function of (2′-5′) A_n: inhibition of (2′-5′) A_n synthetase by heparin and the use of heparin-agarose for partial purification of (2′–5′)A_n synthetase from rabbit reticulocyte lysates. Biochem Int 6: 207–216

Zhang GY, Beltchev B, Fournier A, Zhang YH, Malassiné A, Bisbal C, Ehresmann B, Ehreshmann C, Darlix JL, Thang MN (1993) High levels of 2′,5′-oligoadenylate synthetase and 2′,5′-oligoadenylate-dependent endonuclease in human trophoblast. AIDS Res Huma Retroviruses 9: 189–196

Zhou A, Hassei BA, Silverman RH (1993) Expression cloning of 2-5A-dependent RNase: a uniquely regulated mediator of interferon action. Cell 72:753–765

Zilberstein A, Federman P, Shulman L, Revel M (1976) Specific phosphorylation in vitro of a protein associated with ribosomes of interferon-treated mouse L cells. FEBS Lett 68:119–124

MIX
Papier aus verantwortungsvollen Quellen
Paper from responsible sources
FSC® C105338

If you have any concerns about our products,
you can contact us on
ProductSafety@springernature.com

In case Publisher is established outside the EU,
the EU authorized representative is:
Springer Nature Customer Service Center GmbH
Europaplatz 3, 69115 Heidelberg, Germany

Printed by Libri Plureos GmbH
in Hamburg, Germany